Preface

Of all the aspects of human biology to excite our fascination, few can be more significant than the diversity of human behavior. Even within a culture, our curiosity is captured by those differences that seem to make one person stand out from another — their individual styles of living, their preferences and their beliefs, their strengths and their weaknesses. Such differences are the raw material from which theories of personality are crafted.

Our book tries to analyze the genetic and environmental causes of these differences. Although studies of the genetic and environmental influences are no substitute for good physiological or psychological theories of individual differences, they do provide one important arena in which such models can be tested. For example, can we persist in a "social learning" theory of personality or social attitudes if it turns out that the only detectable effects of parents on their children are genetic rather than social? What are the predictions for the genetic analysis of age-dependent traits of different mechanisms for the acquisition and transformation of information about the world? Is there a relationship between the pattern of genetic and environmental causation of individual differences and the evolutionary and sociobiological significance of the traits for the species? We are still a long way from being able to answer all these questions, but one thing is clear: there is absolutely no point whatever in beginning to speculate about such issues unless we have a clear idea of what is actually happening in the real world of personality and attitude differences.

The main purpose of the book is to give the reader a clearer idea of the state of knowledge, and ignorance, about the causes of individual differences in personality and attitudes, a good feeling for the kinds of data on which such inferences are based, and a sense of the methods of data analysis that are appropriate for answering basic questions about the role of biological and cultural inheritance in human populations. As a consequence, parts of the book are more technical than is often the case, there are many more tables and more than usual caution in circumscribing our less than certain conclusions. By providing much of the original data, we have allowed the reader scope to develop and test his/her own ideas and models and to retrace our steps, at least in some of the simpler cases.

The contract for this book was signed almost fifteen years ago! We were just completing a twin study in London and thought that our results were of sufficient simplicity and interest to justify a book. Why the delay? There are three main reasons.

The first is our growing interest in social attitudes in addition to personality. Initially we were surprised that genes might play any part in the determination of something so obviously "cultural"; then we were pleased that, beyond the contribution of genetic effects, social attitudes did indeed still display all the hallmarks of cultural inheritance; now we are surprised again that the cultural effect may still evaporate into the genetic consequences of assortative mating. Faced with such a consistently developing story, which the reader can reconstruct in the later chapters, any attempt at summary would have been premature.

The second reason for delay has been the rapid explosion of theory and method over the last fifteen years, to which we also have had to devote some time. The publication of Jinks' and Fulker's paper on fitting bio-metrical–genetic models to human behavior was a landmark in 1970. When we first applied these methods to adult personality, the results seemed fairly straightforward and the models described by Jinks and Fulker carried us a long way. The more we became involved in new data, however, the more it became clear that other issues needed further theoretical work, including mechanisms of cultural inheritance, sex limitation, mate selection, develop-mental change, social interaction and trait covariation. Throughout the 1970s, continuing to the present time, we, and many others in the United States whose contributions we cite, recognized the deficiencies of the classical genetic models when applied to human behavior and did our best to develop theoretical models that had explanatory and heuristic value. Most of these ideas do not find their definitive expression in this book, because here we are concerned more especially with the substantive issues of personality and attitudes rather than theory and model-building for its own sake.

The final cause for delay was the completion and publication of other large twin studies, which played such a crucial role in refining some of our early notions based on the smaller sample of London twins and relatives. The large Australian study is still yielding fruit, of which the work described here is only a first sample. We are conscious, even as we write, that new studies are being done that will transcend the ones we describe for their subtlety and creativity.

This work would have been impossible without the financial support of the British Medical and Science Research Councils, the Australian NH &

MRC, The American Tobacco Corporation, NIH, NIAAA and NIMH, and the grants in aid fund of Virginia Commonwealth University. We are indebted especially to Althea Walton, Greg Porter and Judy Silberg for their enormous assistance with preparing and repairing the manuscript.

Richmond, Virginia

Contents

Chapter 1

Another Book on Heritability?

For nearly twenty years the genetic study of human behavior has acquired a medieval flavor in the public forum. Popular discussion has a "scholastic" quality with its concentration on texts, historical material, nuances of expression and writers' credentials. When discussion sinks to this level, new models and new data are largely irrelevant.

Much of the debate has been less than illuminating from a scientific perspective. We think that there are five main reasons for this:

(1) almost exclusive attention paid to intelligence;
(2) restriction of discussion to "heritability" at the expense of other causes of individual differences;
(3) emphasis on reviewing old data rather than presenting original research;
(4) verbal description of data rather than quantitative hypothesis-testing;
(5) small sample sizes and idiosyncratic measurements.

In this book we try to address a number of questions about the causes of individual differences in personality and attitudes in a form that does not sacrifice rigor for didactic simplicity or scholarly doubt for the persuasive power of conviction. Although this makes parts of the book more difficult, the result comes closer to representing what we think and the degree of (un)certainty with which we think it than would be apparent from a more strident account. The principal features of our approach may be presented in contrast with the limitations we listed above.

1.1 PERSONALITY AND SOCIAL ATTITUDES

Most of the "heat" of criticism in the 1970s was directed to the analysis of correlations for IQ. Cognition is an important aspect of human adaptation, but not the only part. There are consistent patterns of individual differences in behavior that emerge in a wide range of studies using quite different instruments from those used to assess intellectual function. These are independent of IQ, yet they are consistent over instruments, occasions, cultures, and even species. In man, they affect how people interact with one another and respond to values of the society to which they belong. These personality variables are invoked to account for liability to some important psychiatric and social disorders, and predict how people respond to certain kinds of drugs and schedules of reinforcement. The background to the description of personality, its measurement, and its practical and scientific importance is summarized in Chapter 2.

Social attitudes are intrinsically interesting for several reasons. First, individual social attitudes commonly change with time. They are thus a monitor of behavioral change, which ought to be affected primarily by the environment. Secondly, they express, however inadequately, the orientation of the individual towards the society in which he lives. Even if the attitudes people express do not correspond exactly to their actual social practices, they represent an individual's willingness to be counted as believer or agnostic, as nationalistic or not, as liberal or conservative. These are the basic currencies that express how an individual views himself in relation to society, how he spends his time, his money and his vote.

Thirdly, social attitudes, perhaps more than any other aspect of behavior, belong to the *human* domain. They represent the interaction between the individual human and the habitat that he has created for himself. They could not exist, in the form we measure them, without religion, politics, law, social problems, and the nuclear family. Because they relate to functions that are "late bloomers" phylogenetically, we might expect them to be especially sensitive to the mechanisms of social learning that characterize the human species. It is when we turn to social attitudes, therefore, that we should find the paradigms of non-genetic inheritance that have so far eluded the behavior geneticist.

1.2 AWAY FROM THE "HERITABILITY HANG-UP"

A paper by Feldman and Lewontin in 1975 accused behavior geneticists of a "heritability hang-up". That is, the over-riding concern of behavior geneticists was to show that behavioral differences were inherited and to estimate how much of the variation we measure is due to genetic causes. A cursory review of the literature prior to 1970, and the subsequent public debate about IQ, may have justified their criticism. Attached to every heritability estimate there remains a "so what?". A heritability estimate does not translate into a prescription for intervention or social change. Indeed, a heritability estimate does not even translate into a selective breeding program for cattle without more knowledge about the kinds of gene action that contribute to the observed genetic differences.

It has been tempting, in the past, to equate "heritability" with "construct validity" or "biological significance". Unfortunately the equation has little foundation. As we shall see, even such a specific trait as a response to an individual questionnaire item may have a genetic component, and yet it would be difficult to construct a separate biological justification for each item. Many of the inherently important questions of population genetics that would help us relate observed genetic variation to the mechanisms responsible for genetic polymorphism have still to be solved.

The chapters that follow provide many instances in which the simple equation "phenotype = genotype + environment" does not do justice to the variety of causes of differences in attitudes and personality. It seems premature to build exotic sociobiological theories to account for certain kinds of genetic polymorphism, developmental process or social interaction until we have understood the basic parameters within which such effects operate. Why should we spend time theorizing about the adaptive significance of parent–offspring interaction before we have shown (a) whether the interaction is independent of genotype, and (b) whether the effects of parent–offspring interaction persist into adult life or whether they merely evaporate when the offspring "leaves the nest"? Different types of social interaction would require different biological explanations. Without the basic parameters to reflect these mechanisms, we cannot begin to theorize constructively.

It will be apparent in many places that our understanding of genetic and environmental effects cannot adequately be represented in the notion of "heritability". At various times we shall consider: how far social interaction between siblings is responsible for creating personality differences (Chapters 5, 6 and 7); how far the family environment creates differences between families (Chapters 6, 7 and 11–16); the causes and effects of assortative mating for personality and attitudes (Chapters 6, 14 and 15); the effects of sex on the expression of genetic and environmental effects (Chapter 5); how genes and environment are organized in their effects on multiple variables (Chapters 10 and 11); the causes of temporal change in attitudes (Chapter 13); developmental changes in the expression of genetic and environmental effects on personality (Chapter 7); and whether or not genes contribute additively to phenotypic differences (Chapters 5, 6 and 7).

Readers who approach the book wanting a purely environmental explanation of human differences will be disappointed and will find reasons to explain away any genetic parameters in our models. That is one reason why we provide the data in as complete a form as possible. On the other hand, anyone who believes that simple additive genetic effects can explain family resemblance for personality and social attitudes will find plenty in the following chapters to prove him wrong.

1.3 PRESENTATION OF ORIGINAL DATA

With a few significant exceptions, our book only uses data that we have either gathered ourselves or have analyzed ourselves from raw data kindly supplied by other investigators. It thus represents a joint research program spanning fifteen years, in which we have tried to address new questions as they arose. Inevitably, the final account is "idealized" in the sense that the story is organized with hindsight and does not necessarily reflect the exact sequence of insight into particular problems or the fact that many early analyses were repeated or improved in the final stages. However, the reader should gain a sense of the close interaction between theoretical developments on the one hand and the realities of data on the other. Sometimes new data were collected, or new analyses conducted because a theoretical problem was recognized that had to be addressed in practice. This is the case with some of the studies of assortative mating described in the later chapters. In other cases, the development of theory was motivated in part by the inability of our original models to explain the findings. This is the case for parts of the treatment of developmental change in gene expression discussed in Chapter 7.

   Wherever possible, we have presented data summaries in the form from which we started the analysis of genetic and environmental effects. The principal exceptions are (1) the analysis of individual items in Chapters 8 and 12 (because there are so many items), and (2) the analysis of extended pedigrees by maximum-likelihood in Chapters 6 and 7 (since that would require tabulating the individual observations). This strategy has resulted in an unusually large number of tables. However, we present the summary statistics so that (i) teachers and students can try some of the analyses for themselves, (ii) critics and researchers can develop their own models, and (iii) the reader can judge for him/herself how far our conclusions follow from the data.

1.4 QUANTITATIVE MODELING AND HYPOTHESIS TESTING

Many models for individual differences and family resemblance result in complex predictions that can only be expressed and tested in quantitative terms. Consider, for example, the predictions made for the correlations between relatives when there are no shared environmental effects. It is intuitively obvious that genetic factors will make identical twins more alike than non-identical twins (see Chapter 3) because identical twins are genetically identical. This much of the genetic argument can be put into words and a simple statistical test of the difference between two correlations. But what does a given difference between the correlations of the two types of twins predict about the resemblance for other kinds of relatives? Are the data consistent with these predictions? If they are not then why not? These questions are inherently quantitative, they require a mathematical model for inheritance and gene expression before they can be tested and they require a statistical approach that enables us to test complex hypotheses. Similar problems apply to the analysis of cultural inheritance. Common sense tells us that the environmental impact of parents on their children creates correlations between parents and children and generates sibling and twin correlations. It takes a quantitative model, however, to show that this type of cultural inheritance has different consequences for different types of biological and social relationship. It also takes a variety of quantitative models to express the predictions of different theories of social interaction between family members. If we restrict our discussion of environmental effects to the simple and intuitively obvious tests of differences between pairs of correlations that we can put into words then we may be able to *detect* family environmental effects, but we shall never come any closer to understanding the *mechanism* of social interaction that created the environmental differences we detect. Chapter 4 introduces some of the basic methods of model-fitting, and subsequent chapters extend these in a variety of ways to encompass more complex hypotheses and intricate data sets.

Our treatment is different from that which normally appears in textbooks. The subject matter is inherently quantitative and statistical. To treat the subject in a purely descriptive manner does not do justice to the scientific questions. It may help the reader to keep separate four quantitative aspects of the subject. These are: (1) the statistical aspect of data summary (the numbers); (2) the way in which we translate our theories of biological and cultural inheritance into models for the statistics that we collect (model building); (3) the statistical principles that we use to estimate parameters and test hypotheses (model fitting); (4) the computer programs and numerical methods that we use to obtain these estimates and statistical tests (number-crunching). Generally, the data summaries rely on statistical methods that are familiar to the advanced student and readily available on most computers. The actual mechanics of deriving the models are conceptually simple but sometimes tedious. It takes more practice to see how to write the model in the first place and to acquire the algebraic "tricks" of derivation. The important issue is to get a "feel" for how we can write and test models for the genetic and cultural effects contributing to family resemblance.

Even if the reader is unfamiliar with all of the statistical methods, it is only really necessary to understand the logical principles behind the inferences that we make. These do not differ from those used in the analysis of any other data set. We did not set out to write a book on statistics or statistical computing. On the other hand, we have tried to provide sufficient background and worked examples for the beginner to understand the basic statistical and numerical issues for the simpler cases (e.g. Chapter 4) and have assumed that the more complex statistical methods of the later chapters will either be familiar to the technical reader or irrelevant to a basic understanding of the logical and substantive issues.

1.5 SUSTAINED STUDY OF SELECTED MEASURES

The history of personality research is littered with small twin studies of large numbers of measures. It is difficult to extract consistent findings from such studies because the sampling errors attached to correlations are quite large and it is not always clear how the individual measures relate to any overall model for the main dimensions of personality. It is impossible to decide whether the results that we get with nuclear families or adoptions, say, are inconsistent with the results from twin studies because the measures differ, the populations differ or because twins differ from non-twins.

Our research is based on a few personality dimensions that have been studied repeatedly with large samples in different populations, or with a number of different types of relationship in the same population. By concentrating on large samples and several kinds of relationship, we hope to be able to test for more subtle features of genetic and environmental determination than are considered in the conventional "heritability study". By looking at repeated samples from the same population and samples from different populations, we shall be able to see whether the mechanisms responsible for individual differences in personality and attitudes generalize over populations.If they do not then we would question the value of such studies unless there were good reasons for predicting a particular pattern of differences, as in the case of secular changes in the causes of educational attainment in Norway, recently described by Heath *et al.* (1985a, b). By focusing on the main dimensions of personality and attitudes, we may not capture all the nuances of individual behavior, but we hope to establish the roots of theory and empirical data on which other ideas can grow.

Chapter 2

Dimensions of Personality

Early genetic studies of personality (see Chapter 3) gave few consistent results. There are probably three main reasons for this: (1) small sample sizes; (2) lack of a coherent personality theory; (3) absence of any systematic theory and methodology to guide data collection and interpretation. In this chapter we outline the elements of the personality theory that guided the development of the measurements that have formed the core of the investigations we describe in our book.

2.1 THEORIES OF PERSONALITY

Even though there may be disagreement about the merits of particular models, there is now a fair amount of agreement among geneticists about the basic approach that is appropriate in the analysis of human variation. The methods will be examined and illustrated throughout the subsequent chapters. They approach the paradigm that Kuhn (1962, 1970, 1974) and others (e.g. Urbach, 1974) consider appropriate for scientific enquiry. There is far less uniformity among psychologists, however, about what are the marks of a good theory of personality. Allport (1937), for example, already discovered over 50 meanings of the term, including quite different and often contradictory definitions. The situation has not improved all that much in recent years, where textbooks of personality present a picture either of benevolent eclecticism or contentious idiosyncrasy. Textbooks of personality frequently present the reader with eponymous chapters devoted to the theories of one particular author each (e.g. Hall and Lindzey, 1970), without any attempt to address substantive issues, to compare different solutions, to look at the empirical evidence for and against a given theory, or try to arrive at overall conclusions. At the other extreme (e.g. Cattell, 1982), authors cite few writers other than themselves and their collaborators, and pay little attention to criticism.

2.2 THE DESCRIPTION OF PERSONALITY

These two attitudes towards personality research are extremes. A good theory of personality should encompass the solid mass of factual material available, and lead to further empirical studies. It might be argued that such a paradigm exists in personality research (Eysenck, 1984). This paradigm in its descriptive aspect, goes back in large part to the ancient Greeks, whose theory of the four temperaments (Choleric, Sanguine, Phlegmatic and Melancholic) was based on solid observation. In its modern form, and based on correlational and factor-analytic methods employed upon the results of self-descriptive questionnaires, ratings by friends and acquaintances, miniature situation studies, experimental investigations, physiological measures, and hormonal and other biochemical assays, this model has transcended the purely descriptive phase of investigation, and has begun to assume a dynamic and causal aspect, relating behavior to fundamental biological factors, whether physiological or hormonal (Eysenck and Eysenck, 1985). The evidence for this statement is reviewed in detail in the book just cited. In this chapter we shall briefly recapitulate the major reasons for suggesting that there does exist a paradigm in personality study, and that this paradigm has both descriptive and explanatory aspects.

We summarize the descriptive aspects of the paradigm, and then outline the various types of tests that such a model must undergo successfully in order to be accepted as being paradigmatic. Three major dimensions of personality emerge consistently as higher-order or superfactors from large-scale factor analyses of matrices of intercorrelations, the elements of which are individual answers to inventory questions, single ratings, or test scores of one kind or another. These three major dimensions have been variously named by different investigators, and there is no agreement on the semantic problem of how best to label them. In this book we shall use the terms suggested by Eysenck and Eysenck (1976), leaving the reader free to substitute other terms should he so desire. The names given to these three superfactors are psychoticism (P), extraversion (E) and neuroticism (N), with ego control, introversion and emotional stability being the opposite ends of the three continua in question. The traits that go to make up each of these three factors are shown in Figures 2.1, 2.2 and 2.3 below; it is the empirically observed intercorrelations between the traits that give rise to the superfactors and legitimate them.

Figure 2.1 shows the traits that go to make up the psychoticism factor, i.e. the figure indicates that a typical high-P scorer is aggressive, cold, egocentric, impersonal, impulsive, antisocial, unempathic, tough-minded and creative (Eysenck and Eysenck, 1976). The term "creative" may seem to stand out from the others as being wrongly chosen, but the evidence for the

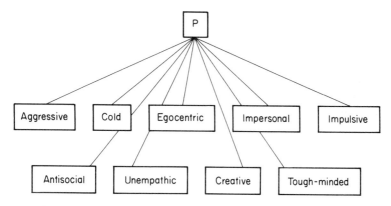

Figure 2.1 Traits correlating together to define the psychoticism dimension (Eysenck and Eysenck, 1985).

proposition that genius and madness are closely allied is quite strong, and the empirical work of Gotz and Gotz (1979a, b) and Woody and Claridge (1977) leaves little doubt about the genuineness of the association (Eysenck and Eysenck, 1976). The distribution of P is very much skewed, with few high scorers and many low scorers.

The introversion dimension is too well known to require much discussion; Figure 2.2 shows the traits characteristic of the typical extravert, indicating that he is sociable, lively, active, assertive, sensation-seeking, carefree, dominant, surgent and venturesome. Introverts, of course, are the opposite in all these respects, and ambiverts are intermediate between these extremes.

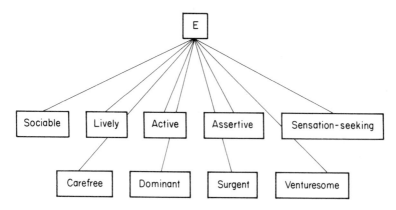

Figure 2.2 Traits correlating together to define the extraversion dimension (Eysenck and Eysenck, 1985).

Ambiverts constitute the majority of the population, with extra-version–introversion showing a fairly normal distribution.

Neuroticism or emotional instability is characterized by such traits as being anxious, depressed, having guilt feelings, low self-esteem, being tense, irrational, shy, moody and emotional (Figure 2.3). Stable people show the opposite traits, and here, too, ambiverts tend to be intermediate, and in the majority, with a continuous, approximately normal, distribution being usually observed in random samples of the population (Eysenck, 1952a).

2.3 TESTING THE MODEL FOR PERSONALITY

2.3.1 Consistency across measurements

For such a model to be widely accepted, clearly a number of conditions must be fulfilled. The first of these is that these three dimensions, in one form or another, should emerge from the great majority of, if not all, statistical studies carried out in this field, and embracing more than just a few restricted traits of personality. A survey of this kind has been undertaken by Royce and Powell (1983), and they concluded their survey by arguing that there were three major dimensions of personality, which they label "emotional stability" (the obverse of neuroticism), "introversion–extraversion", and "emotional independence" (similar to psychoticism in showing lack of trust, lack of cooperativeness, tough-mindedness, lack of affect, dominance and realism). Eysenck and Eysenck (1985) have also surveyed correlational studies of the major instruments used for the investigation of personality, such

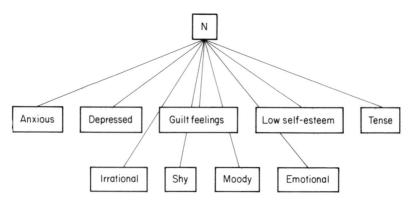

Figure 2.3   Traits correlating together to define the neuroticism dimension (Eysenck and Eysenck, 1985).

as the MMPI, the CPI, the 16PF, and many others. They concluded, that for practically all of these, factors similar to E and N can be observed in factorial analyses, and that in many cases an additional P factor is obvious. We may thus conclude that these three major dimensions or superfactors emerge fairly universally in large-scale studies carried out by American and European psychologists on samples taken from these countries.

2.3.2 Cross-cultural consistency

Two objections might be raised to the acceptance of the paradigm. The first might be that the similarities observed are superficial, depending on subjective judgements of the meaning of individual items or factors defining traits. The second objection might be that what is true of American and European populations might not be universally true and apply in other countries of a very different culture. It thus becomes important to look at cross-cultural comparisons in order to discover whether the same factors could be found in populations whose culture differs very much from the Euro-American.

Recent studies on the use of the Eysenck Personality Questionnaire (EPQ, Eysenck and Eysenck, 1975) in 25 different cultures has suggested an answer to these questions. The EPQ (see Appendix B) is a preferred test for the measurement of P, E and N and it has been widely translated and used in a great variety of countries. In each case, samples of over 500 males and females were tested, factor analyses carried out on the matrices of intercorrelations for males and females separately, and indices of factor comparisons calculated, comparing each country with each other (Barrett and Eysenck, 1984). The results showed a surprising congruity between different nations, with indices of factor comparison nearly always exceeding the 0.95 level, and very frequently the 0.98 level. Thus there are very marked similarities in the personality patterns found in such widely varying countries as Nigeria, Japan, Hong Kong, Brazil, Mainland China, Uganda, Greece and Bangladesh. We may conclude that not only are the superfactors found in many different measuring instruments in the Western world, but we may also conclude that these factors are equally characteristic of third-world countries in Africa and South America, of countries in the Chinese–Japanese culture circle, etc.

2.3.3 Animal models

If these dimensions of personality are so universal within the human species, it might be suggested that evidence for them might also be found among animals. After all, if the major characteristics of human beings are found in interaction with other people, then it would seem that there are three major types of interactions. We may be sociable with others (E); we may be afraid of others (N); or we may be aggressive towards others (P). These three patterns of social interaction are also observed among animals, and the studies of Chamove, Eysenck and Harlow (1972) have shown, on the basis of the analysis of years of careful observation of Rhesus monkeys, that their behavior can indeed be analyzed in terms of these three major factors. It is even possible to discern evidence for these factors in the behavior of non-primate mammals, such as the rat. The work of Eysenck and Broadhurst (1965) and Broadhurst (1975) has demonstrated that tests such as the "open" field may be used with advantage as a measure of emotionality, and extensive work in recent years at the Barcelona Autonomous University (Garcia and Sevilla, 1984) has shown that with suitable alterations the same test can be used for the measurement of extraversion–introversion, using perambulation as a measure of extraversion, and defecation as a measure of neuroticism. Aggression has been measured along many different lines in rats, and presents no problems. The literature has been reviewed in greater detail by Gray (1970, 1973) and by Eysenck and Eysenck (1985).

2.3.4 Developmental consistency

So far, we have considered three tests of the model: support by many different types of questionnaires and ratings, used by different investigators on different samples; universality, i.e. of cross-cultural similarity; and that of animal comparison, i.e. of applying to animals as well as to humans. We now turn to another type of test, namely that of consistency over time. If the variables we are dealing with are truly fundamental then they should also be permanent or semipermanent characteristics of the individual, and longitudinal studies of personality should demonstrate a certain consistency over time. Actually there are two questions that should be carefully distinguished here, as Hindley and Guiganino (1982, p. 127) point out:

> One concerns the extent to which the behavioral characteristics assessed can be regarded as similar in nature at different ages: the issue of what Emmerick (1964, 1967, 1968) and Baltes and Nesselroade (1973) termed continuity versus discontinuity of variables. The other concern extends to whether individuals maintain their relative status across ages on the variables in question.

The first question is dealt with by Eysenck and Eysenck (1985), who conclude that as far as P, E and N are concerned, the junior and senior versions of the EPQ show a sufficient degree of similarity to use them for longitudinal study from early youth to old age. The extensive work of Hindley and Guiganino (1982) is particularly relevant in lending support to this conclusion. Conley (1984) has recently reviewed the literature on longitudinal studies, as well as contributing findings of his own. He uses the formula $nC = Rs$, where C is the observed retest coefficient, R the internal consistency or period-free reliability of the measuring instrument, s the annual stability and n the interval (in years) over which the coefficient is calculated. Comparing the annual stabilities of intelligence and personality traits, particularly P, E and N, he was able to show that these could be estimated at 0.99 and 0.98 respectively. He concluded (p. 11) that: "Intelligence and personality may be regarded as relatively stable characteristics over the length of adult life span." It is important to note that the intervals of the studies summarized by him extended to something like 50 years, i.e. a very lengthy span of time indeed. We shall return to this issue in Chapter 7, when we consider developmental changes in gene expression.

Other authors (e.g. Schuerger *et al.*, 1982; Guiganino and Hindley, 1982; Costa *et al.*, 1983; Eichorn *et al.*, 1981; McCrae and Costa, 1984; and others summarized by Eysenck and Eysenck, 1985) suggest that our model indeed passes this hurdle as well, and that consistency of conduct over many years is characteristic of the three factors P, E and N.

2.3.5 Relationship to psychiatric disorder and social behavior

Another test of the model would be the following. If P, E and N are indeed important aspects of personality then they should have predictable and testable relationships to important areas of social conduct. There is ample evidence that this is indeed so (Wilson, 1981). Personality has been found to be closely related to psychiatric abnormality, in particular neurosis and psychosis (Eysenck, 1971a). Other relationships are discussed in Eysenck and Eysenck (1976). Extraverted types of neurosis (hysteria, psychopathy) may be contrasted with introverted types (anxiety states, phobias, obsessive–compulsive behavior). Physical symptoms of neurosis are more frequently found in extraverted persons, mental symptoms in introverted ones (Eysenck, 1973).

Another area of interest is criminality (Eysenck, 1977; Eysenck and Gudjonsson, 1988). Criminals tend to be high on P, E and N, and there is evidence that individuals with this personality profile indulge in antisocial behavior from quite an early age. These relations are derived from theoretical considerations about the nature of these personality factors, and the evidence suggests that similar personalities commit crimes in communist and third-world countries as do in Europe and the United States (Eysenck, 1985).

A number of studies in Germany have also supported the results originally reported from the United Kingdom and other English-speaking countries (Stellar and Hunze, 1984). Recently there has been growing interest among leading psychiatrists (e.g. Cloninger, 1980) in the capacity of a dimensional model to encompass the known facts on psychopathology.

Research in the sexual (Eysenck, 1976b) and marital (Eysenck and Wakefield, 1981) fields has disclosed many important relations between personality factors P, E and N on the one hand, and sexual attitudes and behaviors and marital happiness on the other. Again these relations were predicted, and the confirmation of these predictions strongly supports the model.

Other relationships that have been discussed in more detail by Wilson (1981) relate to affiliation and personal space, birth order, group interaction and social skills, speech patterns, expressive behavior and person perception, expressive controls, suggestibility, conflict handling, attraction, attitudes and values, recreational interests, occupational choice and aptitude, industrial performance, academic aptitude and achievement, drug use and abuse, and various others. He concludes his survey by stating (p. 239) that:

> Clearly the dimensions of E, N and P . . . have wide explanatory and predictive power across a variety of socially important domains. The suggestion that has sometimes been made to the effect that there are no stable traits which enable us usefully to predict social behavior, is shown to be untenable. Although learning experiences and transient environmental circumstances do have to be considered in predicting what a person will do in a particular situation, so too must their personality be taken into account, or the formula is bound to be incomplete.

2.4 CRITICISMS OF THE MODEL

2.4.1 Criticism of trait theories

So far we have dealt with the model in its descriptive aspects, and such models have sometimes been criticized on a variety of grounds. It has been suggested (e.g. by Mischel, 1968, 1977; Mischel and Peake, 1982) that the study of personality by way of traits is essentially unproductive, and that situations contribute more than do traits to individual behavior. The argument has been discussed in great detail by Magnusson and Endler (1977). A detailed refutation is given by Eysenck and Eysenck (1980), so we do not repeat the arguments here in detail. However, material already surveyed in this chapter demonstrates that Mischel's conception is at best one-sided, and at worst completely mistaken. The issue of person × situations interaction will be considered from a genetic perspective in Chapter 9.

2.4.2 Nomothetic vs. idiographic methods

A rather different type of criticism is that originally put forward by Allport, who argued in favor of an idiographic instead of a nomothetic type of approach to the problem of personality. The usefulness of the nomothetic approach has been demonstrated too often and too incisively to make it possible to disregard the arguments in its favor, whereas there is no evidence that any advance has been made in the idiographic study of personality (for contrary arguments see Runyan, 1982). Indeed, Allport himself, whenever he published empirical data, used the nomothetic rather than the idiographic approach. A more detailed examination of this issue may be found in an earlier publication (Eysenck and Eysenck, 1985).

2.4.3 Inability to explain social behavior

Another frequent criticism of trait psychology is one that has been made in a similar form of the early theory of instincts. It is pointed out that, just as the postulation of an instinct of "sociability" does nothing to explain the phenomena of social behavior, because the instinct is predicated upon the phenomena it is supposed to explain, so similarly a postulation of a trait of "sociability" does nothing to explain social or asocial behavior. In answer, it should be said that while the objection is true, it is irrelevant. Trait psychology is essentially descriptive; it makes no causal claims. To postulate a trait of "sociability" is not to attempt an explanation of sociable behavior: it is merely to claim descriptively that instances of social behavior correlate together over a group of persons, and define and make measurable this particular trait. It might appear a task of supererogation to carry out such work, but this is not so. As Eysenck (1956a) has shown, in an empirical study of the Guilford questionnaire of social shyness, there is not only one type of social shyness, but two quite uncorrelated ones. Some people are lacking in sociability because they do not care for other people (introversion); others behave in an unsociable manner because they are afraid of other people (neuroticism). Thus, because we possess a single term, sociability, to characterize a certain type of behavior, it does not follow that we can postulate a single factor, and careful correlational and factor-analytic work has to be done in order to discover to what extent our semantic habits agree with descriptive facts. Many other examples are given by Eysenck and Eysenck (1985) to illustrate this point. Accurate observation and a well-founded taxonomy of behavior are needed to give us a proper descriptive basis for trait psychology, and it would be quite wrong to rely on simple semantic habits and everyday convenience in our description of personality traits. Thus this particular objection also must be ruled out as not properly applicable to the study of personality traits.

2.5 CAUSAL ASPECTS OF THE THEORY

While a descriptive, taxonomic phase is essential in scientific analysis, and must precede the more causal or dynamic phase, it is not in itself sufficient. We need to ask causal questions, and these, in the personality field, can often be answered in what are often called "reductionist" terms, i.e. in terms that would reduce psychological variables to concepts coming from a more fundamental biological background. Reductionism is often criticized on philosophical grounds, but this would seem to be inappropriate in particular circumstances. It may or may not be the case that all human behavior, including cognition, can be ultimately reduced to physiological or bio-chemical, or physical concepts and theories. Such a belief cannot be supported empirically at the present time. However, certain relationships have been observed between psychological variables, such as those entering into our description of personality, on the one hand, and physiological and biochemical variables, on the other. Thus Eysenck (1967b) has suggested that extraverts are characterized by a low level of cortical arousal, while introverts are characterized by a high level of cortical arousal. He has further suggested that these levels are in part controlled by the ascending reticular activating system, and that these biological differences are causally related to the behavior characteristic of extraversion and introversion respectively (Eysenck, 1981). Similarly, neuroticism has been related to the activity of the limbic system, expressed through the sympathetic and parasympathetic branches of the autonomic system (Stelmack, 1981). Other theories have been prepared that try to account for observed trait variation in terms of the structure, physiology and biochemistry of the nervous system (Gray, 1970; Zuckerman, 1979; Zuckerman *et al.*, 1984).

We need not concern ourselves here with the question of whether such theories are correct or incorrect; they are merely quoted as an example of the possible reduction of factors in the personality field to biological systems that are considered to be more elementary and fundamental.

Physiological variables are not the only ones to be related to the major dimensions of personality. Olweus (personal communication) has shown correlations of 0.5 between adrenaline secretion and both introversion and neuroticism; thus the simple hormonal secretion may account for something like 25% of variation in the aspect of personality usually called "trait anxiety", which is a mixture of introversion and neuroticism. Thus there is evidence of a relationship between behavior and hormone secretion.

Another example of the relationship between personality and biochemical variables is the important effect that the enzyme monoamine oxidase (MAO) seems to exert on sensation-seeking and extraverted behavior. Much work has been done in this field on both monkeys and humans, and quite a strong negative correlation has been established between platelet MAO and extraversion in general, and sensation-seeking in particular. The evidence is surveyed by Zuckerman *et al.* (1984); it is only one of many such relationships that have been studied in some detail in the past. The outcome of these studies leaves little doubt that there are important biological foundations for differences in personality, and this fact by itself suggests that genetic factors might be involved in an important way in causing differences in behavior, with neurobiological structures and secretions mediating this relationship. Early work by Eysenck and Prell (1951) and Eysenck (1956a) gave support to this view. These findings of strong genetic involvement in personality differences contradicted the accepted view (Newman *et al.*, 1937) of little genetic influence on personality, and marks the reawakening of interest in this field. The earlier works have been criticized by Eysenck (1967b).

2.6 HEURISTIC VALUE

The final attribute that justifies giving a model the status of a paradigm is that it is a fertile source of experimental predictions. Deductions can be made from the physiological, causal model that are testable; some of these are themselves physiological, others are experimental along traditional psychological lines. A good review of the former studies that have been carried out in relation to personality can be found in Stelmack (1981), while the latter are reviewed in Eysenck (1967a, 1976a, 1981). The list of such studies is very long indeed, ranging from perceptual variables like sensory thresholds, figural after-effects, pain and sensory-deprivation tolerance, to various aspects of memory, and learning and conditioning. Indeed, it is possible to argue that personality variables interact with almost all the aspects of experimental, social, clinical, educational and industrial psychology that have been studied by empirically minded psychologists, and that this interaction is so close that it accounts for more of the variance, in most studies, than do the so-called main effects.

These, then, are our reasons for adopting the P, E and N model of personality provisionally as our phenotype, for quantitative genetic analysis. It is not necessary for the reader to adopt the personality model that we have outlined above *in toto*. It can be translated into other models, and it should be possible to use our data to effect such a transformation, and to derive some knowledge about the genetics of other systems from the data here given. But it must be doubtful if such a translation is really necessary or advisable in view of the paradigmatic aspects of our model. In due course, the model will be improved, and may even be abandoned. But any better model will need to incorporate at least as much biological, experimental and descriptive evidence, and must offer still more far-reaching theories to mediate between the various aspects of the model. Most personality theorists have been content to look at one particular corner of this whole field, rather than to try and cover its major aspects. Much remains to be studied; it is not claimed for this model that it covers all of personality, but merely that the three dimensions it deals with are important, perhaps the most important, descriptive and causal features of personality. Future research will no doubt add to this array, and may even substitute new, and better, variables for those here singled out for comment. Similarly, future research will undoubtedly look in more detail at the various traits themselves that make up the superfactors along the lines of analysis used for impulsive and sensation-seeking behavior by Eysenck (1983a). In all, it is not claimed that the analysis of personality briefly discussed here is anything but the beginning of the scientific study of personality, and, equally, the discussions of the genetics of personality are in truth only a beginning, laying down in a rough form certain conclusions that future work will undoubtedly refine, and may even reject. Nevertheless, it seems to us that at the moment there is sufficient agreement in the empirical field to justify the claim that the findings are neither statistical artifacts, nor idiosyncratic observations, but that we have here the beginnings of a paradigm that justifies our adopting it as the central focus of our analyses of genetic and environmental effects.

Chapter 3

The Classical Approach: Early Twin Studies of Personality

3.1 HISTORY OF THE TWIN METHOD

3.1.1 Hereditary genius

The first empirical studies of the inheritance of human behavior were described by Francis Galton in *Hereditary Genius* (1869). He attempted to define and measure family resemblance for exceptional ability in a wide range of professions and skills ranging from the law and the church to rowing and wrestling. In these studies, Galton introduced the "proband method" in which "high-risk" families are ascertained through affected individuals. This work showed: (1) that the empirical risk of genius in the relatives of eminent probands was far higher than would be expected for qualities so rare in society as a whole; (2) that the probability that a relative of an eminent proband would himself be eminent declined as the relationship became more remote. These findings led Galton to formulate the first mathematical model for family resemblance, the "Law of Ancestral Heredity", which described the facts of family resemblance quite well but was not based in the more precise understanding of the actual mechanism of inheritance that was to follow from the rediscovery of Mendel's work in the early 1900s.

3.1.2 The history of twins

Galton saw a major problem with his approach. Most individuals who are related biologically are also related socially. Parents provide part of the environment for their children as well as providing their genes. The twin study was proposed as an elegant solution to this difficulty in his book *Inquiries into Human Faculty* (1883). He observes that the objection to statistical evidence in proof of the inheritance of peculiar faculties has always been:

> The persons whom you compare may have lived under similar social condi-
> tions and have had similar advantages of education, but such prominent condi-
> tions are only a small part of those that determine the future of each man's life.
> It is to trifling accidental circumstances that the bent of his disposition and his
> success are mainly due, and these you leave wholly out of account — in fact,
> they do not admit of being tabulated, and therefore your statistics, however
> plausible at first sight, are really of very little use.
>
> No method of inquiry which I had previously been able to carry out — and I
> have tried many methods — is wholly free from this objection. I have there-
> fore attacked the problem from the opposite side, seeking for some new
> method by which it would be possible to weigh in just scales the effects of
> Nature and Nurture, and to ascertain their respective shares in framing the
> disposition and intellectual ability of men. The life-history of twins supplies
> what I wanted.

The logic of the twin method was simple. Identical twins have the same genes and share the same parents. Any differences within pairs of MZ twins must be due to their unique environmental experiences. Non-identical twins have the same parents but different genes. Differences within pairs of non-identical twins must be due to genetic effects *and* their unique environmental experiences. If the differences within pairs of DZ twins are greater than the differences within MZ twins, the excess must be due to genetic effects. Galton's own use of the method was disappointing and confined to a dis-cussion of anecdotal accounts derived from letters sent to him by twins in response to appeal. Nevertheless, his final conclusions show great insight into the dependence of estimates of genetic and environmental effects on the populations and cultures from which they are sampled:

> There is no escape from the conclusion that nature prevails enormously over
> nurture when the differences of nurture do not exceed what is commonly to be
> found among persons of the same rank of society and in the same country. My
> fear is, that my evidence may seem to prove too much, and be discredited on
> that account, as it appears contrary to all experience that nurture should go for
> so little. But experience is often fallacious in ascribing great effects to trifling
> circumstances.

The main weakness of Galton's study was the lack of quantitative beha-vioral data on which to base his analysis. The exploitation of the twin method only began seriously in the 1920s and 1930s with the early develop-ment of behavioral measurements. Foremost among these investigations was the classic study of Newman, Freeman and Holzinger (1937), which supple-mented a sample of twins reared together with 19 pairs of monozygotic twins who had been reared apart. This study, and the many other studies of family resemblance for cognitive abilities, have been reviewed frequently (e.g. Fuller and Thompson, 1978).

3.2 ANALYZING TWIN DATA: THE CLASSICAL APPROACH

The conventional approach to the analysis of twin data addresses two issues: (1) do genetic factors contribute significantly to individual differences? and (2) what are the relative contributions of genetic and environmental effects to variation within twin pairs? These two questions deal with the related statistical issues of *hypothesis testing* (Are genetic effects statistically significant?) and *estimation* (What is our best estimate of the contribution of genetic factors to variation?). Traditional analyses of twin data focus only on estimating and testing the significance of the genetic component. Our subsequent treatment will illustrate the estimation of genetic and non-genetic parameters and testing of complex hypotheses concerning the joint effects of genes and environment.

3.2.1 Analysis of variance

The starting point for the genetic interpretation of twin data is a data summary derived from the pairs of observations made on large numbers of MZ and DZ twin pairs. One convenient approach starts with the nested analysis of variance of each group of twins (see Snedecor and Cochran, 1980) which divides the total variation into that *between* pairs and *within* pairs. If there are N twin pairs and X_{ij} represents the measurement of the jth twin in the ith pair then the analysis computes the mean square between pairs (MSB) from

$$MSB = [\tfrac{1}{2}\sum_i (X_{i1} + X_{i2})^2 - 2N\bar{X}^2]/(N-1),$$

Where $\bar{X}$ is the mean of the $2N$ observations:

$$\bar{X} = \frac{1}{2N} \sum_{ij} X_{ij}$$

The mean square within pairs (MSW) measures the average variation of individual twins around the averages of the pairs to which they belong, and is computed as

$$MSW = \frac{1}{2N} \sum_i (X_{i1} - X_{i2})^2.$$

The analysis is illustrated for neuroticism scores for female MZ and DZ twins comprising part of the large National Merit Twin Study (Loehlin and

Table 3.1 Analysis of variance of neuroticism scores of female MZ and DZ twins from the National Merit Study.

Source	df	Mean square	F	r	Expected MS
Monozygotic					
Between pairs	266	41.9	2.88***	0.4858	$\sigma^2_{WMZ} + 2\sigma^2_{BMZ}$
Within pairs	267	14.5			σ^2_{WMZ}
Dizygotic					
Between pairs	175	34.2	1.58**	0.2236	$\sigma^2_{WDZ} + 2\sigma^2_{BDZ}$
Within pairs	176	21.7			σ^2_{WDZ}

** Significant at 0.01 level; ***, significant at 0.001 level.

Nichols, 1976). The data, kindly made available by Dr R.C. Nichols, are discussed more fully in Chapter 5. The mean squares of the analysis of variance are given in Table 3.1. Separate analyses are conducted for MZ and DZ twins.

The variance ratio $F = MSB/MSW$ may be used to test whether the differences between pairs are statistically significant (see Snedecor and Cochran, 1980). A significant variance ratio indicates that there are genuine differences between pairs of twins that are not explained by sampling alone and implies that members of twin pairs resemble one another more than unrelated individuals paired at random from the population. Both MZ and DZ twins are significantly correlated for their neuroticism scores.

3.2.2 Genetic and environmental components within families

The basic statistical analysis only tells whether there is significant family resemblance in personality. It does not say anything about whether these differences are due to genes or environment. Galton's original argument concentrated on the causes of differences *within* twin pairs, which, in the analysis of variance, contribute to the mean squares within pairs. The mean square within pairs of MZ twins, MSW_{MZ}, is simply a function of environmental effects within families, E_W. The mean square within DZ twin pairs is assumed to be the sum of two components. The first, E_W, reflects the same kinds of environmental influences that make MZ twins differ from one another. The second, G_W, reflects the contribution of genetic effects arising from the segregation within pairs of DZ twins of alleles having different effects on personality. The classical approach does not ask anything about the kinds of gene action contributing to G_W and assumes that genetic and environmental effects are additive and independent (i.e. that there is neither genotype × environment interaction nor genotype–environment correlation — see Chapter 4).

The observed components of variance within MZ and DZ pairs may thus be given expected values in terms of the two parameters of the simple "model":

$$\begin{array}{lll} Observed & Expected & \\ 21.7 = \hat{\sigma}^2_{WDZ} = & E_W + G_W, & (3.1) \\ 14.5 = \hat{\sigma}^2_{WMZ} = & E_W. & (3.2) \end{array}$$

Vandenberg (1966) proposed using the variance ratio (F) test

$$F = \hat{\sigma}^2_{WDZ}/\hat{\sigma}^2_{WMZ}$$

as a test of whether G_W was significantly greater than zero. For these data we obtain $F = 1.50$ for 176 and 267 df, a clearly significant value. Although the F-statistic gives a valid test of significance for the importance of genetic effects over and above the background of random environmental effects, it does not yield a sense of the relative contribution of genetic and environmental factors to individual differences. The pair of simultaneous linear equations (3.1), (3.2) above may be solved to yield estimates of E_W and G_W thus:

$$\begin{array}{l} \hat{E}_W = \hat{\sigma}^2_{WMZ} = 14.5, \\ \hat{G}_W = \hat{\sigma}^2_{WDZ} - \hat{\sigma}^2_{WMZ} = 7.2. \end{array}$$

The units of variation are determined by the units of measurement, so it is more convenient to express the contribution of genetic factors as a *proportion* of the variation caused by both genetic and environmental effects. Holzinger (1929) proposed a statistic that has been used frequently by twin researchers:

$$\hat{H} = \frac{\hat{G}_W}{\hat{G}_W + \hat{E}_W} = \frac{F - 1}{F},$$

which yields $\hat{H} = 0.33$ for our neuroticism data. This statistic is sometimes referred to as the "heritability" estimate, but in fact it does not correspond to any estimate of heritability employed by geneticists elsewhere. Holzinger's H only includes genetic and environmental differences *within* twin pairs and ignores the fact that pairs are also expected to differ from one another because they receive their genes and environments from different parents (see e.g. Jinks and Fulker, 1970).

3.2.3 The intraclass correlation and twin similarity

Just as genetic and environmental differences contribute to differences between twins within a pair, so one pair differs from another because each pair derives its genes and part of its environment from a unique set of parents. The fact that each pair has a unique set of parents makes members of a pair similar to one another and makes one pair different from another. The significant variance ratios reported for MZ and DZ twins indicate that these *between-family* effects are important, but does not directly measure their relative contribution to individual differences. A statistic that is frequently used to summarize data on the resemblance of family members is the intraclass correlation, since it does not depend on the units of measurement and often has a simple intuitive rationale. The intraclass correlation r is appropriate for grouped data that can be analyzed legitimately by the nested analysis of variance. The statistic cannot properly be used with twin data if there are significant effects of birth-order on mean or variance, nor can it be used with unlike-sex twin pairs if there are sex differences in mean or variance.

The intraclass correlation is the proportion of the total phenotypic variance among individuals that is due to factors which differ *between* twin pairs. Another way of saying the same thing is that the correlation measures the proportion of the total variance explained by factors *shared* by twin pairs. The total phenotypic variance in a twin sample is the sum of two components of variance: the component within pairs σ^2_W and the component between pairs, σ^2_B. The intraclass correlation is thus the ratio

$$r = \sigma_B^2/(\sigma_B^2 + \sigma_W^2).$$

The components of variance within MZ and DZ pairs can be obtained directly from the mean squares within pairs MSW (see above). The between-pair components, however, are not equal to the mean squares between pairs. The reason is simple, but may still need explaining. Consider the mean of the ith pair of twins, $\overline{X}_i$. The elementary statistical formula for the variance of an estimate of a mean based on N observations is σ^2/N. In our case, the twins are repeated samples from a "population" with variance σ^2_W, so the variance of a typical pair-mean is expected to be $\frac{1}{2}\sigma^2_W$. Even if no differences are expected between pairs by virtue of their arising from different parents, the average of one pair will differ from another because of these sampling effects alone. So the variance of the pair-means is expected to be $\frac{1}{2}\sigma^2_W$ even if there are no real differences between pairs. Now, if there are differences between pairs over and above the effects of sampling, the variance of the pair-means is $\frac{1}{2}\sigma^2_W + \sigma^2_B$, the extra term σ^2_B being the true variance of pair-means. If it were possible to have families of infinite size (rather than size 2, as in the case with twins) then the contribution of sampling variation would vanish and we would just be left with the true variance of family-means, σ_B^2, which would only reflect differences between pairs without any contribution from factors that create differences within pairs. Our discussion is based on the variances of pair-means. The analysis of variance generates the mean square between pairs, which is not the same thing. A little algebra shows that the mean square between pairs is twice the variance of pair-means, so we have

$$\hat{\sigma}_B^2 = \tfrac{1}{2}(\text{MSB} - \text{MSW}). \tag{3.3}$$

Substituting the mean squares for neuroticism in (3.3), we have, for MZ twins,

$$\hat{\sigma}_{BMZ}^2 = 13.7,$$

and, for DZ twins,

$$\hat{\sigma}_{BDZ}^2 = 6.8,$$

yielding intraclass correlations of 0.486 and 0.224 for MZ and DZ twins respectively.

Just as we expressed the components of variance within pairs in terms of genetic and environmental effects, so we may write expressions for the variance components between pairs as functions of hypothesized genetic and environmental variances between families thus:

$$\hat{\sigma}_{BDZ}^2 = G_B + E_B,$$
$$\hat{\sigma}_{BMZ}^2 = G_B + G_W + E_B.$$

The expectation for DZ twins requires no explanation. Some people are troubled by the inclusion of G_W in the expectation for MZ twins. The difference reflects the mechanism by which the different types of twins are formed. Both MZ and DZ twins derive their genes and some of their environment from unique pairs of parents whose genes and behavior have their own characteristics. These differences contribute to G_B and E_B. However, when DZ twins are produced, each twin is a genetically unique individual obtained by sampling alleles from the parental genotypes. This sampling process (which follows Mendelian processes of segregation and assortment) contributes to G_W in DZ twins. In the case of MZ twins, however, a single sample of parental alleles is generated at fertilization and replicated in two genetically identical individuals. Thus the segregation and assortment that create differences *within* DZ pairs adds to differences *between* MZ pairs.

The total phenotypic variance V_P for MZ or DZ twins is expected to be the sum of four components given that genes and environment act additively and independently:

$$V_P = G_W + G_B + E_W + E_B$$

The intraclass correlations for MZ and DZ twins are then expected to be

$$r_{MZ} = (\hat{G}_W + \hat{G}_B + \hat{E}_B)/\hat{V}_P \qquad (3.4)$$

and

$$r_{DZ} = (\hat{G}_W + \hat{E}_B)/\hat{V}_P. \qquad (3.5)$$

The difference between the correlations is thus an estimate of the proportion of the total variance that is due to genetic factors within families, since

$$r_{MZ} - r_{DZ} = \frac{\hat{G}_W}{\hat{V}_P} = 0.262 \qquad \text{for the neuroticism data.}$$

An alternative form of Holzinger's H, derived from correlations rather than variances, is

$$H = \frac{r_{MZ} - r_{DZ}}{1 - r_{DZ}}. \qquad (3.6)$$

For the example data the value of H is 0.34. The small difference between this value and that derived from the within-pair variances is due to the slight (but not significant) difference in total variance between MZ and DZ twins, which is the denominator in the formula for estimating the correlation coefficients. If the total variances of MZ and DZ twins differ significantly then neither estimate of H is valid, since the simple genetic model predicts that the variances of the two types of twins should be the same if they are sampled from the same population of genetic and environmental effects. This test of the model was often ignored in early analyses. A (two-tailed) F-test comparing the total variances for MZ and DZ twins gives a non-significant value ($F_{533, 351} = 1.009$), confirming that the comparison between correlations is not affected by sampling differences between the distributions of MZ and DZ twins. An alternative to Vandenberg's F for testing the contribution of genetic factors is to test whether the correlation for MZ twins is significantly greater than that for DZ twins. The correlations may be converted into normally distributed z statistics before they are compared to see if they differ significantly (see Snedecor and Cochran, 1980). Since genetic theory predicts an excess of MZ similarity, we conduct a one-tailed test and find that the correlations are significantly heterogeneous ($\chi_1^2 = 9.59$, $P < 0.001$) confirming the potential importance of genetic factors.

3.3 THE SHARED ENVIRONMENT IN THE CLASSICAL APPROACH

The traditional analysis of twin data only looks at differences within pairs. While a significant F may tell us that genetic factors are important and while

H may tell us how large these effects are relative to those of environmental differences within families, these statistics ignore many factors which may have great importance genetically and psychologically. The distinction between E_B and E_W is important theoretically as well as statistically. The "between-families" environmental component will reflect, for the most part, differences in social class, education, parental treatment, prenatal environment, etc., which are shared by members of a pair. Many of the environmental variables considered by sociologists, for example, would contribute to E_B in twins. Similarly, models for familial psychopathology that depend on the environmental transmission of disease from parent to child will create environmental differences between pairs. In contrast, many transient experiences that affect behavior, errors of measurement, accidents, and certain kinds of random life-events, are unlikely to be correlated between relatives and are therefore most likely to be reflected in E_W. These two sources of environmental influence do not exhaust all the issues that might be important in the analysis of twin resemblance. In Chapter 5 we shall consider the effects of social interaction between twins. Such effects will tend to create differences in variance between MZ and DZ twins that invalidate the untested assumptions of the classical approach. Genetic effects within families reflect only a fraction of the genetic differences contributing to variation in a population, and our estimate of G_W may include a variety of different types of genetic effects, including additive, dominant, and epistatic effects, which may be important for our understanding of the genetic polymorphism underlying personality differences (see Chapter 4 for a more detailed discussion of these issues).

Brief experimentation with the expected correlations for MZ and DZ twins reared together will show that we cannot separate G_B from E_B even though we can estimate G_W and E_W without any difficulty. Genetic and environmental effects between families are therefore confounded as long as our study is restricted to twins reared together, unless we are able to make some further assumptions about the relative magnitudes of the parameters. We shall consider how other kinds of data can be used in Chapter 6. For the time being, we note that there is no theoretical justification for any assumption about the relative magnitudes of environmental effects within and between families, because they probably involve quite different mechanisms. However, when mating is random and gene action is additive (see Chapters 4, 5 and 6) the genetic components within and between families are expected to be equal, i.e. $G_W = G_B$.

Given the assumption of random mating and additive gene action, it is possible to estimate the proportion (E_B^*) of the total variation resulting from environmental differences between families:

$$E_B^* = 2r_{DZ} - r_{MZ}. \tag{3.7}$$

No test of significance for E_B will be presented at this stage because better methods will be used for hypothesis testing when we describe our own studies. For the illustrative neuroticism data we find $E_B^* = -0.04$, suggesting a very small contribution from the family environment for these traits. We notice that the estimate of E_B^* is actually slightly negative. Clearly no *variance* can be *truly* negative. However, *estimates* of components of variance may sometimes be slightly (but not significantly) negative by chance alone. *If an estimated component of variance is significantly negative* then we should conclude that *the model is fundamentally wrong,* and have to think again.

3.4 CRITICISM OF THE TWIN METHOD

If the twin method is sound, it provides the single most powerful foundation for the preliminary detection of genetic effects on behavior. The study of nuclear families is of little use because genetic and social causes of resemblance are confounded. It is therefore scarcely surprising that critics of genetic studies of human behavior have directed much of their attention to criticisms of the twin method. The validity of the twin method depends on three main assumptions: (1) that there are only two types of twins, monozygotic and dizygotic and none of the twins presumed to be dizygotic are actually the offspring of different fathers (and so would be half-siblings genetically) or "polar-body twins" which would have an identical genetic complement derived from the mother but different paternally derived genotypes; (2) that twins are representative of the population of genotypes and environments from which they are drawn; (3) that the degree of environmental resemblance for MZ twins is no greater than that for DZ twins (i.e. the ratio E_B/E_W is the same for MZ and DZ twins). Each possible source of error is considered in turn.

3.4.1 Additional types of twins?

There are occasional reports of non-identical twins conceived from different fathers (e.g. Phelan *et al.*, 1982) and "polar-body twins" (e.g. Bieber *et al.*, 1981). Such cases would have to be fairly common before they constituted a significant challenge to the twin method. When large numbers of blood groups are studied in twins and their parents it is possible to compute the likelihood that twins are derived from the same father rather than different fathers. Such studies have provided little indication that a significant proportion of non-identical twin pairs are other than truly dizygotic (see e.g. Elston and Bocklage, 1978).

3.4.2 Are twins typical?

This question has to be answered in two parts. First, we must determine whether the parameters of twin development are comparable with those of non-twins. Then we have to ask whether such differences as are found actually have any impact on the particular traits being studied. For the first part, there is unequivocal evidence that the environmental circumstances of twins before, during and after birth are not typical (see Bulmer, 1970, Chapter 3). Twins have shorter gestation periods than singletons. Even when allowance is made for this fact, they have lower birthweights. The stillbirth rate in twins is twice as high as in singletons and their early postnatal mortality is higher. Twin pregnancies and deliveries show more frequent complications. Do such differences have any lasting effects? Twins are shorter, on average, which could be a long-term effect of retarded fetal growth. Large studies (e.g. Record *et al.*, 1970) have shown a slight but significant deficit in twins' verbal ability scores. The fact that this difference is not demonstrated by the surviving twin whose cotwin died at birth provides some support for Bulmer's contention that "the increased frequency of gross mental retardation in twins is due to their retarded fetal growth, but that the slightly lower intelligence of normal twins is due to differences in their upbringing". Cognitive differences in juvenile twins have also been documented by Hay and O'Brien (1981) and attributed to the unique patterns of social interaction between twins.

3.4.3 Correlations in the environments of twins

Although there are commonly average differences between measurements made on twins and non-twins, there is typically little difference in variance. This finding suggests that the differences that occur on average might therefore not interfere too seriously with the generalizations about the causes of *variation* from the study of twins. The most serious challenge to the twin method is therefore the claim that the environmental correlation of MZ twins is greater than that for DZ twins. The best documentation of this claim is found in the large study of Loehlin and Nichols (1976), who showed that MZ twins are more often dressed alike, have the same friends, share the same room and eat the same food than DZ twins. The large Australian Twin Study of Martin and Jardine confirmed that adult MZ twins have more frequent contact than DZ twins (see Chapter 11). These studies establish a *prima facie* case for doubting the validity of the twin method as a basic tool for the analysis of behavior.

These effects, however, seldom translate into correlates of measured behavior. Loehlin and Nichols, for example, were unable to detect any significant correlation between indices of "environmental" difference within twin pairs and differences in test and rating scores for a wide range of cognitive and personality variables. In spite of correlating six indices of environmental difference with 41 behavioral ratings and measures in MZ and DZ twins (Loehlin and Nichols, 1976, Tables 5.4–5.7), few of the correlations were greater than 0.1, and only 6/492 correlations exceeded 0.2. All six were found in DZ twins.

Similar lack of success attended the attempt by Kendler *et al.* (1986) to predict differences in anxiety and depression scores of twins in the Australian Study as a function of the amount of contact between members of a pair. Although MZ twins communicated more frequently than DZs, there was not the slightest suggestion that twins who communicated with one another more frequently were significantly more alike in their anxiety and depression scores (see Chapter 11).

These studies represent attempts to address the issue by measuring the environment, and have been unable to detect any direct effects of the environment on behavior by correlating behavioral measures with indices of the environment. Three other important strategies have been tried, also without success from the standpoint of identifying clear environmental effects. Scarr (1969a) identified twins whose parents were mistaken about zygosity. If the excess correlation of MZ twins were a function of parental treatment based on beliefs about zygosity then we should expect that DZ children who are believed to be MZ by their parents would correlate more highly than DZ twins whose parents are correct about zygosity. Unfortunately, the samples are small, but in the study the only significant differences between twin correlations were those associated with true zygosity.

Lytton (1977) reported a study of 2½-year-old male twins in which the behavior of parents towards the twins was divided operationally into that initiated by parents and that resulting from a parental response to behavior in their children. He found that MZ twins correlated more highly than DZ twins for parental responses but were no more correlated than DZ twins for "parent-initiated" behavior.

How are these findings to be interpreted? There is a major group of psychosocial measurements gathered on the assumption that they represent indices of the environment. With the exception of measurements of "parent-initiated behavior", these indices are significantly more correlated in MZ twins than in DZs. However, there is no evidence that twin differences for the environmental indices predict differences in psychometric measures. One interpretation is that we have still to measure the right environment for the development of the traits in question, although the current measures encompass many of the domains that were traditionally assumed to be important. The second interpretation is more challenging, and consistent with all the data so far. The alternative model suggests that these so-called "environmental indices" are themselves influenced by genetic factors — that the environments that individuals experience and create are actually functions of genetic differences. Under such a model, the individual has inherent biases of responding which, in part, generates his/her own environment (see Chapter 17).

3.5 TWIN STUDIES OF PERSONALITY BEFORE 1976

Prior to 1976 there were many twin studies of personality. Typically these were small and used tests of unknown reliability and validity. Often the scales are factorially complex and reflect more than one underlying personality dimension. For convenience, however, each of the studies is reported as bearing on the genetic analysis of either "extraversion", "neuroticism" or "psychoticism" depending on the description of the scale (see Chapter 2). We appreciate that such classification may be attempting to force some of the studies into a Procrustean bed.

The sources of data are identified in Table 3.2 with an acronym designat-

Table 3.2 Sources of twin data prior to 1976.

Study	Source	Test
1	Carter (1935)	Bernreuter
2	Newman *et al.* (1937)	Woodworth–Marquis
3	Eysenck and Prell (1951)	Various
4	Cattell (1955)	HSPQ
5	Shields (1962)	Like MPI
6	Vandenberg (1962)	TTS
7	Vandenberg (1962)	Cattell HSPQ
8	Gottesman (1963)	Cattell HSPQ
9	Gottesman (1963)	MMPI
10	Wilde (1964)	ABQ
11	Gottesman (1965)	MMPI
12	Partanen *et al.* (1966)	—
13	Vandenberg (1966)	Stern
14	Reznikoff and Honeyman (1967)	MMPI
15	Owen and Sines (1970)	MCPS
16	Young *et al.* (1971)	PEN
17	Claridge *et al.* (1973)	EPI/16PF
18	Eysenck (1956a)	Various
19	Gottesman (1966)	CPI
20	Scarr (1969b)	Various
21	Nichols (1969)	CPI/VPI
22	Vandenberg (1966)	Myers–Briggs/Comrey

Table 3.3 Twin studies relevant to "neuroticism".

Study[a]	N_{MZ}	N_{DZ}	r_{MZ}	r_{DZ}	H	E_B	Trait
1	55	44	0.63**	0.32*	0.46*	0.01	Neuroticism
2	50	50	0.56**	0.37**	0.30	0.18	Neurotic tendencies
3	26	26	0.85**	0.22	0.81**	−0.41	Neuroticism
4	52	32	—	—	0.38*	—	General neuroticism
4	52	32	—	—	0.36*	—	Nervous tension
5	43	25	0.38**	0.11	0.30	−0.16	Neuroticism
6	45	35	0.10	0.08	0.31	0.06	Stable
7	45	36	—	—	0.69**	—	General neuroticism
7	45	36	—	—	0.52*	—	Nervous tension
8	34	34	0.28	0.38*	0.03	0.48	General neuroticism
8	34	34	0.27	0.32*	—	0.37	Nervous tension
9	34	34	0.47**	0.07	0.45*	−0.33	Depression
9	34	34	0.55**	0.20	0.37	−0.15	Psychastenia
9	34	34	0.47**	0.41**	0.13	0.35	Hysteria
9	34	34	0.39**	0.21**	0.16	0.03	Hypochondriasis
10	88	42	0.53**	0.11	0.17**	−0.31	Psychoneurotic
10	88	42	0.67**	0.34*	0.50	0.01	Psychosomatic
11	82	68	—	—	0.45*	—	Depression
11	82	68	—	—	0.31	—	Psychasthenia
11	82	68	—	—	0.30	—	Hysteria
11	82	68	—	—	0.01	—	Hypochondriasis
12	157	189	0.28**	0.21**	0.06	0.14	Neuroticism
13	50	38	—	—	0.60**	—	Emotional expression
14	18	16	—	—	0.38	—	Depression
14	18	16	—	—	—	—	Psychasthenia
14	18	16	—	—	0.63*	—	Hysteria
14	18	16	—	—	0.57*	—	Hypochondriasis
15	18	24	0.33*	0.26	0.09	0.19	Sleep disturbance
16	17	15	0.61**	0.28	0.48	−0.05	Neuroticism
17	40	45	0.37**	0.23*	0.18	0.09	Neuroticism
17	39	44	0.56**	0.33**	0.34	0.10	Anxiety
17	39	44	0.36*	0.06	0.32	−0.24	Neuroticism
17	39	44	0.37**	0.15	0.26	−0.07	General neuroticism
17	39	44	0.38**	0.06	0.34	−0.26	Free-floating anxiety
17	39	44	0.20	0.10	—	0.00	Nervous tension

[a] Study no. corresponds to source given in Table 3.2.

H computed by original author using the formula in text. Where the author gave only F, we give H as $(F-1)/F$.

* significant at 0.05 level; ** significant at 0.01 level.

Table 3.4 Twin studies related to "extraversion".

Study[a]	N_{MZ}	N_{DZ}	r_{MZ}	r_{DZ}	H	E_B	Trait
1	55	44	0.50**	0.40**	0.17	0.30	Introversion
1	55	44	0.71**	0.34**	0.56	−0.03	Dominance
1	55	44	0.57**	0.41**	0.27	0.25	Sociability
4	52	32	—	—	0.32	—	Surgency
4	52	32	—	—	0.25	—	Adventurous cyclothymia
4	52	32	—	—	0.07	—	Cyclothymia
4	52	32	—	—	0.25	—	Impatient dominance
18	26	26	0.50**	−0.33	0.62**	−0.91	Extraversion
5	43	25	0.42**	−0.17	0.50*	−0.76	Extraversion
6	45	35	0.44**	−0.12	0.46*	−0.68	Impulsive
6	45	35	0.50**	−0.06	0.47*	−0.62	Sociable
6	45	35	0.55**	0.28	0.06	0.01	Reflective
6	45	35	0.55**	−0.06	0.67**	−0.67	Active
6	45	35	0.58**	0.00	0.59**	−0.58	Vigorous
6	45	35	0.61**	0.23	0.20	−0.15	Dominant
7	45	36	—	—	0.31	—	Surgency
7	45	36	—	—	—	—	Adventurous cyclothymia
7	45	36	—	—	0.23	—	Cyclothymia
7	45	36	—	—	—	—	Impatient dominance
8	34	34	0.47**	0.12	0.56**	−0.23	Surgency
8	34	34	0.38	0.20	0.38	0.02	Adventurous cyclothymia
8	34	34	0.19	1.27	0.01	0.35	Cyclothymia
8	34	34	0.21	0.47**	—	0.73	Impatient dominance
9	34	34	0.55**	0.08	0.71**	−0.39	Social introversion
9	34	34	0.24	0.07	0.24	−0.10	Hypomania
10	88	42	0.37**	0.35*	0.03	0.33	Introversion
11	82	68	—	—	0.33*	—	Social introversion
11	82	68	—	—	0.13	—	Hypomania
14	18	16	—	—	0.50	—	Social introversion
14	18	16	—	—	0.39	—	Hypomania
22	40	27	—	—	0.46*	—	Introversion
22	111	90	—	—	0.48**	—	Shyness
19	79	68	—	—	0.49**	—	Sociability
19	79	68	—	—	0.46**	—	Self-acceptance
19	79	68	—	—	0.35*	—	Social presence
19	79	68	—	—	0.49**	—	Dominance
12	157	189	0.51**	0.26**	0.41**	0.02	Sociability
20	24	28	0.83**	0.56**	0.61**	0.29	Need for affiliation
20	24	28	0.86**	0.36	0.78**	−0.14	Friendliness
20	24	28	0.88**	0.28	0.83**	−0.32	Social apprehension
20	24	28	0.93**	0.82**	0.61**	0.71	Likeableness (rating)
21	207	126	0.56**	0.19*	0.46**	−0.18	Extraversion (males)
21	291	193	0.59**	0.39**	0.32*	0.19	Extraversion (females)
15	18	24	0.34*	0.16	0.58*	−0.02	Inhibition–withdrawal
15	18	24	0.45*	−0.38	0.54*	−1.21	Activity level
16	17	15	0.47*	0.07	0.43	−0.33	Extraversion
17	40	45	0.34**	0.29**	0.00	0.24	Extraversion

Table 3.4   *contd*

Study[a]	N_{MZ}	N_{DZ}	r_{MZ}	r_{DZ}	H	E_B	Trait
17	40	45	0.67**	0.25*	0.56**	−0.17	Sociability
17	40	45	0.24*	−0.03	0.26*	−0.30	Impulsivity
17	39	44	0.43**	0.08	0.38	−0.27	Extraversion (factor)
17	39	44	0.49**	0.21*	0.38	−0.27	Cyclothymia
17	39	44	0.27**	0.30**	0.01	0.33	Dominance
17	39	44	0.56**	0.47**	0.00	0.38	Surgency
17	39	44	0.58**	0.30**	0.40	0.02	Adventurous cyclothymia

[a] Study no. corresponds to source given in Table 3.2.
H computed by original author using the formula in text. Where the author gave only F, we give H as $(F-1)/F$.
* significant at 0.05 level; ** significant at 0.01 level.

ing the tests used. The data for the three main categories of test are given in Tables 3.3–3.5. Wherever possible, we give correlations, sample sizes, Holzinger's H values, and Vandenberg's F with its associated significance level. Our estimate of the proportion of variance attributed to E_B is also appended. Frequently the results were only tabulated in terms of F ratios and heritabilities, so we are unable to reconstruct the original correlations. Where correlations were given without H values we calculated Holzinger's H based on the correlations. Where F ratios were given alone we calculated H from $(F-1)/F$.

Most of the correlations are based on fewer than 100 twin pairs and many based on fewer than 50. A correlation based on 100 pairs has a standard error of approximately 0.1, and one based on 50 pairs has a standard error of about 0.15. It is therefore scarcely surprising that the correlations fluctuate widely. Small sample sizes notwithstanding, almost all the F ratios exceed unity, though many are not significant. As a consequence, most of the estimates of H are also positive.

The tables give the impression of great heterogeneity. Some studies confirm a genetic component to personality differences, others do not. Some give large estimates of E_B^*, others do not. With results like these, it is asking a lot to find patterns in the data. There are likely to be two main reasons for the variation. One is the sampling variation attached to estimates obtained from small studies. The other is genuine heterogeneity between findings for different populations and tests.

Table 3.5 Twin studies related to "psychoticism".

Study[a]	N_{MZ}	N_{DZ}	r_{MZ}	r_{DZ}	H	E_B	Trait
1	55	44	0.44**	−0.14	0.51**	−0.72	Self-sufficiency
4	52	32	—	—	0.32	—	Tender-minded
6	45	36	—	—	0.03	—	Tender-minded
8	34	34	0.55**	0.47**	0.06	0.39	Tender-minded
9	34	34	0.57**	0.18	0.50*	−0.20	Psychopathic deviate
9	34	34	0.59**	0.19	0.42	−0.21	Schizophrenia
9	34	34	0.44**	0.18	0.05	−0.08	Paranoia
8	34	34	0.60**	0.05	0.56**	−0.30	Self-sufficiency
11	82	68	—	—	0.39*	—	Psychopathic deviate
11	82	68	—	—	0.33*	—	Schizophrenia
11	82	68	—	—	0.38*	—	Paranoia
14	18	16	—	—	0.35*	—	Psychopathic deviate
14	18	16	—	—	0.39	—	Schizophrenia
14	18	16	—	—	0.44	—	Paranoia
13	50	38	—	—	—	—	Dependency needs
22	111	90	—	—	0.13	—	Dependence
22	111	90	—	—	0.22	—	Empathy
22	111	90	—	—	—	—	Hostility
19	79	68	—	—	0.32*	—	Socialization
12	157	189	0.25**	0.16*	0.18	0.07	Aggressiveness
21	207	126	0.53**	0.15	0.44**	−0.23	Socialization (males)
21	291	193	0.55**	0.49**	0.14	0.43	Socialization (females)
21	220	137	0.37**	0.14	0.27*	−0.09	Aggression (males)
21	296	197	0.43**	0.31**	0.18	0.19	Aggression (females)
15	18	24	0.45*	−0.10	0.16	−0.65	Aggressivity
17	39	44	0.68**	0.25	0.57**	−0.18	Tender-minded
17	39	44	0.34**	0.05	0.31*	−0.24	Paranoid trend
17	39	44	0.39**	0.01	0.38	−0.37	Self-sufficiency

[a] Study no. corresponds to source given in Table 3.2.
H computed by original author using the formula in text. Where the author gave only F, we give H as $(F-1)/F$.
* significant at 0.05 level; ** significant at 0.01 level.

3.6 POWER AND SAMPLE SIZE

The issue of sample size has been addressed repeatedly in theoretical studies (Eaves, 1972; Eaves and Jinks, 1972; Klein *et al.*, 1973). Martin *et al.* (1978) conducted detailed power calculations to find out how large samples have to be to obtain reliable results in studies of twins reared together. The answer will depend on the hypothesis to be tested, the test statistic chosen, the relative proportions of MZ and DZ twins, and the population values of the parameters. The study assumed that a model-fitting approach would be used to

Table 3.6   Total number of pairs required for 95% power of rejection of false hypotheses at 5% level when "true" model is (a) $E_W E_B$, (b) $E_W V_A$.

(a) True model $E_W E_B$

True model	False model									
	E_W p_{MZ}					$E_W V_A$ p_{MZ}				
E_W E_B	0.1	0.3	0.5	0.7	0.9	0.1	0.3	0.5	0.7	0.9
0.1 0.9	22	22	22	22	22	33	45	67	119	385
0.3 0.7	36	36	36	36	36	73	115	164	278	854
0.5 0.5	69	69	69	69	69	298	325	430	696	2 055
0.7 0.3	191	191	191	191	191	1 491	1 229	1 485	2 289	6 534
0.9 0.1	1 718	1 718	1 718	1 718	1 718	20 904	13 508	15 119	22 534	62 948

(b) True model $E_W V_A$

True model	False model									
	E_W					$E_W E_B$				
E_W V_A										
0.1 0.9	66	45	34	28	23	388	118	63	40	36
0.3 0.7	108	74	57	46	38	886	313	208	186	303
0.5 0.5	212	145	110	89	75	2 181	852	640	670	1 344
0.7 0.3	588	402	306	247	207	7 026	2 914	2 356	2 683	5 955
0.9 0.1	5 284	3 615	2 748	2 216	1 857	68 016	28 982	24 232	28 784	66 800

decide between a number of hypotheses about the genetic and environmental causes of variation (see Chapters 4 and 5). This is the method that makes the most efficient use of the data. To conduct the power study, they assumed that variation in a trait is caused by a particular combination of factors, and determined how many twin pairs would be needed to reject incorrect models for twin resemblance.

Table 3.6 shows what happens when the causes of variation in a population are due entirely to environmental effects (E_B and E_W) and an investigator attempts to fit models that ignore the effects of the environment between families. The critical question here is "How many twin pairs would be needed to be 95% certain that we would know we had fitted a 'wrong' model when we ignored the family environment?" The number of twin pairs required is tabulated for various proportions of the two environmental components, a number of values for the proportion p_{MZ} of MZ pairs in the sample, and for two alternative "wrong" models. The second wrong model ignores the shared environment and assumes that individual differences are caused entirely by within-family environmental effects (E_W) and additive genetic effects (V_A, see Chapter 4). Suppose, for example, that we measure a trait in a population in which 30% of the variation is due to family environmental effects and we construct a sample that comprises 30% MZ pairs (p_{MZ} = 0.3). If we studied 1229 twin pairs then we would be 95% certain of showing that the "V_A, E_W" model could not explain the data. This figure assumes that the twins are all sampled from the same population and measured with the same test.

Table 3.7 considers the more complex case of sampling a population in which all three factors contribute to variation: the environment within and between families and additive genetic effects. We asked how many pairs would be needed to reject three false hypotheses about the causes of individual differences. The first false model ignored both genetic and family environmental effects (the E_W model). The second (E_W, E_B) model ignored genetic effects (i.e. it assumed that all twin resemblance was due to the shared environment) and the third (E_W, V_A) ignored the shared environment and assumed that all family resemblance was genetic.

Suppose that we sample a population in which the proportions of variance explained by E_W, E_B and V_A are 0.3, 0.3 and 0.4 respectively and had 30% MZ twin pairs in the sample. Table 3.7 shows that we would be 95% sure of rejecting the first model (the model that assumes no family resemblance) with only 54 twin pairs. This is a relatively trivial hypothesis, however. It would require 660 pairs to be 95% sure of correctly rejecting a purely environmental hypothesis and 718 pairs to reject a model that assumed all differences between pairs to be genetic. The precise results depend on the relative proportions of the three components of variance. These calculations, however, show that very large studies are needed if we are to have much confidence in our ability to resolve even comparatively simple hypotheses about the causes of family resemblance. Indeed, the entire sample size available from the twin studies on multiple variables prior to 1976 is only barely enough to meet the demands of hypothesis testing. It is therefore not surprising that the results of these early studies are very variable even if we were to ignore sources of real heterogeneity between them.

3.7 EARLY TWIN STUDIES: A "META-ANALYSIS"

Without access to all the raw data, it is impossible to conduct any statistical analysis of these data that is completely satisfying. Many of the correlations are based on subtest scores computed from the same data, so that the correlations for different primary factors are not independent. As an approximate guide to the consistency of these studies, we have nevertheless conducted an analysis of the heterogeneity of the correlations for each broad psychological

Table 3.7 Total number of pairs required for 95% power of rejection of false hypotheses at 5% level when the "true" model is $E_W E_B V_A$.

True model			False model E_W p_{MZ}					False model $E_W E_B$ p_{MZ}					False model $E_W V_A$ p_{MZ}				
E_W	E_B	V_A	0.1	0.3	0.5	0.7	0.9	0.1	0.3	0.5	0.7	0.9	0.1	0.3	0.5	0.7	0.9
0.1	0.1	0.8	57	42	33	27	23	417	126	68	44	42	4 126	5 086	6 943	11 315	33 124
	0.3	0.6	43	35	30	26	23	497	152	85	59	65	417	521	725	1 207	3 625
	0.5	0.4	33	30	27	25	23	656	208	123	94	124	134	173	245	417	1 285
	0.7	0.2	27	25	24	23	22	1 226	422	277	248	414	61	81	117	204	646
0.3	0.1	0.6	89	67	53	44	38	1 070	381	257	235	400	6 835	6 899	9 031	14 553	42 733
	0.3	0.4	63	54	47	42	37	1 798	660	466	455	848	645	718	966	1 583	4 715
	0.5	0.2	47	44	41	39	36	4 931	1 915	1 449	1 542	3 191	200	242	335	558	1 685
0.5	0.1	0.4	162	125	101	85	74	3 136	1 234	940	1 002	2 065	11 204	9 230	11 458	17 929	51 800
	0.3	0.2	102	92	84	77	72	10 167	4 124	3 268	3 567	7 990	997	963	1 233	1 963	5 732
0.7	0.1	0.2	382	313	265	229	202	15 017	6 265	5 110	5 892	13 278	16 512	11 720	13 720	20 814	58 596

category on the assumption that all the correlations are independent. Such a "cavalier" approach to the data is only defended by the prospect of "better" analyses of new data in later chapters.

For each dimension of personality we only include studies for which correlations were available. Following Bartlett (see Snedecor and Cochran, 1980) we converted each correlation coefficient to a normally distributed "z" value, with its associated variance. A pooled weighted average z may then be obtained, and the variance of the observed zs around their average value tested for heterogeneity by chi-square. Within a given personality dimension, the analysis was performed separately for MZ and DZ twins. Because of the large number of correlations and the relatively small sample sizes of each, an iterative procedure was followed in which the small bias in each expected correlation is allowed for in the analysis.

The results of these analyses are summarized in Tables 3.8 and 3.9. With the exception of the "neuroticism" data on DZ twins, all the sets of correlations are significantly heterogeneous. The standard errors of the pooled z values, however, are calculated from a statistical model that assumes that all the correlations are sampled from a single population with a single correlation. The presence of significant heterogeneity implies that tests differ significantly (for example in their reliability) or that twin populations differ significantly (for example in the relative contributions of genetic and environmental effects).

The heterogeneity chi-square reflects the sampling of tests and twin populations as well as the sampling of individual pairs within studies that would occur even if the same test were used and the same twin population sampled in every case. The heterogeneity chi-square is the ratio of the sum of squares of differences in z values between samples to the variance in zs predicted from statistical theory alone. The standard error appropriate to the

Table 3.8 Summary statistics from analysis of twin studies prior to 1976.

| | Personality dimension | | | | | |
| | Neuroticism | | Extraversion | | Psychoticism | |
Statistic	MZ	DZ	MZ	DZ	MZ	DZ
Number of rs	22	22	36	36	15	15
Pooled z	0.476	0.220	0.591	0.245	0.497	0.234
s.e. (z)	0.032	0.033	0.024	0.025	0.026	0.029
χ^2	48.76	11.01	90.45	79.82	25.89	36.31
df	21	21	35	35	14	14
Corrected s.e.	0.049	0.024	0.038	0.038	0.035	0.047
Pooled r	0.443	0.216	0.531	0.240	0.460	0.229
Total df	960	895	1807	1569	1486	1167

Table 3.9   Summary genetic analysis of pooled correlations.

Statistic	Personality dimension		
	Neuroticism	Extraversion	Psychoticism
$z_{MZ} - z_{DZ}$	0.256	0.346	0.263
s.e. (difference)	0.055	0.054	0.059
df	42	70	28
t	4.65	6.41	4.46
P			
H	0.290	0.383	0.300
E_B	−0.011	−0.051	−0.002

z based on pooling heterogeneous *independent* samples (strictly speaking, our samples are not independent) is $\sigma_h = \sigma_z(\chi^2/d)^{1/2}$ where σ_z is the theoretical standard error and χ^2 is the heterogeneity chi-square based on d df. Except in the case of the neuroticism data from the DZ samples, these corrected standard errors are all greater than the theoretical errors. Had the correlations been homogeneous, we could have used the pooled error variance (Table 3.8) based on the total numbers of twin *pairs* in the global sample — a very large number. However, we may now only compare the average value of the correlation with the variance based on the differences *between* studies. The corrected standard error of the pooled z thus has the same df as the heterogeneity chi-square (see Table 3.8).

The pooled correlation for MZ twins is higher than that for DZ twins, averaged over all tests. Is this difference significant? A simple test compares the average value of z for MZ twins with that for DZ twins, using the pooled error corrected for heterogeneity. In Table 3.9 we give the difference between the pooled zs for the three personality dimensions, together with its standard error obtained by adding the sampling variances of the pooled zs, i.e.

$$\sigma_d = (\sigma^2_{hmz} + \sigma^2_{hdz})^{1/2}.$$

Using a one-tailed t-test assuming independent samples, we find that the average correlation for MZ twins significantly exceeds that for DZ twins for all three groups of personality test variables, *even when we make allowance for the heterogeneity of different studies with respect to the sampling of tests and twin populations*. More refined analyses cannot be attempted without the original data or more appropriate summary statistics.

The pooled correlations for MZ and DZ twins may be substituted in the formula (3.6) for Holzinger's H to yield an average estimate of the contribution of genetic factors to variation *within* twin pairs. Similarly, we can use

(3.7) to give an estimate of the shared environmental contribution, given that mating is random and gene effects are additive.

The results for measures grouped as indices of "neuroticism" and those that we have assumed to be related to "psychoticism" are very similar. Approximately 30% of the variation within families could be attributable to the segregation of genetic differences within families. In both cases the estimated contribution of E_B^* is so small as to be negligible. Although the values are slightly negative, they are unlikely to differ significantly from zero, bearing in mind that the standard errors of the original pooled correlations are around 0.03.

Holzinger's H is larger for the extraversion measures than for scales loading on the other two dimensions because the difference between MZ and DZ correlations is greater for the extraversion measures. Although we do not provide a test of significance, it is worth noting that the standard error of a difference between typical pooled heterogeneous zs in Table 3.8 is around 0.055. The difference between the zs of MZ twins for the "neuroticism" and "psychoticism" scales is about 0.02, clearly not significant, whereas the MZ correlation for extraversion exceeds those for the other two factors by about 0.10. The excess resemblance for extraversion in DZ twins compared with the other two factors is rather less (about 0.02).

The estimate of E_B^* for extraversion is -0.051 from the pooled data, less than zero, scarcely significant if at all, but more negative than the values for neuroticism and psychoticism. If such a finding turns out to be generally true, it would betoken a significant difference in the basis of variation in extraversion compared with the other personality dimensions. A negative estimate of any component of *variance*, if it is statistically significant, points to failure of the model under which estimates of genetic and environmental components have been computed. Among the factors that might lead to such an effect are non-additive genetic effects, competitive social interactions based on genetic differences, and greater environmental correlation of MZ twins. These issues will be examined in more detail in Chapter 5.

3.8 SUMMARY

The existing literature on personality provides a *prima facie* case for the contribution of familial effects to variation in personality, but the low DZ correlation (about 0.25 in the studies reviewed here) suggests that about 75% of the variation in typical personality differences is due to effects for which individuals differ *even when they share the same parents*.

Several problems have emerged, which will be addressed in subsequent chapters.

(1) Different studies have used different measures. Although our "meta-analysis" suggests that there are some significant general trends over and above the heterogeneity of individual studies, differences between populations have been confounded with those between measures. The problem is compounded by the fact that measures have often been taken "off the shelf" without being related to any testable theory. Our subsequent analyses consider a common set of variables in a number of populations to generate the elements of a model against which other studies may be evaluated.

(2) Sample sizes have been small. The correlations summarized in Tables 3.8 and 3.9 involve between 900 and 1800 pairs overall, although these numbers are inflated by the fact that a number of different scores are included from the same twin pairs. Such small numbers make it impossible to test any but the most trivial hypothesis, and certainly do not allow us to test for many of the more subtle effects that are of great biological and psychological interest. Most of the studies we consider in the following chapters are much larger. Even then, they are tantalizingly small for some purposes.

(3) The models underlying the analyses of these data have either not been specified or have simply been "dull". The emphasis has been on detecting genetic variation and estimating "heritability" at the expense of testing alternative hypotheses. We shall identify areas where the simple model of independent and additive effects of genes and environment breaks down. Among issues to be examined are the additivity of gene action, the effects of cultural inheritance, the causes and consequences of assortative mating, developmental changes in gene expression, and sibling interaction.

(4) Statistical methods have been inappropriate or inefficient. Three areas in which better statistical methods are helpful are: (i) providing tests that can reject a "wrong" model; (ii) making best use of the data (i.e. giving the most powerful tests and smallest standard errors for parameter estimates); and (iii) testing more complex hypotheses than the presence or absence of a genetic component.

(5) There has been very little attempt at replication across samples from the same population or from different populations. Few studies of family resemblance in behavior have tried to use a consistent set of measures in a number of populations, or tried to evaluate the robustness of findings within a population over different measures of the same psychological construct. Although we cannot present replications for all our findings, we describe many studies for which results have now been replicated.

(6) Although twins are a convenient starting place for behavior–genetic studies, any model for individual differences that makes sense of twin data alone, or only adoption data, or nuclear-family data, has limited value. In Chapter 6 we shall consider how far the models that work well with twin data can encompass the results for other types of kinship, including the parents and children of twins, extended kinships, spouses, the spouses of twins, and some adoption data.

Chapter 4

Introduction to Model Fitting

4.1 WHAT'S WRONG WITH THE OLD WAY?

The classical methods used in Chapter 3 are simple and require only the flimsiest intuitive grasp of how genes and environment affect the correlations between relatives. They have been used for almost half a century. They have established a strong case for the contribution of genetic differences to personality. There are four main reasons for needing a better approach.

(1) The statistical method does not specify or test any explicit model for individual differences. Estimates of genetic and environmental parameters will be biased if the model is wrong. The classical model, for example, will give the wrong answer if there are social interactions between twins (see Chapters 5 and 6). A method is needed that tests the assumptions behind a particular model.

(2) The method only works with twins. Most genetic and environmental effects of interest cannot be estimated efficiently and without bias simply by comparing two variances or correlations. The method used in the classical twin studies will therefore not generalize to estimating multiple parameters from complex datasets and testing compound hypotheses. We need a method that can also handle adoptions, extended families and other kinds of relationships (see e.g. Chapter 6).

(3) The classical approach only considers genetic causes of family resemblance. A more flexible method must allow for the effects of social interaction and the shared environment.

(4) The parameters of the classical model for twin data (G_W, E_W etc.) are based only on intuition and do not reflect any more specific theory of genetic causation or non-genetic transmission. If this approach were to be extended to other relationships then new parameters would have to be invented to explain every correlation between relatives.

One of the main goals of science, the search for simplicity in the midst of apparent complexity, would be defeated.

In this chapter we outline and illustrate the modeling approach to the study of family resemblance that will feature in almost everything that follows. The method allows us to test complex hypotheses about genetic *and* non-genetic causation by using many different types of relationship simultaneously and gives great flexibility to the investigator to express alternative theories of causation in a form that allows the derivation of their consequences for the correlation between relatives.

A method that meets these criteria is inherently self-correcting. If a complex unique explanation is required for every data point then a simple model will fail. If a genetic model is too simple for the data then it will be shown to be false. No method can guarantee the truth, but it can exclude theories that are manifestly wrong.

Abstract ideas about causation are crystalized in a model for the statistics that can be derived from data on various sets of relatives. The "model" represents the bridge between theory and data, by translating the "semantic" components of a theory (Torgersen, 1958) into "syntactic" definitions so that the rules of logical and mathematical inference can be used to deduce new predictions about the results of empirical studies. Without such a quantitative model, it is impossible to know, for example, what parent–offspring correlation to expect from a knowledge of the correlations between twins. The model forces us to look more closely at the data by making us expect particular quantitative patterns. If these patterns occur then we may conclude that our model receives some support; if they clearly do not then our model is obviously wrong and some better alternative must be found. Figure 4.1 summarizes the place of the model in diagramatic form.

It is convenient to distinguish two important parts of the modeling process: model *building* and model *fitting*. The stage of model building consists in deciding how the causes of variation can be expressed in a mathematical form. The stage of model fitting consists of estimating the parameters of a model and deciding whether it fits the actual data. Each is considered separately.

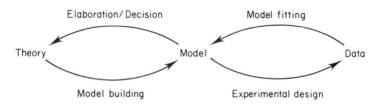

Figure 4.1   The place of the model in the analysis of individual differences.

4.2 MODEL BUILDING

The way in which models are developed will become clearer by looking at the examples in the following chapters. However, we do not begin building models for personality differences in a vacuum. We are guided by the cumulative experience of quantitative and behavioral genetics over the last eighty years. Many of the most informative studies have been conducted in plants and animals, which are far more amenable to genetic and environmental manipulation than man. This body of research suggests some of the broad features that models for human differences may encounter.

4.2.1 Genotype and phenotype

The first basic distinction that we need to make is that made originally by Johannsen (1909) between *genotype* and *phenotype*. He observed that certain kinds of differences were transmissible between generations and could be modified by selection within a population derived by crossing and recrossing pure breeding-strains. On the other hand, even though pure breeding-strains were not uniform for many characteristics, such within-strain differences could not be transmitted to subsequent generations or modified by selection. The discrimination between transmissible differences that were available to artificial selection and differences that were neither transmissible nor selectable led to the distinction between those characteristics of the organism that were expressed and measurable (the "phenotype") and those that influenced the phenotype but were capable of alteration by selective breeding (the "genotype").

This basic idea can be represented by a (linear) model in which the phenotype of the *i*th individual (P_i), expressed as a deviation from the average value of the population, is the sum of a "genotypic effect" G_i and an environmental effect E_i. Thus

$$P_i = G_i + E_i \tag{4.1}$$

Since we can only measure the phenotype of an individual directly, there is an infinite number of genetic and environmental effects that can satisfy the equation for each individual, so that neither the genetic nor the environmental effects can be identified statistically for any individual and there would be as many equations like (4.1) as there are individuals in a sample. However, *if genetic and environmental effects are independent* (see Section 4.2.3 below) the phenotypic *variance* V_P is the sum of the variances of the genetic effects G and the environmental effects E:

$$V_P = G + E. \tag{4.2}$$

We now have only one equation, but still two unknown parameters, *G* and *E*, for which there remains an infinite number of solutions unless we have data on the covariances between relatives to which *G* and *E* make different contributions (see Section 4.2.5 below).

4.2.2 Genotype × environment interaction

The elementary model above makes a number of very strong assumptions which may not generally be true. Among the foremost of these is the assumption that there is no *"genotype × environment interaction"* (G × E). Although G × E is often interpreted in purely statistical terms, it is more important to understand what it means in genetic terms. The above model assumes that genes only contribute to the "average" expression of an individual's trait. One way of visualizing the idea is to think of a number of genotypes replicated many times (as is possible with many other species) and raised in a number of different levels of some environmental treatment (Figure 4.2). Each line represents the response of a given genotype to changing levels of environment. In Figure 4.2 the genotypes differ in trait value on average. These are the "genotypic" effects, represented by "G" in the model. Furthermore, there is an average trend of trait value with level of environment that is the same for all genotypes. These environmental differences correspond to the "E"s in the model. The important point, however, is that all the genotypes have identical response curves relating trait value to environment. The lines are parallel, indicating an absence of G × E interaction.

In careful studies of other species, however (see e.g. Mather and Jinks (1982) and references therein for many examples), the response to changes in

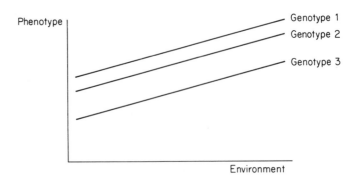

Figure 4.2 Constant sensitivity of genotypes to environmental influences in the absence of genotype × environment interaction.

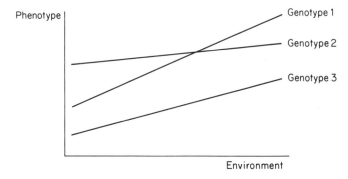

Figure 4.3 Genetic differences in sensitivity to environment creating genotype × environment interaction.

the environment is not the same for all genotypes. That is, there is G × E interaction. One possible form of G × E is illustrated in Figure 4.3. In the illustration the response curves are not parallel. Such interactions would have great significance in man because they could represent the mechanism underlying certain forms of physiological and psychiatric disorder. Genetic differences in sensitivity to sodium, for example, might account for the etiology of hypertension. Genetic differences in sensitivity to environmental stress may underlie the origin of depression and anxiety. An important result of such genetic mechanisms is that they do not lead to the familial aggregation of disease unless there is also family aggregation of the environment.

A number of papers have considered aspects of the theory and analysis of G × E in man (e.g. Jinks and Fulker, 1970; Eaves *et al.*, 1977; Eaves, 1982, 1984; Lathrop *et al.*, 1984), but most of the practical insights come from the study of other species in which greater genetic and environmental manipulation is possible. Several important results have been summarized by Mather and Jinks (1982).

(1) The genes that affect sensitivity to the environment are often quite different from those affecting average trait value. That is, it is possible to select artificially for high or low sensitivity to the environment without affecting average response.

(2) The kind of gene action (e.g. effects of heterozygosity) shown by genes affecting sensitivity to the environment may be quite different from that of genes affecting average values.

(3) Sensitivity to different environmental variables may be mediated through different sets of genes.

In animals and plants the data required to construct diagrams like Figure 4.3 can be collected experimentally. In man, however, this is not so easy, since genotypes cannot be replicated (except in the case of monozygotic twins) and environments cannot easily be specified or controlled. However, there is a number of situations in which G × E can be detected. Some of these are reviewed by Eaves *et al.* (1977) and Eaves (1982, 1984). If the effects of the environment can be measured then the model for family resemblance can be expressed, conditional on the environmental measurements of the family members. That is, the contribution of genetic factors to the correlation between relatives can be expressed as a function of the environment. Similarly, if a specific genetic marker can be identified then the sensitivity to the environment can be measured in MZ twins of known genotype (cf. Martin *et al.*, 1987). Generally, the analysis of G × E in man requires that we are lucky enough to specify the genetic effects, or the environmental effects, or both, although it is theoretically possible with data on large numbers of separated twins to detect G × E interactions between unspecified genetic and environmental factors (Jinks and Fulker, 1970; Eaves *et al.*, 1977).

The practical importance of G × E lies in the implication that people may differ in their sensitivity to the same environmental change ("One man's meat is another man's poison"). G × E is important theoretically because it represents the genetic control of adaptation to changing environmental circumstances and is expected to affect the dynamics and equilibria of populations under selection. Trojack and Murphy (1981) have examined some of the theoretical aspects of homeostatic responses to perturbations in the environment. At present, we are far from an adequate analysis of G × E in man. The reason lies not so much in the want of theory and methods as in the shortage of adequate data. Until very recently, behavior geneticists have been content to measure traits in kinships without attempting to quantify the environment. Epidemiologists, on the other hand, have tried faithfully to measure the environment but ignored the genes. As long as the environment can be specified, it may be possible to detect genetic effects on response to the environment (Eaves, 1984; Kendler and Eaves, 1986), given large enough samples of appropriate kinships, so that G × E may be examined.

4.2.3 Genotype–environment covariance: the genetic environment

Behind (4.2) lies the second important assumption that genetic and environmental effects are uncorrelated so that the total phenotypic variance is simply the sum of two variance terms: that due to genes (G) and that due to environment (E). Under many circumstances of great practical and theoretical interest this will not be the case, but there will be genotype–environment covariance (CovGE). If there is CovGE then we modify (4.2) to include the covariance of genes and environment:

$$V = G + E + 2CovGE. \tag{4.3}$$

We shall consider some of the causes of CovGE in man later, but the biometrical–genetical literature on plants and animals (see e.g. Mather and Jinks, 1982) provides important clues to its interpretation. In general, CovGE will arise whenever the environment is "caused by" the genotype of the individual or one of his relatives. The classical case is that of genetic maternal effects. In rats, mice and *Nicotiana rustica* there is evidence that the maternal genotype has a direct environmental impact on the phenotype of her offspring. If the genes that create the maternal effect also exert a consistent effect on the offspring phenotype directly (what Haley *et al.* (1981) call the "one-character model" of maternal effects) then there will be CovGE because the environment created by the maternal genotype is correlated with the effect of her offspring's own genotype on the offspring phenotype. Other patterns of CovGE will arise if the genotypes of other relatives form a salient part of the individual environment (see e.g. Eaves, 1976a, b).

4.2.4 The number of genes

The success of Mendel's early experiments was due to four factors: his use of pure breeding-lines; his selection of clear-cut, all or nothing differences; his (opportune?) focus on multiple characteristics that were influenced by genes widely separated on the genome; and his careful study of extremely large samples. His concentration on discrete "segregating" characters such as tall or short peas was crucial to his ability to uncover the basic laws of inheritance, which have subsequently been understood at the cellular and molecular level (Sutton, 1903; Watson and Crick, 1953). However, such "all-or-nothing traits" are only a small part of the measurable variation in any species. Most phenotypic variation in species from microorganisms to man is continuous. That is, individuals can be measured but not categorized at the phenotypic level. Part of our inability to assign individuals to "genetic" categories may be due to the "smoothing effect" of the environment, but detailed studies of variation in other organisms suggests that this is not the whole story.

At about the time Mendel had published his *Experiments in Plant Hybridization*, Francis Galton was turning his attention to the laws of heredity in Man (for a more detailed discussion of Galton's contribution see e.g. Burt, 1962; Forrest, 1974; Pearson, 1973). Galton assembled an impressive collection of data on a wide variety of anthropometric variables ranging from stature and span, through sensory acuity and physical strength. The amount of data he gathered was so great that it is only in the last few years that researchers have begun to analyze it with the aid of computers (McClearn and Johnson, 1982).

The main impact of Galton's work was the demonstration that continuous variables, such as stature, were correlated between relatives and that the degree of correlation increased with the closeness of familial relationship. Galton was apparently unfamiliar with Mendel's work, but conflict ensued at the turn of the century when the results of Mendel's hybridization experiments were publicized by Bateson and Punnett (1908). Apparently there was a conflict between "Mendelian" inheritance, which applied to the kinds of discrete characters that Mendel had examined, and "Galtonian" inheritance, which applied to continuous characters. This was not resolved immediately.

A central protagonist in the debate was Galton's biographer and successor, Karl Pearson. Pearson pointed out the inconsistency, as he thought, between Mendel's results and those of Galton and himself, working in conjunction with Alice Lee (Pearson and Lee, 1903). The discrepancies were several. Apart from the fact that Mendel's characters were discontinuous and Galton's were continuous, Pearson showed that Mendelian inheritance predicted different correlations between relatives from those found for Galtonian traits. He also showed that Mendel's theory predicted different variances *within sibships* as a function of their average trait value, although no such "heteroscedasicity" was apparent for Galtonian traits. The conflict was frequently odious, especially on Pearson's part. Resolution, however, was achieved as a result of experimental and theoretical developments in the first twenty years of this century. On the experimental side, two studies had shown that many of the principles of Mendelian inheritance could be demonstrated for continuous characters in plants. In 1909 Nilsson-Ehle showed that several Mendelian factors in wheat resulted in the same difference between red and white grain. In crosses in which more than one factor were considered together, he showed that the *degree* of redness was apparently a function of the *number* of "red-producing" factors. Similarly, East (1915) showed that there was a great increase in variability for a continuous trait (corolla length in *Nicotiana longiflora*) in the progeny of a cross between two F_1 plants derived from true-breeding parents (see Mather and Jinks, 1982). These results paralleled the segregation observed by Mendel in the progeny of crosses between F_1 hybrids in his experiments. East and Nilsson-Ehle both recognized that their findings were consistent with the cumulative effects of many genes, each with small effect on a continuous phenotypic variable, and thus laid the one cornerstone of the "polygenic" theory of continuous variation.

The second major step was a theoretical one, taken by Fisher (1918). In a classical paper, he showed how the biometricians' own data could be reconciled with Mendelian inheritance. There were three main aspects to Fisher's argument. First, following the experimental results of East and Nilsson-Ehle, Fisher adopted the polygenic model in which a large number of genes (infinitely large in Fisher's analysis) of individually small effect were responsible for continuous variation and family resemblance in man. Secondly, Fisher questioned the generalization drawn from Mendel's experiments that all hybrids for a given character would resemble one or the other parent (i.e. he relaxed the assumption of complete dominance). Thirdly, he allowed in various ways for the effects of assortative mating, i.e. for the genetic consequences of the fact that spouses were correlated for many traits such as stature.

Allowing for "incomplete dominance" was an important component of Fisher's theory, because Pearson had already shown that strictly Mendelian inheritance would only produce a correlation of 1/3 between parent and offspring, a value below that observed for stature and, in Pearson's judgement, in clear contradiction to the "Mendelian" model. "Assortative mating" was potentially important for two reasons: it produced a correlation between the genetic effects of spouses and generated correlations within individuals between Mendelian factors that were otherwise independent functionally and even unlinked genetically. The correlation between genetic effects of spouses led to an increase in the correlation between relatives and, because of the resulting correlation of genetic effects within individuals, generated a marked increase in the phenotypic variance of a heritable trait over that expected in a randomly mating population. Fisher's ability to represent the elements of polygenic inheritance in terms of a mathematical model built on the principles laid down by Mendel enabled him to predict very well the pattern in the correlations between different kinds of relatives for stature using the same data that Pearson had used to develop his criticism of the widespread application of Mendelian principles to continuous variation. Fisher's paper was thus one of the major intellectual triumphs in the early history of genetics in this century.

The seventy years since Fisher's seminal paper have seen the detailed analysis of the consequences of Mendelian inheritance for continuous variation. A major finding has been the recognition that a large number of loci are responsible for continuous variation and that they are distributed throughout the genome. Support for this view comes from the biometrical–genetic study of continuous traits in fungi (Caten, 1979), *Drosophila* spp (e.g. Breese and Mather, 1957, 1960) and *Nicotiana rustica* (Jinks and Towey, 1976) and from population geneticists (e.g. Lewontin, 1974), who have maintained that "fitness" is necessarily a polygenic trait and that models that assume otherwise are utterly inappropriate.

4.2.5 Gene action and interaction

In the previous chapter we defined two empirical genetic components G_W and G_B to reflect the contribution of genetic effects to variation within and between families. The parameters are intuitively sensible, but do not reflect any precise idea of how genes behave in kinships or are expressed in phenotypes. Indeed, they do not even require that we take any notice of Mendel! Such empirical parameters are only satisfactory for a limited set of relationships. If we are to develop more general models, it will be necessary to redefine them in terms of gene effects and the frequencies of increasing and decreasing alleles.

The extensive biometrical studies of plants and animals show that all the principles of Mendelian inheritance for discontinuous traits extend completely to the polygenic case. Thus, for example, the kinds of gene action displayed by individual loci and the types of interaction shown between pairs of loci are also manifest in polygenic systems. As a result, the genetic variance G is not adequately represented by one parameter but rather by several. Biometrical studies have shown that G can be further subdivided into components due to differences between homozygotes (additive genetic effects), heterozygous effects (dominance deviations) and interactions between different genes (epistatic effects). Thus, in principle, we may partition the genetic variance as follows:

$$G = V_A + V_D + V_{A \times A} + \ldots, \qquad (4.4)$$

where V_A denotes the additive genetic variance, V_D the dominance variance, and the series of terms $V_{A \times A} + \ldots$ the various forms of interaction between homozygous and heterozygous effects.

There are several detailed derivations of the expectations of genetic parameters, including those of Mather and Jinks (1982) based on Mather's (1949) notation, Falconer (1981) and Bulmer (1980). Mather (1949) considered the contribution of a single locus with two alleles to a continuous phenotype (see Figure 4.4).

The midpoint between the two homozygotes on the scale is m. The additive deviation between midpoint and the increasing homozygote is d (the decreasing homozygote is thus $-d_a$ units from m). The heterozygous deviation h_a measures the displacement of the heterozygote from the midpoint. If $h_a = 0$ then the heterozygote is intermediate and there is said to be "no dominance". If $h_a = d_a$ then there is complete dominance for increasing phenotypic effect (at this locus). If $h_a = -d_a$ then the heterozygote corresponds to the decreasing homozygote and there is complete dominance for decreasing trait expression. Intermediate values of h reflect various levels of incomplete dominance for increasing or decreasing trait values. The effects and frequencies of the three genotypes may be summarized thus:

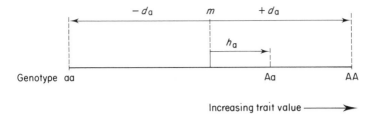

Figure 4.4 Effect of a single locus on a continuous phenotype (after Mather and Jinks, 1982).

genotype:	AA	Aa	aa
effect:	$m + d_a$	$m + h_a$	$m - d_a$
frequency:	u_a	$2u_a v_a$	v_a^2

The subscript "a" is introduced to distinguish the frequencies and effects at the A/a locus from those of other loci.

Mather (1949) shows that the contribution of such a locus to the genetic variation in a randomly mating population is $\sigma_d^2 + \sigma_h^2$, where

$$\sigma_d^2 = 2u_a v_a [d_a (v_a - u_a) h_a]^2,$$
$$\sigma_h^2 = 4u_a^2 v_a^2 h_a^2.$$

If there are many similar loci, each with their own additive and dominance effects and allele frequencies, then it may be shown (Fisher, 1918; Mather, 1949) that their individual contributions to the genetic variance may be added together as long as the loci are independent (i.e. in linkage equilibrium). Under these circumstances, we have

$$\left.\begin{aligned}
V_A &= \tfrac{1}{2}D_R = 2\sum_a u_a v_a [d_a + (v_a - u_a) h_a]^2, \\
V_D &= \tfrac{1}{4}H_R = 4\sum_a u_a^2 v_a^2 h_a^2,
\end{aligned}\right\} \tag{4.5}$$

where summation is over all loci. We use the notation V_A and V_D of Falconer (1981) for the additive and dominance variance components because it has been widely used in the literature in human quantitative genetics. Mather (1949), however, defines a parameter D_R ($=2V_A$) in representing the contribution of additive genetic effects and H_R ($=4V_D$) to represent dominance contributions. The difference results from the fact that Mather defines his parameters in terms of the differences observed between inbred lines and crosses derived from them, whereas Falconer begins, as we have done, with the genetic variance in a randomly breeding population. The difference is only a matter of scale, and translation between the two notations is straightforward.

The additive component of genetic variance in randomly mating populations does not comprise just additive genetic effects ("d"s) except in the case of the $u = v = \frac{1}{2}$ at every locus since a term in h appears in (4.5). The dominance component, however, will consist only of dominance effects in the absence of epistasis. This means that the presence of a statistically significant dominance component will still indicate significant non-additivity of genetic effects. Mather and Jinks (1982, p.217) show graphically how the allele frequencies affect the relative magnitudes of the additive and dominance components for the contribution of a single locus.

The definitions of the additive and dominance components are still more complex if there are epistatic interactions between loci. Also, additional components are required to specify the effects of epistasis fully. Mather (1974, see Mather and Jinks, 1982, p.221) gives the expectations for the covariances of relatives amended for epistasis. In practice, however, epistatic interactions, for the most part, are highly correlated with dominance in the expectations, so that independent estimation of dominant and epistatic effects is likely to be almost impossible in man when there are other effects such as cultural inheritance and assortative mating. Heath *et al.* (1984) have claimed that genetic non-additivity at loci affecting dermatoglyphic pattern intensity is consistent with a model of additive × additive epistasis rather than dominance. We shall assume that inclusion of a dominance parameter in our model is sufficient to capture any significant genetic non-additivity.

Heritability

The distinction between additive and dominance effects has resulted in the definition of two different heritability coefficients, which have sometimes been confused by psychologists in the past. The "broad" heritability includes all sources of genetic variation in the numerator and all sources of variation, genetic and environmental, additive and non-additive, in the denominator. In terms of the components of gene action, the broad heritability is

$$h_b^2 = \frac{V_A + V_D}{V_A + V_D + E}.$$

The ratio that finds greater application in plant and animal breeding, however, is the narrow heritability

$$h_n^2 = \frac{V_A}{V_A + V_D + E}.$$

The biological importance of the narrow heritability lies in the fact that only additive genetic effects (and additive × additive epistatic interactions) contribute to the resemblance of parents and offspring, and are thus the only effects relevant to predicting the response to selection in the short term. These effects determine how much of the variation in a population can be "fixed" by selection. However, many other factors, such as the number and linkage relationships between loci, are required to predict the intermediate and long-term response to selection (see e.g. Mather and Jinks, 1982, p. 339; Lewontin, 1974), so the application of such formulae to support artificial improvement of the human species is naive and irresponsible (Jacquard, 1984).

Covariances between relatives

Having defined the genetic contribution to the total variance in terms of gene effects and frequencies, it is relatively straightforward, though tedious, to show that the same basic components may be used to represent components of variance, or covariance, between relatives (see e.g. Mather and Jinks, 1982). Expected covariances for some of the more commonly encountered relationships are given in Table 4.1.

Under the simple model we have assumed here, the expected component of variance between families is equal to the expected covariance between relatives. Thus, for example, the genetic part of the component of variance between DZ twin pairs is

$$\sigma^2_{BDZ} = \tfrac{1}{2}V_A + \tfrac{1}{4}V_D.$$

Since the total variance is $V_A + V_D$, the genetic component of the variance within DZ pairs is

$$V_p - \sigma^2_{BDZ} = \tfrac{1}{2}V_A + \tfrac{3}{4}V_D.$$

Table 4.1 Expected covariances between relatives under additive–dominance model with random mating in the absence of cultural inheritance.

| | Contribution to covariance | |
Relationship	V_A	V_D
MZ twin	1	1
DZ twin/sibling	$\frac{1}{2}$	$\frac{1}{4}$
Parent–offspring	$\frac{1}{2}$	0
Half-sibling	$\frac{1}{4}$	0
Avuncular	$\frac{1}{4}$	0
Grandparental	$\frac{1}{4}$	0
First cousin	$\frac{1}{8}$	0
Adopted	0	0

4.2.6 The forgotten environment

The main focus of theoretical and experimental studies of continuous varia-tion in plants and animals has been the elucidation of genetic contributions to the phenotype. When the environment is important, it has either been the subject of experimental manipulation (as in the case of G × E experiments), genetic manipulation (as in the studies of genetic maternal effects), or ran-domized so that its effects are not confounded with genetic effects on the resemblance between relatives. Fisher's (1918) paper, for example, treats the environment as a random variable that does not contribute to correlation between relatives.

If environmental effects are uncorrelated with genotypes then we may simply add the contributions of E_W and E_B to the appropriate variance com-ponents to obtain expectations in terms of both genetic and environmental effects. The expectations of variance components for MZ and DZ twins reared together are given in Table 4.2.

Our parameters E_B and E_W correspond to Jinks and Fulker's parameters E_2 to represent the contribution of the shared environment to differences *between* families and E_1 to represent the contribution of the environment to differences *within* families. Analogous parameters, σ^2_{be} and σ^2_{we} were employed by Cattell (1960) for the same purpose. It is easy to think of human examples of these two kinds of environment. Effects of diet, education, parental treatment, etc., which are shared by members of the same family, will be part of E_B. Individual experiences, chance happenings, accidents, etc. will create environmental differences within the family and will contribute to E_W. When considering behavioral variables that may be subject to fluctua-tions over time and errors of measurement, we recognize that environmental effects within the family may also be a function of such changes. Thus the same individual measured on two occasions will not have exactly the same phenotype, so that pairs of individuals from the same family will have diffe-rent scores simply because of fluctuations due to measurement error. In some

Table 4.2 Expected components of variance and expected mean squares for MZ and DZ twins reared together.

	Variance component			
Statistic	V_A	V_D	E_W	E_B
Between MZ pairs	1	1	0	1
Within MZ pairs	0	0	1	0
Between DZ pairs	$\frac{1}{2}$	$\frac{1}{4}$	0	1
Within DZ pairs	$\frac{1}{2}$	$\frac{3}{4}$	1	0

of the later analyses we shall attempt to determine how much of E_W can be explained by such short-term fluctuations and how much is due to lasting environmental differences within the family (Chapters 5 and 9).

Psychologists are more likely to be interested in the family environment E_B because many mechanisms of social interaction and cultural inheritance of concern to developmental, clinical and social psychologists will create differences in the environments *between* families. For example, parental rearing attitudes, insofar as they are consistent over time, will create an environment for behavioral development that is shared by siblings in the same family. Psychiatric disorders in parents might create an adverse environment shared by all their children living in the same home. Insofar as attitudes of children are influenced culturally by the attitudes of their parents, we expect environmental effects common to siblings in the same family.

As long as our analysis is confined to twin data (e.g. in Chapter 5), our two environmental parameters will be sufficient to specify many of the sources of environmental effects on personality and attitudes. However, although our definition of of E_B suffices for many applications in this book, it is inappropriate for all relationships and social interactions. Just as genetic effects can be represented more parsimoniously by specifying the mechanism of inheritance and gene action, so the environment can be modeled more effectively if the sources and target of social interaction can be specified. Many models of cultural inheritance (e.g. Cavalli-Sforza and Feldman, 1973a, b; Eaves, 1976a, b; Rice *et al.*, 1980) assume that the environment shared by siblings is caused by aspects of the parental phenotype or environment. Under these conditions, the environmental correlation between relatives will depend on the type of relationship (see Chapters 6, 7 and 16).

Similar findings emerge for the environmental correlations between twins when there are sibling interactions based on genetic differences. In Chapter 5 and elsewhere we shall consider models for cooperation and competition between siblings in the attempt to detect their effects on measurements of behavior.

4.3 MODEL FITTING

4.3.1 Criteria for a "good" model

(a) Consistency

Consistency is a basic requirement of any model. The implications of an assumption must be traced through every part of the dataset. We should not assume something in one part of the analysis and assume something else in another. The requirement of consistency would be violated, for example, if we tried to fit a model in which we specified an environmental component in the correlation between parents and offspring but did not deduce a corresponding environmental correlation between siblings. This example is obvious, but many are not so straightforward and might involve a failure to impose necessary constraints on parameter values that are implied by a particular model. The dangers of inconsistency are greatest when we do not begin with a causal theory, but try to introduce *ad hoc* parameters into a model to explain particular anomalies in the data.

(b) Parsimony

Insistence on simplicity, or parsimony, is the only safeguard against building castles in the air. It is always tempting to squeeze the last drop of inspiration from a set of data by adding in "interesting" parameters, but as the amount of squeezing increases, the precision and generality of prediction decreases. Simple models are most informative. Their parameters have the smallest standard errors. Their predictions are most sweeping. When models get very complicated they are probably missing the point. This basic principle of parsimony was expressed by the fourteenth-century dissident British philosopher, William of Occam, in the maxim "It is vain to do with more what can be done with fewer." In practice, if a simple model fits the data then it is always easy to devise a more complex model that will fit as well or even better, but the more complex model adds nothing to our understanding. If two or more models fit the data, therefore, we use "Occam's razor" and adopt the simpler until it is disproved by a more powerful study. There will be several cases where we have to modify the conclusions of earlier studies in the light of more extensive data.

(c) Goodness of fit

We reject any model that does not fit the data. This criterion is important because models that do not fit cannot predict anything. The estimation of Holzinger's *H* from twin data, for example, assumes a model that it does not test. Very large values of *H* could be produced if the correlation between DZ twins were very small, even negative. Yet such small DZ correlations, allied to moderate MZ correlations, could indicate something fundamentally wrong with the model and preclude the estimation of *H* altogether.

(d) Statistical significance of parameters

The use of statistical significance as a basis for including parameters in a model is the guarantee that models are developed cautiously and rationally. If, as a basic rule, we do not elaborate parameters unless there are good statistical reasons then there will certainly be occasions when we shall miss effects that are really there, but we shall also avoid the worse error of ascribing substance to mere chance. Occasionally, this rule has been applied foolishly, and researchers have "grubbed around" in their data in search of significance. The rules of inference that we use, however, only apply to decisions that were agreed upon *a priori*. They are likely to be grossly misleading if the model-fitting exercise degenerates into fitting every possible combination of parameters and "picking the best".

4.3.2 Rationale of model-fitting methods

The criteria for a "good" model lead to a number of features that we should expect from a model-fitting procedure: (1) simultaneous estimation of several parameters from multiple data points (e.g. correlations); (2) a test of whether or not the model fits the data; (3) standard errors of parameter estimates. In addition to these, we may also want a method that (4) makes the most efficient use of the data (i.e. gives estimates that are as precise as possible) and (5) allows for the fact that different parts of the data are known with different precision.

Several statistical methods satisfy many or all of these criteria. They all depend on defining a function of the model parameters and data that increases as the model gets worse. By choosing the model and parameter set that minimizes this "loss function", we find the model that gives the best fit to the data. A careful choice of loss function can yield estimates that have desirable statistical properties including those specified above. This approach is familiar to anyone who has ever done a regression analysis, because the regression slope is chosen to minimize the sum of squared deviations between the data points and the predicted regression line. In this application the residual sum of squares is the "loss function". If the model fits and the errors are normally distributed then we can obtain estimates of the standard errors for the intercept and slope. In the examples that we consider, the data points are seldom individual, independent observations. Rather, they are variances, mean squares, covariances, or the correlations between relatives. In regression analysis the parameters are the intercept and slope(s) in the regression equation. In our case they are parameters describing the causes of variation or the pattern of correlations between different relationships.

Many of the model-fitting methods are now available in computer packages such as GLIM (NAG, 1985; Nelder and Wedderburn, 1972), SAS

(1985) and LISREL (Jöreskog and Sörbom, 1981) or can be programed in FORTRAN using existing algorithms for numerical optimization such as those of the Numerical Algorithms Group (NAG, 1982).

In our applications we have typically used one of two loss functions: the (weighted) residual sum of squares or minus the natural logarithm of the likelihood. In most cases that we consider, the two approaches give answers that are almost identical, and the choice between them is largely a matter of convenience. In this chapter we consider two simple applications of the method of weighted least squares so that a reader unfamiliar with the approach will be able to experiment with some of the other data in the book.

4.3.3 Weighted least squares applied to twin mean squares

Our first illustration applies the method of weighted least squares to the neuroticism mean squares from the National Merit Study given in Table 3.1. A model for the mean squares that includes additive and dominance effects and environmental effects within and between families is given in Table 4.3. Our task is to estimate the parameters of such a model from the observed mean squares. In this case the analysis was conducted with a program written for the SAS "MATRIX" procedure (SAS, 1982, 1985b). The program is reproduced in Appendix A and may easily be modified for other models and datasets.

It turns out that all four parameters cannot be estimated simultaneously since, although there are four mean squares and four parameters in the "full" model, the total variances of MZ and DZ twins are expected to be equal, leaving only three estimable combinations of parameters (see Chapter 5).

A simpler form of the model, which is important theoretically, is that which sets both V_D and E_B to zero. That is, we estimate only two parameters, V_A and E_W, from the four mean squares. This model excludes all non-genetic sources of family resemblance and assigns all genetic differences to purely additive effects. That is, the phenotype for neuroticism is assumed to be unaffected by cultural inheritance within the family. Purely additive gene

Table 4.3 Expected mean squares for twins reared together.

Statistic	Expected mean square			
	V_A	V_D	E_W	E_B
Between MZ pairs	2	2	1	2
Within MZ pairs	0	0	1	0
Between DZ pairs	$\frac{3}{2}$	$\frac{5}{4}$	1	2
Within DZ pairs	$\frac{1}{2}$	$\frac{3}{4}$	1	0

action may imply that there is no strong correlation between the phenotype for neuroticism and fitness.

(a) The model and expected mean squares

The observed mean squares from Table 3.1 may be represented by the vector **x**, with element x_i being the ith mean square. In our example, there are four mean squares. We wish to fit a simple model that includes the two parameters V_A and E_W. The (presently unknown) parameters are denoted by the vector θ The expected values for each mean square, $E(x_i)$, are obtained by multiplying each parameter by a coefficient and summing over parameters:

$$E(x_i) = \sum_j a_{ij}\theta_j.$$

The above procedure can be expressed much more simply in matrix notation, as in the computer program given:

$$E(\mathbf{x}) = \mathbf{A}\theta.$$

(b) Specifying the loss function

The loss function should have three properties: (1) it should measure the current discrepancies between the observed mean squares and the values predicted from the parameters of the model; (2) it should give greater weight to data points that are known more accurately; (3) it should be related to some known statistical distribution so that we can devise a statistical test to determine whether the final value of the loss function is greater than might be expected by chance alone given that the model fits. If the original test scores are normally distributed then it turns out that the following weighted sum of squared deviations has these desirable properties:

$$s^2 = \sum_i w_i[X_i - E(X_i)]^2,$$

given an appropriate choice for the weights w_i corresponding to the mean squares.

(c) Choice of weight

Fisher (1935) suggested that an appropriate weight is the amount of information I about the statistic. The amount of information is the inverse of the variance of the observation. Using the amount of information has intuitive appeal since it means that data points that are known less precisely (i.e. that have greater sampling variance) receive smaller weight. It may be shown

(Kendall and Stuart, 1977, Vol. 1) that the variance of a variance (or mean square) is equal to

$$\sigma_V^2 = \frac{2\{E(V)\}^2}{d},$$

where $E(V)$ is the expected values of the variance, and d is the corresponding degree of freedom. Unfortunately, we do not have the expected value of the variance unless we know the parameter values, and since the latter are precisely what we are trying to estimate, it would seem that the problem is insoluble. If the model is any good, however, the observed values of the mean squares should be reasonably close to their expected values, so we shall use the observed mean squares to generate weights that will serve as trial values. The trial weights may then be refined when the parameters have been estimated.

(d) Minimizing the loss function

In matrix notation, the loss function is

$$s^2 = (\mathbf{x} - \mathbf{A}\theta)'\mathbf{W}(\mathbf{x} - \mathbf{A}\theta)$$

(cf. Appendix A). The parameter values that minimize s^2 *could* be found by trial and error, but this process would be tedious and relatively inaccurate. However, the function will be a minimum for parameter values that set the first derivatives of s^2 to zero with respect to θ. Thus we may write the so-called "normal equations"

$$\frac{\partial s^2}{\partial \theta} = 0$$

and solve for θ_1 and θ_2. Application of elementary rules for differentiation of s^2 with respect to the unknown parameters yields normal equations that may be expressed most easily in matrix form:

$$\mathbf{A}'\mathbf{W}\mathbf{A}\theta - \mathbf{A}'\mathbf{W}\mathbf{x} = 0 \tag{4.6}$$

Since the elements of $\mathbf{A}$ and x are known (the coefficients of the parameters in the model and the observed mean squares) and $\mathbf{W}$ is being given approximate values derived from the observed mean squares, we have enough information to solve for the parameters. We rewrite the above expression as

$$\mathbf{A}'\mathbf{W}\mathbf{A}\theta = \mathbf{A}'\mathbf{W}\mathbf{x} \tag{4.7}$$

and multiply both sides by the inverse of $\mathbf{A}'\mathbf{W}\mathbf{A}$ to yield the solution:

$$\hat{\theta} = (\mathbf{A}'\mathbf{W}\mathbf{A})^{-1}\mathbf{A}'\mathbf{W}\mathbf{x}. \tag{4.8}$$

(e) Results

The above procedure may be followed numerically using the data given in Table 3.1 and the computer program in Appendix A. We are given the vector of mean squares x with their associated degrees of freedom. Substitution of the observed mean squares in the formula for the amount of information yields the first approximation to the diagonal weight matrix:

$$
W = \begin{bmatrix}
0.07576 & 0.0 & 0.0 & 0.0 \\
0.0 & 0.63496 & 0.0 & 0.0 \\
0.0 & 0.0 & 0.07481 & 0.0 \\
0.0 & 0.0 & 0.0 & 0.18688
\end{bmatrix}.
$$

The matrix of coefficients is

$$
A = \begin{bmatrix}
E_W & V_A \\
1 & 2 \\
1 & 0 \\
1 & 1.5 \\
1 & 0.5
\end{bmatrix}.
$$

By multiplying matrices, we obtain the normal equations

$$
\begin{bmatrix}
0.972405 & 0.357168 \\
0.357168 & 0.518070
\end{bmatrix} = \begin{bmatrix}
18.9949 \\
12.2138
\end{bmatrix},
$$
$$
A'WA = A'Wx,
$$

yielding the solution

$$
\begin{bmatrix}
1.3771 & -0.9494 \\
-0.9494 & 2.5848
\end{bmatrix} \begin{bmatrix}
18.9949 \\
12.2138
\end{bmatrix} = \begin{bmatrix}
14.562 \\
13.536
\end{bmatrix},
$$
$$
(A'WA)^{-1}A'Wx = \hat{\theta}.
$$

The current parameter estimates may now be substituted in the expectations to yield the expected mean squares:

$$
\begin{bmatrix}
41.63 \\
14.56 \\
34.87 \\
21.33
\end{bmatrix} = \begin{bmatrix}
1 & 2 \\
1 & 0 \\
1 & 1.5 \\
1 & 0.5
\end{bmatrix} \begin{bmatrix}
14.562 \\
13.536
\end{bmatrix},
$$
$$
E(x) = A\theta.
$$

whence we obtain the weighted residual sum of squares:

$$
s^2 = [x - E(x)]'W[x - E(x)]
$$
$$
= 0.06657.
$$

Table 4.4   Iterative weighted least-squares estimation of genetic and environmental components of variance from MZ and DZ twin mean squares.

Iteration	s^2	$\hat{E}_W$	$\hat{V}_A$	$\sigma^2_{\hat{E}_W}$	$\sigma^2_{\hat{V}_A}$	$\text{Cov}_{\hat{E}_W \hat{V}_A}$
1	0.06657	14.562	13.536	1.377	2.585	−0.949
2	0.06621	14.564	13.543	1.382	2.596	−0.957

The current value of s^2 has been computed using the observed mean squares as trial weights, but the amount of information should properly be based on the expected mean squares. The above procedure is thus repeated, substituting the expected values $E(\mathbf{x})$ in the expression for the elements of the weight matrix, to yield new estimates, and a new (smaller) s^2. Since the estimates will now change somewhat, so will the expected mean squares, and consequently the procedure may be repeated a third time and so on, until two successive values of s^2 differ by a suitably small amount. In the appended SAS program we stopped the iterative procedure when two successive values of s^2 differed by less the 0.01. Table 4.4 presents the results of the iterative weighted least-squares analysis for the example data. Convergence is rapid and, for this case in which the model fits well, we find that the parameter estimates after the first iteration are close to the final values.

(f) Interpreting the results

The matrix $\mathbf{A'WA}$ at the final solution is the expected information matrix and equal to one half of the matrix of second partial derivatives of the loss function s^2 at the minimum. If our original mean squares are normally distributed, i.e. if samples are large enough, then the minimum value of s^2 follows the chi-square distribution when the model fits. The df of chi-square are obtained by subtracting the number of parameters (2), from the number of raw mean squares (4). This chi-square provides an overall test of whether the model fits the data, since, under the null-hypothesis that the residual effects are due to chance alone, the chi-square should not be significant. From tables we find that our chi-square is not significant ($P > 50\%$), so we have no reason to reject our simple model.

When the model fits, the inverse $(\mathbf{A'WA})^{-1}$ assumes considerable importance because it is the covariance matrix of the parameter estimates. By taking the square roots of the diagonal elements, we may generate standard errors for the parameter estimates. Thus, taking the final value of the covariance matrix from Table 4.4, we have

$$\sigma_{\hat{E}_W} = 1.18,$$
$$\sigma_{\hat{V}_A} = 1.61.$$

The off-diagonals give the covariances between the estimates, which may be used to decide how reliably the experimental design can separate the parameters of the model and to generate standard errors of other functions of the parameters including "heritability" ratios (see e.g. Eaves, 1969; Young *et al.*, 1980). In our example the correlation between the estimates of V_A and E_W is

$$0.9568/\sqrt{1.3818 \times 2.5960} = 0.5052.$$

The correlation between the estimates is not very high, compared with that between estimates of additive and dominance genetic effects (see Chapter 5).

The units in which behavior is measured are usually arbitrary, so it often suffices to express the estimated components of variance as proportions of the total. In our example the best estimate of the phenotypic variance is

$$V_A + E_W = 13.54 + 14.56 = 28.10.$$

The proportions of variance attributable to additive genetic effects and environmental differences within families are 0.49 and 0.51 respectively, suggesting a moderate contribution of inherited factors to the observed variation in neuroticism in this sample.

4.3.4 Fitting a simple non-linear model to correlations

The above model represented expected mean squares in twins as linear functions of genetic and environmental parameters. Many of the more complex models that involve assortative mating or cultural inheritance, however, are not linear. A consequence of the non-linearity is that the first derivatives of the loss function (the normal equations (4.6) above) are not linear functions of the unknown parameters, so that, for a given set of weights, there is not a direct solution for the unknown parameters. We illustrate the principles of fitting a simple non-linear model involving a single genetic parameter to the correlations derived from the mean squares for neuroticism.

The model is a simple path model (see Chapter 6) for the correlations between twins and is another form of the V_A, E_W just fitted to the mean squares. Path models express the correlation between the relatives as regression functions of latent variables. In our simple example we assume that the only latent variable responsible for family resemblance is the genotype and the regression of the (standardized) neuroticism score on the (standardized) latent variable is the "path coefficient" h (see Chapter 6 for more details). Genetic factors thus contribute a proportion h^2 of the total variance. The remaining portion of the variance, $e^2 = 1 - h^2$, is due to random environmental factors. The parameter h^2 is thus the same as V_A expressed as a proportion of the total variance, and e^2 is analogous to E_W in this model.

Since the genotypes of MZ twins are completely correlated, the correlation of MZ twins is also expected to be h^2, in the absence of additional environmental correlation. The genetic effects of DZ twins have a correlation of 0.5 when gene effects are additive, so the DZ correlation is expected to be $\frac{1}{2}h^2$. The observed correlations are 0.4858 and 0.2236 for MZ and DZ twins respectively.

We have two correlations, and one parameter (h), so we might expect that the method of weighted least squares could be used to give a single best estimate for h and to test the model. That is indeed so, but there are some differences in procedure as we are now working with correlations rather than mean squares and because the model now is non-linear in terms of the parameter of interest (h).

In fitting models to mean squares, we assumed that samples were large enough for the mean squares to be approximately normal. The normal approximation is very much poorer for correlation coefficients (see e.g. Snedecor and Cochran, 1980), especially when the true value of the correlation is not zero. The z-transformed correlation

$$z = \tfrac{1}{2}\ln\left(\frac{1+r}{1-r}\right)$$

is normal, however, given that the original observations are normal, with variance equal to $1/(N-3)$ for the product moment correlation and $1/(N-1.5)$ in the case of the intraclass correlation. The z-transformed correlation has the additional advantage that the variance does not depend on the expected correlation (unlike the variance of a mean square), so that iteration of the weights is not required in seeking a weighted least-squares solution.

However, a model that can be linearized easily for the expected correlations is not linear in the zs, with the result that we still require an iterative procedure for its solution. We now need to minimize

$$s^2 = [z - E(z)]'W[z - E(z)]$$

with respect to the unknown parameter(s). The general numerical solution of complex minimization problems is not discussed in detail here (but see Chapter 6 for some additional considerations). For a more technical account of the problems and alternative methods see a general text on optimization (e.g. Greig, 1980). Most of these methods, however, are built on a well-known simple iterative procedure for the numerical solution of equations due to Newton. Newton's method for solving an equation of the form $f(x) = 0$ (our "normal equations" are of this form) relies on the fact that, for a given trial value x_0, the tangent to the curve $f'(x_0)$ can be drawn and extrapolated to give the value of x_1 for which $f(x_1) = 0$. A new tangent, $f'(x_1)$, may then be drawn and extrapolated as before, and so on until a satisfactory solution is obtained. Some experimentation with functions of different shapes will show that the method: (1) will converge on the right solution if the function is quadratic; (2) will not necessarily find the "global" minimum if the function has local minima; (3) will not necessarily converge if the function is "convex" (i.e. bulges upward on part of the curve). The most likely danger that we shall encounter is (2), but this can be avoided by careful choice of initial values and by starting from different places.

The same method can be extended to the solution of equations in more than one unknown. In more complex problems, constraints may be required *ex-hypothesi* (see Chapter 6) or may be required to prevent intermediate values "wandering" to nonsensical solutions.

Fortunately, there are many commercially available subroutines for minimizing functions of this type, which differ in their robustness, flexibility, the type of function being minimized, their ability to handle constraints and how much user programming is required. In this book we shall use many different routines partly as a result of the continual improvement of such software in the last decade. To estimate h in our example, we employed the commercially available software package TK! Solver (Software Arts, 1982) for the IBM Personal Computer. Specification of general genetic models is tedious with the current version of TK! Solver but the interactive features of the program make it a useful teaching device for simple models and small numbers of correlations.

Figure 4.5 gives the graph of the weighted residual sum of squares s^2 for values of h between 0.1 and 0.9. This crude figure suggests a minimum value close to zero when h is approximately 0.7. Using a step size, d of 0.01, we computed numerical values for the first derivatives of s^2 with respect to h by the method of central differences. For a given value of h, we compute s^2, and the values of s^2, at $h - d$ and $h + d$, f^- and f^+ respectively. For a given step size, an approximation to the first derivative is

$$f'(h) = (f^+ - f^-)/2d.$$

Figure 4.6 gives the graph of the first derivatives for values of h between 0.65 and 0.85. As far as the eye can judge, a value of $h = 0.7$ gives a first derivative close to zero. The trial-and-error procedure, though instructive, is imprecise.

We require the h that sets $f'(h)$ to zero to a satisfactory approximation. Newton's method requires that we draw the tangent to the curve, and extrapolate to $f'(h) = 0$. The tangent to the curve relating $f'(h)$ to h is the second derivative $f''(h)$ of s^2 with respect to h, and is used in the actual numerical extrapolation. A numerical approximation to $f''(h)$ is

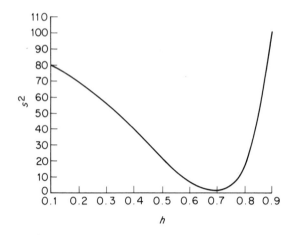

Figure 4.5 Graph of residual sum of squares s^2 for different values of genetic parameter h.

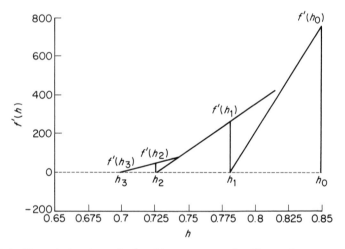

Figure 4.6 First derivative of s^2 with respect to h, illustrating extrapolation procedure to search for best-fitting value of the genetic parameter h.

$$f''(h) = (f^+ + f^- - 2s^2)/d^2,$$

where s^2 is evaluated for a particular h.

In order to extrapolate from the current trial value to the next value of h, we subtract from the current value a correction:

$$\delta_0 = f'(h_0)/f''(h_0).$$

Figure 4.6 also illustrates the first steps of Newton's method applied to the solution of the equation $ds^2/dh = 0$, employing numerical approximations to the first and second derivatives. Starting with a trial value of $h = 0.85$, the algorithm gave a $f'(h) < 10^{-6}$ after six iterations for $h = 0.694 \pm 0.02$. The progress of Newton's method may be followed in Table 4.5. The weighted sum of squares of the residual zs was $s^2 = 0.065$, which is approximately distributed as chi-square for 1 df since we have estimated one parameter from two correlations. As we should expect, *ceteris paribus*, the model fits as well as the similar biometrical model fitted the mean squares earlier. The proportion of variance in the phenotype that may be attributed to genetic differences between individuals is $h^2 = 0.48$, and the proportion due to environmental factors, given that the model is correct, is $e^2 = 1 - h^2 = 0.52$. Again, these values are close to those obtained by the alternative method starting with mean squares and estimating the components of variance, V_A and E_W.

4.4 SUMMARY

This chapter has not exhausted all the nuances of model fitting, but has served to introduce some of the basic ideas so that a reader who is unfamiliar with the approach can appreciate the concepts without necessarily having to understand all the numerical and statistical details later on. Although other

Table 4.5 Progress of Newton's algorithm for estimation of a path coefficient from twin correlations.[a]

Iteration	h	h^2	$f(h) = s^2$	$f'(h)$	$f''(h)$
0	0.850000	0.7225	42.52	772.09789	11 674.821
1	0.783867	0.6144	10.49	274.54812	4 742.601
2	0.725977	0.5270	1.12	70.77849	2 573.542
3	0.698475	0.4879	0.08	8.45325	1 985.434
4	0.694217	0.4819	0.07	0.15634	1 909.424
5	0.694135	0.4818	0.07	0.00008	1 907.997

[a] First and second derivatives of s^2 computed numerically using a step size of 0.01 (see text).

models may be more complicated and the numerical methods more technical, the basic principles remain. The same approaches that we have described can be extended to the estimation of multiple parameters, the analysis of multivariate data, and parameter estimation using different statistical criteria. The model fitting approach is not fool proof, but it is a distinct improvement over earlier methods used in behavior genetics because: (a) it makes explicit assumptions that may otherwise be hidden; (b) it forces us to put our ideas down on paper in a form that can be tested; (c) it makes the best possible use of the information contained in a set of data; (d) it gives a "goodness-of-fit test" that can be used to decide whether a model has any value before interpreting the parameters; (e) it provides estimates of standard errors that can be used for deciding which factors contribute significantly to individual differences.

Chapter 5

Adult Twin Studies of the Major Personality Dimensions

The twin study plays a central role in the development of our model for individual differences in personality. Initially we want to understand the outcome, rather than the process, of personality development, so we look first at adults. We also wish to explore the most general features of individual differences before examining the minute details, so it is reasonable to start building a model for the major dimensions of personality rather than for more specific traits. We believe that this starting point provides the strongest theoretical core from which we can move on to address the important questions of development, specific aspects of behavior, the relationship between personality and psychiatric disorders, and the generalizability of our findings over other kinds of biological and social relationship. In this chapter we therefore describe the results of applying the methods that we have described to four large bodies of personality data on adult identical and fraternal twins. For convenience these are designated the "London Study" (see e.g. Eaves and Eysenck, 1977), the "US Study" (see e.g. Loehlin and Nichols, 1976), the "Swedish Study" (Floderus-Myrhed *et al.*, 1980) and the "Australian Study" (see e.g. Martin and Jardine, 1986).

These studies have several features in common. First, they are very large and allow us to resolve certain alternative hypotheses with reasonable confidence. Secondly, they have included measures of the major personality dimensions or items from which scores on these dimensions may be derived. Thirdly, they have been conducted by different investigators in different populations. Fourthly, the data are available in a form that allows us to use model-fitting methods.

Chronologically, the US data were collected first, followed by the London data, the Swedish data and the Australian data. However, we first began to apply our methods to the London data, and it was in the light of our experience with this sample that we interpreted the findings of the US and Swedish data. Finally, the Australian Study allowed a unique opportunity to test the models that had emerged from our earlier analyses. Since this order best reflects the way in which our own understanding has evolved, we shall consider the London Study first, followed by those from the US and Sweden, and concluding with the results of the Australian Study.

5.1 THE LONDON DATA

5.1.1 Ascertainment

The backbone of the London data is derived from twins' responses to the Eysenck Personality Questionnaire obtained in the early 1970s. The Questionnaire (see Appendix B) provides raw scores on the three main dimensions of Eysenck's personality theory: psychoticism (P); extraversion (E) and neuroticism (N); together with scores on a "lie" scale (L) designed to identify subjects responding in a "socially desirable" manner. Important ancillary data come from a slightly earlier study using the "PEN", a precursor of the final EPQ, and a slightly more extended form of the EPQ administered as part of a study of sexual attitudes (Martin and Eysenck, 1976). In later chapters we shall consider the additional information about genetic and environmental determination of personality that comes from juvenile twins, their parents, spouses, adoptees and extended kinships.

The collection of the twin data began in summer 1971. The twin subjects were volunteers who had agreed to participate in the Volunteer Twin Registry established from the Department of Psychology at the Institute of Psychiatry in London. The registry was established originally in 1969 by Dr John S. Price with the aid of a grant from the Medical Research Council. A small number of pairs had cooperated in earlier studies, but significant enhancement of the registry followed with the opportunity to contact 155 pairs of identical twins who took part in a David Frost television program broadcast by Rediffusion Limited in March 1968. From these 155 pairs, 102 pairs subsequently completed and returned an early form of the Eysenck Personality Inventory. With MRC support, the registry was expanded by public appeal.

A further 300 pairs were recruited following publicity in the *Daily Mail*. Subsequent appeals and publicity in the press, and on radio and television, had secured a total of 1261 pairs by June 1971. The structure of the registry at this time is given in Table 5.1.

Registries of this kind require continual recruitment to replace pairs lost by attrition. In order to maintain cooperation in subsequent studies, twins who did not return questionnaires were not pursued strongly. The 1977 audit of returns from the twin registry yielded the response rates given in Table 5.2.

Approximately two-thirds of the individuals approached returned ques-

Table 5.1 Twins enrolled on the Institute of Psychiatry Twin Registry in 1971.

Children (under the age of 12 years)			
	MZ	DZ	Total
Male	88	81	169
Female	97	111	208
Unlike-sex	—	157	157
Total	185	349	534

Juniors (12–16 years)			
	MZ	DZ	Total
Male	22	16	38
Female	26	41	67
Unlike-sex	—	35	35
Total	48	92	140

Adults (16 years @)			
	MZ	DZ	Total
Males	67	42	109
Females	222	155	377
Unlike-sex	—	101	101
Total	289	298	587
		Total	1261

Table 5.2 Response rates from Twin Registry in 1971.[a]

	Cooperation of pair members			
Twin type	Both cooperate	One only cooperates	Neither cooperates	Total (pairs)
MZ female	244	66	104	414
DZ female	153	47	48	248
MZ male	94	33	50	177
DZ male	64	31	50	145
DZ male/female	97	28	94	219
Total	652	205	346	1203

[a] Individuals over 16 years returning questionnaire mailed in 1976.

tionnaires, and data were available on slightly fewer than 50% of complete pairs. Our pairwise response rate should be compared with 64% achieved in the Australian Registry (Martin and Jardine, 1986), 76% in the Swedish Study (Floderus Myrhed *et al.*, 1980) and 72% in the US sample (Loehlin and Nichols, 1976).

If twins responded independently to the survey then we should expect 473.2 pairs to be concordant for cooperation, 562.6 pairs to be discordant and 167.2 pairs to be concordant for non-cooperation. Our data thus show a highly significant excess of agreement between twin pairs on whether or not to participate in the study. The concordance for cooperativeness is apparently highest in DZ twins of unlike sex, largely owing to an excess of non-cooperative pairs.

The table immediately reveals two sources of bias that are common to studies of twin volunteers. There is a marked excess of females and monozygotic twins. The excess of MZ twins occurs in spite of a known 2 : 1 excess of DZ twins in the British population (see e.g. Bulmer, 1970). Identical twins tend to display greater interest in being twins, and, in the early days of the registry, there was a conscious effort to recruit MZ twins. Obviously, our sample is unrepresentative by zygosity and sex. It also has a marked excess of respondents in the upper socio-economic levels. The implications of these factors for the generality of our findings are difficult to assess. Biases in ascertainment by sex may not be a problem if the males and females themselves are representative of their respective populations. Similarly, bias in favor of MZ twins is not necessarily a problem unless the twins sampled are not representative of the parent twin populations for the traits under investigation. Of greater concern are those biases that might lead to errors in the estimation of variances and correlations for the traits in question. Thus if personality and social class are significantly associated then the fact that our twin sample is biased with respect to SES could lead to underestimation or overestimation of the population variance in personality, and, since SES is familial, we should expect such biases to be reflected in the between-family components of variance. The correlations between social class and scores on the main dimensions of personality, however, are small (Eysenck and Eysenck, 1969), so we do not expect sampling biases with respect to SES to affect our analysis seriously.

Of equal concern are the factors leading to significant pairwise association in cooperativeness, especially if concordance in cooperation reflects similarity in personality. In our sample there is little evidence that MZ twins are more concordant in cooperating than DZ twins. Among MZ female twins, 84% of pairs are concordant for cooperation or non-cooperation. This compares with 81% in female DZ twins. For male twins the rates are very similar: 81% and 79% respectively (cf. Table 5.2). There is therefore no evidence that MZ twins are more concordant in responding than DZ twins, or that males tend to differ in concordance from females. The fact that zygosity does not affect concordance in responding eliminates genetic factors as one of the main determinants of participation in the study and gives us more confidence that our findings have some general value.

5.1.2 Zygosity determination

The analysis of variation in twin samples into genetic and cultural components presupposes that an accurate determination of zygosity can be made. No approach is completely foolproof, since even a pair of non-identical twins might, by chance alone, be identical for alleles at a large number of loci. When the number of loci is sufficiently large, however, the chances of this occurring are fairly remote, so that twins identical for all or many loci may, with a high degree of confidence, be judged to be monozygotic. The most reliable method of zygosity diagnosis is therefore, to determine the genotype of each twin at a large number of loci (see e.g. Bulmer, 1970). This approach requires drawing blood from the twins and is clearly impossible on the scale required for studies like ours, so alternatives have to be found. When twins can be interviewed face-to-face, careful observation is often adequate to establish zygosity in all but a few of the most doubtful cases. When twins merely participate by mail, a simple reliable criterion for diagnosis by questionnaire is needed.

A method that has been used widely involves administration of a "zygosity questionnaire" to the entire sample and conducting a validity check on a subset using blood-typing to establish zygosity by the most commonly accepted reliable method. An early study by Cederlof et al. (1961) decided on the zygosity of 145 twin pairs on the basis of two questions concerning physical similarity when growing up and being confused in childhood. Zygosity based on questionnaire responses agreed with blood-typing results in 98.6% of these pairs, but a further 55 doubtful cases of their original sample of 200 pairs could not be diagnosed by their questionnaire. Nichols and Bilbro (1966) found they could correctly determine the zygosity of 93% of their twin sample on the basis of a questionnaire concerning physical similarity. Martin and Martin (1975) showed disagreement between diagnoses based on blood groups in only one out of 47 twin pairs, and subsequently showed that the discrepant case arose because of the parents' having been misinformed by their physician. Comparable results have been reported for a veteran twin sample in the USA (Jablon et al., 1967). Magnus et al. (1983) conducted a discriminant analysis on 207 twin pairs for whom zygosity had been determined both by blood groups and questionnaire. They found that combining the questionnaire data of both twins led to a misclassification rate of 2.4%. If questionnaire data from only one twin were used then the misclassification rate was increased to 3.9%.

In our sample, zygosity was validated for a subset of 178 like-sex twin pairs (Kasriel and Eaves, 1976) who visited the Institute of Psychiatry between March 1970 and September 1972. Prior to the study, the twins had individually completed questionnaires that included the following two questions relating to similarity:

(1) "In childhood, were you frequently mistaken by people who knew you?"

(2) "Do you differ markedly in physical appearance or coloring?"

Blood samples obtained from the twins were tested initially with 18 anti-sera at the MRC Blood Unit at the Lister Institute, and in doubtful cases a further five antisera were used. The responses of the twins to the questionnaire are cross-tabulated by zygosity on the basis of blood-typing in Table 5.3.

The data show that twins classified as likely to be monozygotic on the basis of blood groups almost always (92/94 pairs) give identical responses to both questions. Both agree that they were alike in appearance and confused for each other in childhood. By contrast, only five out of 84 dizygotic pairs agreed that they were both confused for one another and alike in appearance. The remaining DZ twin pairs showed every conceivable response pattern to the two questions (cf. Table 5.3). On the basis of these data, a criterion for zygosity diagnosis by questionnaire was established in which all pairs where both twins stated they did not differ in appearance *and* were mistaken for one another as children would be treated as monozygotic. Any kind of disagreement between twins about the two questions was taken as indicating they were dizygotic. On the original population, this criterion would lead to the misclassification of 5% of DZ twins as MZ and 2.5% of MZ twins as DZ. Of the 178 pairs studied, therefore, only seven (3.9%) would be misclassified by this criterion. Since our criterion was based on the same data, however, we should expect the actual errors of diagnosis on new data to be slightly greater. Generally, it is accepted that the questionnaire approach to zygosity diagnosis is correct in about 95% of possible cases. The

Table 5.3   Validation of zygosity diagnosis in London sample.

		MZ		DZ		
Question 1[a]	Question 2[b]	Female	Male	Female	Male	Total
Both state	Both alike	36	56	1	4	97
confusion	Disagree	1	—	1	2	4
	Both unalike	—	—	3	2	5
Disagree	Both alike	1	—	2	1	4
	Disagree	—	—	2	4	6
	Both unalike	—	—	2	6	8
Neither states	Both alike	—	—	3	—	3
confusion	Disagree	—	—	2	3	5
	Both unalike	—	—	18	28	46
Total		38	56	34	50	178

[a] In childhood were you frequently mistaken by people who know you?

[b] Do you differ markedly in physical appearance and coloring?

precise effect of such errors of diagnosis on the genetic analysis are difficult to assess. If errors are random, such that they merely cause MZ and DZ twins to be mixed up in both directions, then they will tend to lead to underestimation of the genetic component and an overestimation of non-genetic family resemblance. Estimates of the genetic contribution may be inflated if the most similar DZ pairs are misclassified as MZ. In general, the effect will depend on the causes of misclassification, but with a small error rate the effects are expected to be small.

5.1.3 Transformation of the raw scores

The scoring procedure adopted for the EPQ, in common with many personality tests, consists in summing the number of keyed responses to the items loading on each factor in turn. Most published validation studies employ the raw scores in computing means, correlations, etc. When we study family resemblance, however, there is no reason to suppose that the raw scores are the best indicator of the underlying trait nor that their pattern of inheritance should be especially straightforward. Geneticists therefore commonly employ a transformation prior to genetic analysis in order to minimize the complexity of non-additive effects for a particular scale. Both psychometricians and geneticists agree that there is nothing sacred about a particular scale of measurement. A good scale is one that can be used to maximize predictive power in terms of as few parameters as possible. Lord and Novick (1968) observe:

> If we construct a test score by counting up correct responses (zero-one scoring) and treating the resulting scale as having interval properties, the procedure may or may not produce a good predictor of some criterion. To the extent that this scaling process produces a good empirical predictor the stipulated interval scaling is justified . . . If a particular interval scale is shown empirically to provide the basis of an accurately predictive and usefully descriptive model, then it is a good scale and further theoretical developments might profitably be based on it. Thus measurement (or scaling) is a fundamental part of the process of theory construction.

Similarly, Mather and Jinks (1982, p. 63) argue:

> The scales of the instruments which we employ in measuring our plants and animals are those which experience has shown to be convenient to us. We have no reason to suppose that they are specifically appropriate to the representation of the characters of a living organism for the purposes of genetical analysis. Nor have we any reason to believe that a single scale can reflect equally the idiosyncrasies of all the genes affecting a single character . . . The scale on which the measurements are expressed for the purposes of genetical analysis must therefore be reached by empirical means. Obviously it should be

one which facilitates both the analysis of the data and the interpretation and use of the resulting statistics.

Thus a scale chosen to represent behavioral differences is often a matter of convenience rather than any belief that it represents the underlying biological or psychological process exactly. A different scale may be better for genetic analysis and uncover simplicity that is lost in the raw data. We belabor the issue of scaling because our genetic analyses of the personality dimensions are not conducted on the raw P, E, N and L scores, but rather on scores transformed to remove most of their undesirable statistical properties.

Nowhere is the problem of scale more apparent than in the analysis of the P scale of the EPQ. The "P" items generally have low endorsement frequencies in the P direction. As a result, the distribution of raw P scores is very highly skewed. There are also large sex differences in mean and variance (Table 5.4).

Early attempts to fit a model for twin resemblance in P foundered miserably because of sex interactions in both genetic and environmental effects and especially because of $G \times E$ interaction. Jinks and Fulker (1970) proposed a test of certain types of $G \times E$ interaction which involves correlating the means of each identical twin pairs with the corresponding within pair standard deviations (or absolute intrapair differences). In the absence of shared environmental effects, applying this test to monozygotic twins reared together detects genetic effects on sensitivity to the environment that are correlated with genetic effects on average trait value. The method does not detect genetic control of sensitivity to the environment mediated by genes independent of those affecting average trait value (Jinks and Fulker, 1970; Eaves *et al.*, 1977). These interactions are theoretically important because they mean that environmental effects on personality do not operate equally on all individuals but that some people are inherently more sensitive than others to their environment. If $G \times E$ interaction produces a correlation between means and standard deviations for pairs of monozygotic twins then

Table 5.4   Summary statistics for raw P scores of EPQ.

	Males	Females
N	318	770
Mean	3.365	2.131
Minimum	0	0
Maximum	14	11
Median	2.657	1.675
Mode	2	0
Variance	8.485	4.184
Skewness	1.174	1.246

it implies that the genes that generate the average level of a trait also mediate its sensitivity to the environment. For example, if monozygotic twins with high average neuroticism scores (i.e. that have the "neurotogenic genotype") are also more variable in their neuroticism scores then it is implied that the development of neurotic behavior may result from genes that promote neurotic behavior also making individuals more sensitive to environmental stress. Similar considerations apply to measures of other personality traits or physiological variables presumed to contribute to disease liability.

We computed the "mean-absolute difference" regressions for the raw psychoticism scores summarized in the previous table, as suggested by Jinks and Fulker, and found a correlation of 0.47 for female MZ twins and 0.50 for males in the London sample. When the linear correlations are computed for the extraversion and neuroticism scales, in contrast, these are close to zero. The absence of a linear relationship for E and N, however, does not imply the absence of systematic genotype × environment interaction. There is indeed a highly significant *quadratic* regression of absolute intrapair differences on pair means for MZ twins. The within-pair differences are large in the middle of the distribution and small in the tails, implying that intermediate levels of these traits are far more sensitive to the environment than values at either extreme.

If these results really mean that there is G × E interaction for personality then they would have enormous implications for our understanding of personality and psychiatric disorder. Certainly, on the raw scales there is G × E and if we decide to use the raw scales as our instruments for predicting change or the effects of treatment, then we cannot ignore its effects. However, the finding leaves two questions unanswered: (i) If there is G × E interaction then what kind of environmental effects are the genes interacting with? (ii) Is the scale on which interaction is detected a simple (linear) function of the hypothesized underlying dimensions of behavior that they purport to measure?

We attempted to answer both these questions at once for all the scales of the EPQ. We supposed that a significant portion of the variance within MZ twin pairs was due to errors of measurement. That is, even in the most ideal and controlled environment, the same individual does not complete a questionnaire in the same way on two occasions. He distributes his "keyed" responses over somewhat different answers each time. Such variation is inherent in the process of measurement and is not likely to be explained in terms of environmental factors that are readily accessible to measurement. Under these circumstances, there will be a relationship between the mean raw score on a scale and the variance (or standard error) within twin pairs. Thus we have a model for responses to items of a given scale that is very simple but works quite well in practice. Each subject is presumed to have his own predisposition to endorse items of a particular scale in a keyed direction. The subject's predisposition is measured by his probability p of a keyed response. Subjects are expected to differ in their p values. It is assumed (unrealistically) that a subject distributes his keyed responses randomly and independently over each item in the scale on each occasion and that the probability of endorsement is constant for all items within a scale.

If these strong assumptions are correct then the error variance in the raw score R on an N-item scale for a subject of predisposition p follows from the binomial distribution thus:

$$\sigma_R^2 = Np(1 - p).$$

Given these assumptions, it follows that the error variance will be highest for intermediate trait values ($p = \frac{1}{2}$) and lower for extremes ($p \rightarrow 0$ or 1). If the intrapair variance for MZ twins is due in large part to measurement error then we should expect the quadratic "mean-difference" regression that is actually observed for traits in which test scores range over much of the scale. Since, in the case of the P scale, very few subjects endorse more than one or two items, there are very few subjects with high raw scores, so that the overall mean-variance trend has a strong linear component since $R = Np$ approximately for small p.

These considerations lead us to expect that standard transformations will eradicate most of the mean-variance regression if the main source of $G \times E$ interaction is a correlation between true score and measurement error. In the case of the P scale, for which p is small, the square-root transformation should remove most of the regression. In the case of E, N and L, for which a much broader range of scores are represented, we expect the angular (arcsine) transformation to remove the mean-difference association.

The specified transformations indeed remove the mean-variance correlations for the raw scores in all the EPQ data sets to which we have direct access. The square-root transformation, for example, reduces the mean-difference correlation to -0.11 in female twins and -0.02 in males. Examples of other transformations have been examined by Martin and Jardine (1986), but are no better than those based on these simple theoretical criteria.

A transformation based on a strong model for measurement error removes almost all the $G \times E$ interaction for these tests, so the interaction need not be interpreted in terms of interaction between genes and environmental factors of long-term importance. In our analyses we therefore work with the transformed personality measures rather than the raw scores.

Alternative scales may also be considered. For example, the "latent-trait" model (see e.g. Birnbaum, 1968) results in maximum-likelihood estimates of trait values under a more general model relating the probability of response to particular items to subjects' trait values and item parameters. The items of the EPQ are not sufficiently numerous and variable in endorsement frequency to allow for good estimation of subjects' trait values over the entire range. Indeed, we attempted to estimate latent trait scores from a 76-item neuroticism questionnaire and found that the resulting scores were much *more* variable in the extremes because we had so little information about the trait values of extreme subjects. The endorsement frequencies of neuroticism items appear to cluster around intermediate values. Clearly, all the psychometric problems of personality measurement are far from solved by existing tests. In Chapter 11 we employ latent trait theory in an attempt to test whether variation in liability to depression is caused by one gene or many.

5.1.4 Age correction

Adult personality scores on the EPQ scales change significantly with age. The ages of family members are correlated and the ages of twin pairs are perfectly correlated. Therefore the regression of test scores on age is expected to inflate the resemblance of family members and, in the case of twins, lead to overestimation of the variation between pairs. If the regression and age distributions are the same for MZ and DZ twins then the excess between pair variation should be the same for MZ and DZ twins and should lead to a pattern of twin differences resembling that caused by the family environment. For this reason, therefore, the mean squares (or twin correlations) have to be corrected for the average trend in personality with age.

Typically, age correction is achieved by regressing test scores on age and assigning to each individual an "age-corrected" score, which is the deviation of his observed test score from the value predicted from knowledge of his age and the best-fitting age regression. Such correction is only strictly appropriate if the individuals in the sample are unrelated and hence uncorrelated in age. In data with variable family structures the maximum-likelihood method (see Chapter 6) can be applied to estimate age regressions as part of the model-fitting analysis. With twin data, however, the perfect correlation in age means that the sums of squares between pairs can be corrected directly for age regression within each twin group. Since members of a pair are identical in age, the within-pair variances are not affected by age differences. We let the scores of the first and second twins of the ith pair be t_{i1} and t_{i2} respectively, and $\overline{T}$ be the sample mean. We let x_i be the age of the pair and $\overline{X}$ be the mean age of the sample. The sum of squares between twin pairs, uncorrected for age, is

$$SS_b = \tfrac{1}{2}\sum_i (t_{i1} + t_{i2})^2 - 2N\overline{T}^2.$$

The sum of squares of age (between pairs) is

$$SS_a = 2\sum_i x_i^2 - 2N\overline{X}^2,$$

and the between pairs sum of products of age and scores is

$$SP_{ab} = \sum_i (t_{i1} + t_{i2})x_i - 2N\overline{T}\,\overline{X}.$$

The estimated regression of twin pairs on age is thus

$$b = SP_{ab}/SS_a.$$

The sum of squares due to regression of pairs on age is

$$SS_r = (SP_{ab})^2/SS_a,$$

so that the sum of squares between pairs corrected for the regression on age is

$$SS'_b = SS_b - SS_r.$$

The uncorrected sum of squares between pairs has $N - 1$ df, where N is the number of twin pairs. The corrected sum of squares has one fewer df, $N - 2$, to allow for the fact that the linear age regression has been estimated from the twin-pair data. In most of our analyses of twin pair data, the analyses of variance tables contain the sums of squares corrected for age in this way. Non-linear regressions may be computed if required, and additional reductions in the sum of squares between pairs accompanied by further reductions in the df.

5.1.5 Correction for sex differences

The effects of sex may be of two kinds. There may be average sex differences, reflecting effects common to all members of a sex, and "person × sex" interactions resulting from sex differences in genetic and environmental effects. A genetic example would be the case of sex limitation in which some genes are expressed in males and others in females. The effects of such interactions are not removed from the data by typical corrections for sex but can be the subject of informative genetic analysis.

The method of correction for the average effects of sex is similar to that employed for age correction. Test scores of individuals are expressed as deviations from the mean for their sex rather than the mean of the population. In the case of like-sex twins correction of the analyses of variance for sex is unnecessary since separate analyses of variance are conducted for each subgroup of twins so that variation within each sex and zygosity subgroup is automatically expressed around the subgroup means. In contrast, pairs of unlike sex require special treatment. Sex differences inflate estimates of the variance within unlike-sex twin pairs. Therefore, in the analysis of unlike-sex pairs we remove that part of the within-pair sum of squares that is due to the average sex difference. A full analysis of variance of unlike-sex twin pairs would identify the following sources of variance:

Source	df
Age regression	1
Between pairs	$N - 2$
Sex	1
Age × sex	1
Residual	$N - 2$
Total	$2N - 1$

The residual item contains effects due to genetic segregation and environmental differences within families. In practice, we have found no age × sex interaction in our samples of unlike-sex twins, so that the tabulated analyses of variance are only corrected for the overall age effect within these groups and the sum of squares for age × sex interaction is thus pooled in the residual item. For comparison with the other twin groups, and to reflect the theoretical perspective on which the subsequent analysis is based, we label the residual term for unlike-sex pairs "within pairs" in tables.

The possibility of genotype × sex interactions raises an important question that is only partly resolved by using the analysis of variance to summarize twin data. The analysis of variance assumes that the same error variance applies to each observation in the sample. In the case of unlike-sex pairs this means that the within-sex variances have to be homogeneous across sexes. This will not necessarily be the case if there is sex limitation. If there is sex limitation, then the analysis of variance is not the ideal data summary and the variances and covariances of the first and second twin are to be preferred as sufficient second-degree statistics. However, the computations required by fitting the models to covariance matrices are somewhat heavier (see e.g. Chapter 6), and our analyses of twin data typically focus on mean squares.

5.1.6 Data summary

The analyses of variance of the transformed P, E, N and L scores for each group of adult twins in the London sample are presented in Table 5.5. The linear regression on age is removed from the sums of squares between pairs, and the average effect of sex is removed from the sum of squares within unlike-sex pairs.

Table 5.5 Mean squares for transformed P, E, N and L from the London adult twin sample (EPQ).[a]

Zygosity	Sex	Statistic	df	Mean squares			
				P	E	N	L
Monozygotic	Female	Between pairs	231	327	109	105	130
		Within pairs	233	138	40	43	42
	Male	Between pairs	68	457	143	129	160
		Within pairs	70	138	32	43	52
Dizygotic	Female	Between pairs	123	388	108	78	132
		Within pairs	125	191	75	68	52
	Male	Between pairs	45	291	79	72	58
		Within pairs	47	194	54	62	62
	Unlike	Between pairs	66	346	75	87	106
		Within pairs	67	221	53	62	74

[a] Corrected for age and sex; multiplied by scale factor to remove decimals.

The within-pair variances of MZ twins reflect the joint effects of measurement error and long-term environmental differences within families on variation in the four EPQ scales. If there is a genetic component in personality differences then the variances within DZ pairs are expected to exceed those within MZ pairs of the same sex. This prediction is borne out for almost all variables in both sexes. In the majority of cases the mean squares between pairs significantly exceed the corresponding mean squares within pairs, indicating significant correlations between the personality scores of twins. Although the data within pairs imply a genetic component to family resemblance, more rigorous analysis of the genetic and social determinants of differences between families is achieved by the methods of model fitting discussed in Chapter 4.

5.1.7 Fitting models without sex limitation

We consider a number of alternative hypotheses for the causes of variation in each personality dimension. Table 5.6 gives the contributions of five theoretically important parameters to the mean squares between and within pairs for MZ and DZ twins, in the absence of sex differences in genetic and environmental effects.

The parameters are the additive genetic variance V_A, the dominance genetic variance V_D, the environmental variance E_W, *within* families, the environmental variance E_B *between* families and the genotype–environmental covariance C_A due to competition or cooperation between siblings based on additive genetic difference.

Table 5.6 Expectations of mean squares assuming no sex differences in genetic and environmental effects.[a]

Zygosity	Sex	Mean square	Expectation				
			V_A	V_D	C_{AS}	E_W	E_B
MZ	F	B	2	2	4	1	2
		W	.	.	.	1	.
MZ	M	B	2	2	4	1	2
		W	.	.	.	1	.
DZ	F	B	$1\frac{1}{2}$	$1\frac{1}{4}$	3	1	2
		W	$\frac{1}{2}$	$\frac{3}{4}$	-1	1	.
DZ	M	B	$1\frac{1}{2}$	$1\frac{1}{4}$	3	1	2
		W	$\frac{1}{2}$	$\frac{3}{4}$	-1	1	.
DZ	M/F	B	$1\frac{1}{2}$	$1\frac{1}{4}$	3	1	2
		W	$\frac{1}{2}$	$\frac{3}{4}$	-1	1	.

[a] Sibling interaction based on genetic-dominance effects omitted, not all parameters may be estimated simultaneously.

Several important points should be considered in studying the expectations given in the table. First, although there are ten mean squares and only five parameters, all five parameters cannot be estimated simultaneously because the expectations for males and females are identical under the general model. Also, the effects of dominance cannot be estimated simultaneously with those of the family environment as long as we have only twins reared together. Secondly, the coefficients specified for the additive genetic component assume mating to be random. Assortative mating based on genotype or phenotype will inflate the variance between families by a factor which depends on the amount and type of assortative mating. If there is assortative mating then Fisher (1918) showed that the genetic variance between siblings is increased by

$$V_{AM} = \frac{A}{2(1-A)}V_A ,$$

where A is the correlation between the additive genetic deviations of spouses. The value of A will depend on the contribution of genes affecting personality to the phenotype on which mate selection is based, the intensity of mate selection and the mechanism of familial transmission (see Heath and Eaves, 1985). If assortative mating is based directly on the measured phenotype for personality then $A = h^2\mu$, where h^2 is the narrow heritability in the assortatively mating population and μ is the marital correlation between mates. We shall explore the effects of assortative mating in later chapters (e.g. Chapters 6 and 16). At present it is sufficient to note that our estimate of the shared environment from twins reared together is biased upwards by the genetic effects of assortative mating if we assume mating to be random when it is not. The estimate of the between-family component $\hat{E}_B$ is thus $E_B + V_{AM}$. Similarly, the estimate of the additive genetic variance $\hat{V}_A$ will be smaller than the true genetic component in the assortatively mating population, but will approximate the value that would be obtained if the population under investigation were subsequently mated at random.

Although we are unable to resolve the four parameters V_A, V_D, E_W and E_B simultaneously with just twins reared together, we may nevertheless, fit and interpret a number of simpler versions of the model. Eaves (1970) has shown that if we ignore dominance and estimate V_A, E_W and E_B then our parameter estimates will be biased by a function of any dominance effects that have been discounted. Thus if dominance is wrongly ignored then we shall obtain

$$\hat{E}_W = E_W,$$
$$\hat{E}_B = E_B - \tfrac{1}{4} V_D,$$
$$\hat{V}_A = V_A + \tfrac{3}{2} V_D.$$

Dominance therefore tends to reduce the estimate of the family environmental component and increase our estimate of the additive genetic contribution. If the estimate of E_B should ever be significantly negative then this is possibly due to non-additive genetic effects. Conversely, if we ignore E_B and estimate the three remaining parameters then we obtain

$$\hat{V}_A = V_A + 3E_B,$$
$$\hat{V}_D = V_D - 2E_B,$$
$$\hat{E}_W = E_W.$$

In both cases the estimates of E_W are unbiased. The above relationships enable us to translate parameter estimates of one model into those of the other, if necessary.

A third consideration relates to the effects of competition and cooperation. Such social interactions have two effects on family resemblance and individual differences: (1) they create an additional source of environmental variation; and (2) they generate genotype–environment covariance if some of the genes generating the social interaction also affect the phenotype directly. In order to detect sibling interactions based on genotype, we may either compare the statistics for individuals reared at different densities (e.g. twins and singletons) or we may compare individuals of differing degrees of genetic relationship at the same density (e.g. MZ and DZ twins). The former design detects sibling interactions as differences in total variance (Eaves, 1976b) due to both the environmental and genotype–environmental components of sibling interaction. The twin design cannot resolve the environmental component of sibling interaction from other genetic and environmental effects, but the genotype–environmental covariance resulting from sibling interaction may produce differences in the total variances of MZ and DZ twins and aberrant patterns of twin resemblance. When a model involving genetically based sibling interactions is fitted to data on twins, therefore, the estimate of V_A will contain both the direct effects of genes on the phenotype *and* the indirect (environmental) effects of the cotwin's genes. A single parameter V_A is used to represent both these effects when they are confounded. The genotype–environment covariance parameter C_A', however, is not confounded with the additive genetic variance, and is represented separately in the model.

We fitted six basic models to the twin data for each of the four variables. These are described separately.

(1) E_W only

A model that includes only environmental differences within families predicts no family resemblance and therefore no significant differences between twin pairs. All variation would occur within pairs, and the within-pair variances should be the same for MZ and DZ twins. Examination of the raw mean squares and almost every twin study in the past suggests that this model is utterly inappropriate. It is included to provide a baseline against which the statistical importance of subsequent gains can be assessed.

(2) E_W and E_B only (the "environmental model")

This model assumes that genetic factors play no role in individual differences but that environmental differences between families account for the significant resemblance between twins. The second model is theoretically important because it promises to identify those variables for which the causes of family resemblance are purely cultural. Such variables would be valuable markers for testing more subtle hypotheses about the mechanism of cultural inheritance in man. Our test of this model is equivalent to showing that the correlations of MZ and DZ twins do not differ significantly and that the within-pair variances are the same for both types of twin. Our preliminary examination of the mean squares suggests that this model is unlikely to fit the EPQ data very well.

(3) V_A and E_W only (the "simple genetic model")

The theoretical importance of this model lies in the radical assumption that all family resemblance is genetic and that there is no non-genetic similarity between twins. The model tests whether parental influence and correlated learning experiences are affecting personality in the twins. If the model fails then we may need to consider shared environmental effects, genetic dominance, assortative mating or sibling interaction as alternative hypotheses — with or without genetic factors. The model makes very strong predictions about the similarity of other types of family members and suggests that, apart from the idiosyncratic differences in environmental experiences within families, all the measured variation is genetic. This model challenges many common suppositions about the role of the family in child development. In so far as certain psychiatric disorders represent extremes of the distribution of normal personality, it supports a biological model for psychiatric disease in preference to a cultural model. It embodies the assumption that the MZ correlation is exactly twice the DZ correlation (cf. Chapter 3). It does not assume that all the variation is genetic, because the model includes a parameter E_W representing environmental effects within families. It is termed the "simple genetic" model because it is assumed that genes alone are responsible for the correlations between relatives.

(4) E_W, E_B and V_A (the "genotype–cultural" model)

This model includes both genetic and cultural components of family resemblance and is the most complicated model that can be fitted to twin data without taking sex limitation into account. If this model fits better than both of the previous models then we conclude that the trait is affected by both genetic and environmental factors and that the relevant environmental influences are correlated between family members. Before this model is accepted, the parameter estimates should all be significantly greater than zero. The model will improve the fit to any twin data set in which the correlation of DZ twins is significantly less than the MZ correlation *and* significantly greater than half the correlation of MZ twins.

(5) E_W, V_A and V_D (the "dominance" model)

Model 5 recognizes that the genetic component of variation may comprise both additive and dominance effects. The variation due to dominance is represented by V_D in the model. Ideally, we should like our model to include both E_B and V_D, but they cannot be resolved with only data on twins reared together. We have to be content with a comparison of results for models 4 and 5, using the sign and significance levels of the parameters in the attempt to decide whether failure of the simpler models was due to dominance or the family environment. A significantly negative value of dominance in model 5 may be taken as evidence in favor of model 4 (or the effects of sibling

cooperation). A significant negative estimate of E_B in model 4 may support
the hypotheses of dominance or sibling competition. The dominance model
is likely to improve the fit to any data set in which the DZ correlation is
significantly less than half the MZ correlation. However, as Martin *et al.*
(1978) have shown in simulation studies, the amount of dominance has to be
very large in order to be detectable even by quite large twin studies.

(6) V_A, C_A and E_W (the "competition–cooperation" model)

The effects of competition and cooperation are represented in this model in
the simplest possible form. The parameter C_A represents the genotype–environment covariance generated if the same genes have a direct effect
on the trait and an indirect effect on the environment of a cotwin. If the genes
responsible for sibling interaction differ from those responsible for the direct
effects on the trait then the effects of sibling interaction will remain
undetected as long as the study employs twins alone. However, comparison
with singletons may then permit further resolution of the direct and indirect
effects of genes (see Chapter 7). A negative C_A implies competitive interactions, since genes that increase trait expression in one twin of a pair are
generating an environment that *reduces* trait expression on the other twin. A
positive parameter value supports the hypothesis of cooperation or mutual
reinforcement, since genes that increase the trait value of one twin create an
environment for the cotwin that tends to *increase* the trait expression. In the
presence of competition or cooperation, estimates of the additive genetic
variance derived from twin data alone will be confounded with the environmental effects created by social interaction generated by genetic differences.
In practice, it may be difficult with twin data alone to resolve the effects of
competition from those of dominance, since both lead to an increase in the
ratio of MZ : DZ correlations. Similarly, the effects of cooperation may be
hard to distinguish from other environmental sources of twin resemblance,
which are represented by E_B in the model. A recent paper by Carey (1986)
presents another model for sibling interaction which allows the effects of
competition and cooperation to develop over time. We restrict our analysis
to the model for interaction based on genetic effects originally formulated by
Eaves (1976b).

5.1.8 Method

The method of weighted least squares described in the previous chapter was
used to fit the six models to the four sets of mean squares. The goodness-of-fit
chi-squares may be used to decide which models could explain the mean

squares and to help decide which parameters should be included or dropped from the model. If we have two models, one which contains p parameters and one in which k of these parameters have been set to zero *ex hypothesi* then the loss function s_r^2 for the "reduced" model with $p - k$ parameters will be greater than that for the "full" model, s_f^2 with p parameters. Given that our statistical assumptions are correct, the difference between these two chi-squares $d = s_r^2 - s_f^2$ is itself a chi-square for k df. If this chi-square is significant then the k parameters cannot legitimately be omitted from the model, for to do so makes the fit significantly worse. If the difference is not significant then there is some justification for adopting the simpler model because the data do not justify the additional parameters of the full model.

5.1.9 Results

5.1.9.1 *Psychoticism*

The model-fitting results for the transformed P scores are given in Table 5.7. The model that assumes no familial correlation in P scores fails badly (model 1), and that which assumes only environmental causation barely fits ($\chi^2_8 = 14.9$, $P = 0.06$). It is tempting to conclude that an environmental model is just adequate for psychoticism. However, the fit of the V_A, E_W model is very much better, having a goodness-of-fit chi-square for 8 df very close to its expected value ($P = 0.52$). The issue is resolved by the results for models 4, 5 and 6, since adding dominance, the family environment or competition to the model yields no improvement over model 3, which

Table 5.7 Results of fitting models to psychoticism scores of adult London twins assuming no sex differences in genetic and environmental effects.

Model	Parameter estimates					Goodness-of-fit		
	E_W	E_B	V_A	V_D	C_A	χ^2	df	P
1	260***	—	—	—	—	85.8	9	<0.0001
2	165***	96***	—	—	—	14.9	8	0.06
3	139***	—	122***	—	—	7.2	8	0.52
4	142***	20	101**	—	—	6.7	7	0.46
5	142***	—	160**	−40	—	6.7	7	0.46
6	139***	—	122***	—	<1	7.2	7	0.41

Note: Significance of parameter estimates tested by comparing with theoretical error. Such standard errors only apply for models that fit the data. Significance levels as follows:

* $P < 0.05$; ** $P < 0.01$; *** $P < 0.001$.

assumes only additive gene action and environmental effects within families. There is no evidence whatever of competition or cooperation ($C_A < 1$ and not significantly different from zero). The non-significant negative estimate of V_D in model 5 precludes any significant contribution of genetic non-additivity to psychoticism, and the non-significant positive estimate of E_B under model 4 suggests that the family environment's role in individual differences is too small to be detected in these data. Since model 3 fits very well and cannot be improved markedly by adding the most important alternative parameters, we conclude that variation in psychoticism is best explained by a very simple model in which genes act additively and the environmental causes of variation differentiate between individuals *within* the family rather than between families.

A model that "fits" is not necessarily "right", because effects that are large enough to be practically important might nevertheless be missed by chance alone. In Chapter 3 we discussed the power studies of Martin *et al.* (see Tables 3.6 and 3.7). Our sample of adult twins comprises 543 twin pairs, of which approximately 50% are MZ pairs. Table 3.6 shows that our study is clearly large enough to reject the purely environmental model ("E_B, E_W") when the MZ correlation is about 0.5 and due entirely to genetic factors. Similarly, our study has an excellent chance of rejecting the E_W, V_A model when the trait is entirely environmental with MZ correlations of around 0.5. However, when the MZ correlation is partly genetic and partly environmental, discrimination is much more difficult. From Table 3.7, we find that a sample would have to be approximately twice the size (1233 pairs) to be 95% certain of detecting a between-family environmental component that explained 30% of the total variance when 20% was due to genetic effects and the remainder explained by environmental effects within families was due to genetic factors.

Under the E_W, E_B, V_A model, the parameter estimates and their standard errors are as follows:

parameter	estimate	standard error
E_W	141	11
E_B	20	32
V_A	100	37

The total variance is thus $141 + 20 + 100 = 261$. Dividing each parameter estimate by the total variance yields the proportional contribution of the three sources to individual differences in psychoticism as measured by the transformed P scale thus: 54% within family environmental effects; 8% between-family environmental effects, 38% additive genetic effects. Taking into account the standard errors, however, we recognize that the contribution of the family environment could differ greatly from the estimated value.

We have dwelt on the issue of power and sampling variation at this early stage because it is important to have a sense of what family studies can and cannot achieve. We are reasonably confident, on the basis of our study, that genetic factors contribute almost as much to variation in P scores as the environment unique to each individual in the family. On the other hand, although we have no convincing evidence that the environment shared by family members plays any role whatever, as much as 30% of the total variation could be due to E_B, and yet not be detected with any certainty in the London sample. Subsequent analyses will show that the effects of the family environment can be detected for other variables but not for psychoticism.

5.1.9.2 Extraversion

The results for extraversion (Table 5.8) are quite similar to those for P. The E_W model is rejected soundly, along with the model that assumes all family resemblance is due to the shared environment (model 2). Once again, the third model, which assumes only additive genetic effects and within-family environmental causes, gives a very good fit. Little improvement follows the addition of the family environment (model 4), dominance (model 5) or competition (model 6). Thus it is fairly certain that the family environment does not contribute significantly to variation in extraversion. An important ambiguity is revealed when we examine the genetic parameters, however, since the inclusion of dominance in the model yields an estimate of the dominance variance that is larger than the additive component. Furthermore, in model 5 neither genetic component differs significantly from zero. The following matrix gives the variances of the parameter estimates for model 5

Table 5.8 Results of fitting models to extraversion scores of adult London twins assuming no sex differences in genetic and environmental effects.

Model	Parameter estimates					Goodness-of-fit		
	E_W	E_B	V_A	V_D	C_A	χ^2	df	P
1	78***	—	—	—	—	96.8	9	<0.0001
2	50***	28***	—	—	—	33.7	8	<0.0001
3	40***	—	39***	—	—	11.9	8	0.15
4	38***	−11	52***	—	—	10.2	7	0.18
5	38***	—	16	24	—	10.2	7	0.18
6	38***	—	46***	—	−4	10.1	7	0.18

Note: Significance of parameter estimates tested by comparing with theoretical error. Such standard errors only apply for models that fit the data. Significance levels as follows:

* $P<0.05$; ** $P<0.01$; *** $P<0.001$.

on the diagonal, their covariances in the upper triangle and their correlations in the lower triangle:

$$
\begin{array}{c}
 \\
E_{\mathrm{W}} \\
V_{\mathrm{A}} \\
V_{\mathrm{D}}
\end{array}
\begin{array}{ccc}
E_{\mathrm{W}} & V_{\mathrm{A}} & V_{\mathrm{D}} \\
\left[\begin{array}{ccc}
9.22 & 10.27 & -17.02 \\
0.17 & 387.73 & -390.02 \\
-0.28 & -0.98 & 409.72
\end{array}\right]
\end{array} .
$$

The exceptionally high negative correlation (-0.98) between estimates of the additive and dominance components of genetic variation in these data is characteristic of attempts to resolve additive and dominance effects with twin data. We may obtain an estimate of the total genetic variance as $\hat{V}_{\mathrm{A}} + \hat{V}_{\mathrm{D}} = 15.9 + 24.0 = 39.9$. The variance of the estimate is

$$
V(\hat{V}_{\mathrm{A}}) + V(\hat{V}_{\mathrm{D}}) + 2 \operatorname{Cov}(\hat{V}_{\mathrm{A}} \, \hat{V}_{\mathrm{D}}) = 387.73 + 409.72 - 2 \times 390.02 = 17.41.
$$

The square root of this value gives the standard error of the estimate of the total genetic variance as 4.17. Thus the total genetic variance is known with remarkable precision and differs very significantly from zero although we are unable to resolve the genetic variance into its additive and non-additive components and show that either is significant by itself. The problem is caused mainly by the high correlation between coefficients of the additive and dominance components in the model for twin data (cf. Table 5.6) and constitutes the major weakness of the classical twin design as a tool for resolving the components of genetic variance.

In the absence of statistical evidence for the unique contribution of dominance, we assume that the genetic variation is largely additive, but recognize that a large dominance effect might remain undetected against the background of genetic additivity.

On balance, our assessment of the London data for extraversion leads us to propose the same basic model as that for psychoticism: additive gene action; within-family environmental effects; little or no effect of the family environment.

5.1.9.3 Neuroticism

Neuroticism shows a very similar pattern of results to extraversion. Both environmental models fit very poorly (Table 5.9) and the additive genetic model fits very well indeed without the addition of dominance, the family environment or competition. The case for shared environmental effects is very weak, since the estimate of E_{B} from fitting model 4 is significant and negative. It is unlikely that the negative E_{B} results from competition since the small negative estimate of C_{A} in model 6 is not significant and the change in chi-square over model 3 is trivial. However, the estimate of V_{D} in model 5 is

Table 5.9 Results of fitting models to neuroticism scores of adult London twins assuming no sex differences in genetic and environmental effects.

	Parameter estimates					Goodness-of-fit		
Model	E_W	E_B	V_A	V_D	C_A	χ^2	df	P
1	75***	—	—	—	—	67.6	9	<0.0001
2	53***	22***	—	—	—	20.9	8	0.0075
3	44***	—	30***	—	—	5.2	8	0.73
4	42***	−16*	49***	—	—	2.5	7	0.93
5	42***	—	−2	34*	—	2.5	7	0.93
6	43***	—	38***	—	−4	4.0	7	0.78

Note: Significance of parameter estimates tested by comparing with theoretical error. Such standard errors only apply for models that fit the data. Significance levels as follows:

* $P<0.05$; ** $P<0.01$; *** $P<0.001$.

just significant and associated with a marginally significant improvement in fit over model 3. The negative estimate of the additive genetic component in model 5 is not unusual when attempting to resolve additive and dominance effects in twin data, and stems from the very high correlation between estimates of these two parameters (see above). Our preliminary study of neuroticism therefore provides substantial support for the contribution of genetic factors and the environment within families, but gives no significant indication that the shared family environment plays any role. Although there is a hint of dominance, we remain cautious about interpreting the estimates of additive and dominance parameters too stringently because of their large sampling errors and the high correlations between them.

5.1.9.4 The Lie scale

The results for the lie scale are the most enigmatic of the four sets considered here. Although the environmental models both fail, the fit of the various genotype–environmental models is far from compelling. While the observed and expected mean-squares do not differ significantly under model 3, the chi-square is significant at the 10% level (Table 5.10) and none of the additional parameters make any marked improvement.

Inspection of the data (Table 5.5) shows substantial difference between sexes in the relative magnitudes of the MZ and DZ correlations. This may indicate a different mechanism of family resemblance in males and females. Further discusion of the lie scale is thus postponed for the analysis of sex differences in genetic and environmental determination.

Table 5.10 Results of fitting models to lie scores of adult London twins assuming no sex differences in genetic and environmental effects.

Model	Parameter estimates					Goodness-of-fit		
	E_W	E_B	V_A	V_D	C_A	χ^2	df	P
1	88***	—	—	—	—	119.6	9	<0.0001
2	51***	37***	—	—	—	21.7	8	0.0056
3	43***	—	45***	—	—	13.9	8	0.09
4	44***	9	35**	—	—	12.9	7	0.07
5	44***	—	64**	−18	—	12.9	7	0.07
6	44***	—	39***	—	3	12.9	7	0.07

Note: Significance of parameter estimate tested by comparing with theoretical error. Such standard errors only apply for models that fit the data. Significance levels as follows:

* $P < 0.05$; ** $P < 0.01$; *** $P < 0.001$.

5.1.10 Models incorporating sex-limited effects

In Table 5.11 we present the elements of a model for sex limitation in the expression of genetic and environmental effects. Initially the term "sex limitation" was reserved for inherited traits, such as baldness, in which autosomal genetic differences are expressed only in one sex. However, just as biometrical genetics has broadened the definition of dominance to include various degrees of intermediate dominance, so we may use the term "sex limitation" to refer to any case in which the effects of autosomal loci on a continuous trait differ between the sexes. Baldness would be an extreme case in which the genetic effects were only expressed in one sex. Sex-limited effects may be of two kinds: (1) the same genes may contribute to variation in both sexes, but the magnitude of their effects is different in males and females ("scalar sex limitation"); (2) some genes expressed in one sex may not be expressed in the other ("non-scalar sex limitation"). Our use of the term "sex limitation" covers sex differences in the scale of gene action *and* sex differences in *which* autosomal genes actually affect a trait. For simplicity we also use the term "sex limitation" to include the analogous process in the environmental determinants that would arise if males and females were exposed to different salient cultural environments. This more general use of the term causes no problem because it will always be clear from the particular model whether we are speaking of genetic effects, environmental effects, or both.

The basic principles of modelling all types of sex limitation, genetic and environmental, are the same. We consider the genetic aspects first and confine ourselves to the treatment of additive gene action because the treatment

Table 5.11 Expectations of twin mean squares under V_A, E_W, E_B model when genetic and environmental effects vary over sexes.[a]

Zygosity	Sex	Mean square	V_{AM}	V_{AF}	V_{AMF}	E_{WM}	E_{WF}	E_{BM}	E_{BF}	E_{BMF}
MZ	F	B	.	2	.	.	1	.	2	.
		W	.	.	.	.	1	.	.	.
MZ	M	B	2	.	.	1	.	2	.	.
		W	.	.	.	1	.	.	.	.
DZ	F	B	.	$1\frac{1}{2}$	.	.	1	.	2	.
		W	.	$\frac{1}{2}$	.	.	1	.	.	.
DZ	M	B	$1\frac{1}{2}$	.	.	1	.	2	.	.
		W	$\frac{1}{2}$	.	.	1	.	.	.	.
DZ	M/F	B	$\frac{1}{2}$	$\frac{1}{2}$	$\frac{1}{2}$	$\frac{1}{2}$	$\frac{1}{2}$	$\frac{1}{2}$	$\frac{1}{2}$	1
		W	$\frac{1}{2}$	$\frac{1}{2}$	$-\frac{1}{2}$	$\frac{1}{2}$	$\frac{1}{2}$	$\frac{1}{2}$	$\frac{1}{2}$	-1

[a] Not all parameters may be estimated simultaneously.

of dominance deviations requires no new principles. As long as we confine our analysis to like-sex pairs, we simply use the same expectations as those in Table 5.6 for the classical autosomal case, except that we now define a separate additive component for each sex: V_{AM} for males and V_{AF} for females. The two components measure the effects of all genes which contribute to differences in either sex. Genes contributing to V_{AM} may or may not contribute to V_{AF} and *vice versa*. However, the expectations for unlike-sex relatives are more complex because genes that affect the trait in males may not do so in females and *vice versa*. The covariances of unlike-sex relatives only reflect genes whose effects are expressed in both sexes. We define V_{AMF} to represent this covariance term. This parameter only occurs in the expectations of covariances (and hence mean squares) for unlike-sex relationships. If genes affecting males are distinct from those affecting females then V_{AMF} will be zero, even though V_{AM} and V_{AF} may be highly significant. These circumstances could also arise, theoretically, if the genes affecting the trait were expressed in both sexes but the directions of individual gene effects were inconsistent over sexes.

We may define the parameter r_{AMF} to represent the consistency of additive genetic effects in males and females:

$$r_{AMF} = V_{AMF}/(V_{AM}V_{AF})^{1/2}.$$

Although the value of this correlation should lie between -1 and 1, the components of variance estimated from the linear model may not satisfy this range constraint. We know of no case in which the failure to impose this constraint has led to serious errors in deciding between models of personality in twins.

The same approach may be used to specify sex-limited effects on the between-family components of environmental variation. The model in Table 5.11 defines E_{BM}, E_{BF} and E_{BMF} in a form analogous to the sex-limited genetic components. The within-family environmental components never contribute to the covariances between relatives, whatever their sex, so there is no need to define a covariance term E_{WMF} for the within-family environment. However, since the magnitude of E_W may differ between males and females, separate variance components are included for these effects in the model (Table 5.11).

5.1.11 Allowing for sex limitation

It is not possible to fit every possible combination of the parameters specified in Table 5.11. We considered two major aspects of the sex-limitation model: genetic and environmental. In one model we assume no family environmental effects but allow for sex differences in E_W and V_A. The model is thus the sex-limited analogy of model 3 presented in the previous analyses of the personality dimensions. Since the model involves five parameters, the goodness-of-fit chi-square has $10 - 5 = 5$ df. This chi-square may be subtracted from that obtained for model 3 to yield a chi-square test for improvement in fit having 3 df since the more general, sex-limitation, model has three additional parameters. The second model we tested is the sex-limited version of the E_B, E_W model employed previously (model 2 in Tables 5.7–5.10). This model also yields a chi-square having 5 df for testing goodness of fit and may be compared against model 2 for the same trait to see whether sex limitation of environmental effects is a necessary feature of any model for personality differences.

The results for the genotype–environmental model are given in Table 5.12 and those for the purely environmental model in Table 5.13.

The findings for the effects of sex limitation of the additive genetic component and within-family environment are remarkably consistent over all four traits. For P, E, N and L the improvement in fit resulting from the sex-limitation model is marginal and never approaches significance. The data provide no evidence that the four aspects of personality are better explained by a model in which different genes and environmental effects contribute to variation in males and females. The values of r_{AMF} are instructive. They all exceed 0.7, confirming a high degree of communality between genetic effects on the two sexes. Indeed, the fact that no improvement is achieved by the sex-limitation model points to the four correlations being no different from their upper bound of unity in the London data. However, the standard error of r_{AMF} is expected to be large (since a value of 0.7 is not significantly

Table 5.12 Improvement in V_A, E_W model when allowance is made for sex differences in genetic and environmental effects in adult London twin data.

Parameter	Trait			
	P	E	N	L
E_{WM}	135***	31***	43***	49***
E_{WF}	139***	42***	45***	40***
V_{AM}	144***	43***	35***	38***
V_{AF}	116***	37***	29***	47***
V_{AMF}	112*	33*	26	30?
r_{AMF}	0.866	0.814	0.807	0.713
Goodness-of-fit χ_5^2	6.5	8.9	4.5	12.3
P	0.26	0.11	0.48	0.03
Improvement in fit χ_3^2	0.7	1.3	0.7	1.7

Note: Improvement in fit is judged by subtracting the goodness-of-fit chi-square under this model from that obtained when sex-limited effects are deleted from the model (cf. Tables 5.7–5.10). Significance levels as follows:

? $P<0.10$; * $P<0.05$; ** $P<0.01$; *** $P<0.001$.

Table 5.13 Improvement in E_B, E_W model when allowance is made for sex differences in environmental effects in adult London twin data.

Parameter	Trait			
	P	E	N	L
E_{WM}	162***	40***	50***	56***
E_{WF}	157***	52***	52***	45***
E_{BM}	119***	35***	28***	32***
E_{BF}	97***	27***	22***	43***
E_{BMF}	56*	16*	13?	15?
r_{BMF}	0.521	0.528	0.527	0.416
Goodness-of-fit χ_5^2	10.1	29.7	18.7	12.2
P	0.07	<0.0001	0.0022	0.03
Improvement in fit χ_3^2	4.8	4.0	2.2	9.5

Note: Improvement in fit is judged by subtracting the goodness-of-fit chi-square under this model from that obtained when sex-limited effects are deleted from the model (cf. Tables 5.7–5.10). Significance levels as follows:

? $P<0.10$; * $P<0.05$; ** $P<0.01$; *** $P<0.001$.

different from unity), so there could be a large effect of sex limitation that would remain undetected in a study with relatively few unlike-sex twin pairs. The London data, however, give little or no reason to suppose that any genes exercise effects specific to either sex.

Allowing for sex differences in the family-environmental component (Table 5.13) gains little in comparison with the E_B, E_W model already considered. The sex-limited environmental model fails in every case and only gives a significant improvement for the Lie scale ($\chi^2_3 = 9.5$, $P < 0.05$). Thus the data imply some form of heterogeneity across sexes for the "Lie" scale, but it is not explained adequately either in terms of a "purely" genetic model or a purely environmental model.

We attempted to account for the anomalies in the results for the Lie scale by adding a parameter for family environmental effects to the expectations of female like-sex twins. The results of this analysis are summarized in Table 5.14.

The results indicate a significant improvement in fit, which confirms that females are especially sensitive to their social environment as far as their scores on the Lie scale are concerned. However, the modification does not remove all the anomalies. The estimates of V_{AF} and V_{AMF} do not differ significantly from zero, yet the estimate of r_{AMF} is 1.19, consistent with the expression of the same genes in both sexes. Thus, although we can be certain that there are genetic effects in males and effects of the family environment in females, we are unable to decide whether there are any genetic effects in females and whether there is sex-limited gene expression.

5.1.12 Conclusions from the adult London twin data

At the risk of drawing sweeping generalizations prematurely, we summarize

Table 5.14 Sex differences in the causes of variation in EPQ lie scores: effects of adding a family environment parameter to expectations for female twins.

Parameter	Estimate
E_{WM}	49.1
E_{WF}	42.4
V_{AM}	38.1
V_{AF}	17.1
V_{AMF}	30.5
E_{BF}	28.9
χ^2_4	7.9
$P\%$	<10
r_{AMF}	1.19

the results of our analysis so far because they provide a framework for interpreting subsequent studies.

(1) There is evidence for family resemblance in P, E, N and L because the E_W model fails badly in every case. We conclude that personality differences are not caused entirely by the unique environment of the individual within the family.

(2) A model that excludes genetic factors but incorporates the environment shared by family members does not fit very well. A model that tries to explain family resemblance in genetic rather than environmental terms generally fits much better. There is thus strong support for a *genetic* component of family resemblance in personality from these twin data.

(3) A model that allows for both genetic factors and the family environment is consistently better than a model that includes only the environment but little better then a model that assumes all family correlations to be genetic. Thus there is strong support for a genetic component, but very little support for the effects of the family environment.

(4) Including genetic dominance in the model does not improve the fit. However, estimates of the dominance parameter may be numerically large and remain not significant. The simplest model therefore favors only additive genetic effects, but a substantial amount of dominance might remain undetected.

(5) Allowing for sex-limited expression of the genes affecting P, E and N does not improve the fit. The same genes seem to operate in both sexes and contribute equally to variation in males and females.

(6) The effect of the environment within families is approximately the same in both sexes.

(7) There is little evidence that sibling competition or cooperation contributes to the development of adult personality differences.

The conclusions provide a perspective against which other studies may be evaluated. They have quantifiable weaknesses, which we hope other studies may remedy. The model-fitting approach is conservative and self-correcting. It avoids drawing conclusions that are beyond the statistical power of the study, but prevents us adopting a theory that is clearly inconsistent with the data. The conclusions that we have reached so far are strong, simple, testable and, above all, replicable.

5.2 THE US STUDY

Loehlin and Nichols (1976) administered the California Personality Inventory (CPI) as part of an extensive investigation of 850 pairs of like-sex twins ascertained as a result of their participation in the National Merit Scholarship Qualifying Test (NMSQT). Although the sample has a very restricted range for ability, the correlations between measures of ability and personality are slight, so selection for high ability is not expected to bias the results for personality. The sample is unusual for uniformity in age, since the test is administered to high-school juniors. The ascertainment procedure is described by Loehlin and Nichols (1976, Chapter 2). The CPI is a 480-item questionnaire, yielding scores on a large number of criterion-keyed scales. Loehlin and Nichols (p. 12) comment that "for most of our CPI scales, then, our samples appear to be reasonably representative of the portion of the population from which they come." In addition to the usual CPI scales, scores were generated from the item responses on the dimensions of extraversion and neuroticism using *a priori* scoring keys supplied by Eysenck.

The scores were included on a datatape kindly supplied by Dr Robert Nichols. We conducted a model-fitting analysis of the extraversion and neuroticism scores (Eaves and Young, 1981). The raw extraversion and neuroticism scores were significantly skewed such that there was a linear relationship between the means and intrapair standard deviations for identical twins. An optimization procedure was employed that devised the best transformation of the form $x' = \ln(a + bx)$ for removing the skewness, where x is the raw score and a, b are constants to be determined. To the nearest integer, Young found that skewness was minimized by $a = 300$, $b = -1$ for neuroticism and $a = 30$, $b = -1$ for extraversion. The resulting transformation has no theoretical basis, but is merely selected for convenience. Since the group is homogeneous with respect to age, correction for age was deemed unnecessary. Thus the mean squares between pairs are based on $N - 1$ df. The analyses of variance for the twins in each sex and zygosity group are given in Table 5.15. The lack of unlike-sex pairs in the sample restricts the range of hypotheses that might be tested, especially hypotheses concerning sex differences in the expression of genetic and environmental effects. Thus, although it is possible to compare the magnitudes of variance components between the sexes, we shall be unable to distinguish between "scalar" and "non-scalar" forms of sex limitation (see p. 97 above). Indeed, even if the estimates of V_A and E_B are the same in males and females, it is theoretically possible that different genes and environments may still contribute in each sex. We fitted three of the basic models listed above to the mean squares for extraversion and neuroticism. These were: the "simple genetic" model (model 3); the "environmental" model (model 2) and the

Table 5.15 Mean squares for transformed neuroticism and extraversion data from the NMSQT twin study.[a]

Twin type	Item	df	Mean square	
			Extraversion	Neuroticism
MZ female	Between pairs	266	13.2	41.9
	Within pairs	267	2.9	14.5
MZ male	Between pairs	178	11.7	53.8
	Within pairs	179	3.2	14.0
DZ female	Between pairs	175	10.6	34.2
	Within pairs	176	5.6	21.7
DZ male	Between pairs	110	9.5	37.8
	Within pairs	111	6.5	24.6

[a] Computed by Young *et al.* (1980) from data kindly made available by Dr Nichols. Raw scores have been transformed (see text).

"genotype–cultural" model (model 4). We did not fit model 1, because there is clearly significant resemblance between twins and we can deduce the results for the dominance model from those for model 3 from the relationships given above (p.88).

The analysis was conducted in two ways. In the first case the same parameter values were assumed to apply in both sexes. Since there are eight mean squares, the goodness-of-fit chi-square will have 8 − 2 = 6 df for models 2 and 3 and 8 − 3 = 5 df for model 3. In the second case the parameters were allowed to have different values in both sexes. Thus for model 3 two separate estimates of V_A and E_W were obtained, making four parameters in all and resulting in a chi-square for goodness-of-fit having 8 − 4 = 4 df. Comparing the fit of models that assumed parameters common to both sexes with models in which each sex had unique parameter values provides a test of heterogeneity of genetic and environmental effects across sexes. For example, if we denote the chi-square obtained under model 3 with parameters constrained to be the same in males and females by χ^2_A and let χ^2_B be the chi-square obtained when V_A and E_W differ between sexes then the difference in chi-square,

$$\chi^2_C = \chi^2_A - \chi^2_B,$$

is itself a chi-square for 2 df that may be used to test for heterogeneity over sexes.

The results of the model-fitting analysis are presented in Tables 5.16 and 5.17 for neuroticism and extraversion respectively. The parameters are presented for sexes considered jointly ("both sexes") and for each sex considered separately. The overall fit of the model that allows for sex differences in parameter values may be judged by adding up the chi-squares and their asso-

Table 5.16 Model-fitting results for NMSQT neuroticism data.[a]

Data set	Model	Parameter estimates			Fit	
		E_W	V_A	E_B	χ^2	df
Both sexes	2	17.6***	—	124.0***	31.48***	6
	3	14.4***	15.6***	—	5.34	6
	4	14.2***	17.9***	−2.2	4.88	5
Females only	2	17.4***	—	10.7***	11.02***	2
	3	14.6***	13.5***	—	0.06	2
	4	14.5***	14.5***	−0.9	0.01	1
Males only	2	18.1***	—	14.8***	15.49***	2
	3	14.1***	18.7***	—	0.88	2
	4	13.9***	23.3***	−4.3	0.41	1

[a] The models are identified by the numbers employed in the text.
Significance level: *** $P < 0.001$.

Table 5.17 Model-fitting results for NMSQT extraversion data.

Data set	Model	Parameter estimates			Fit	
		E_W	V_A	E_B	χ^2	df
Both sexes	2	4.2***	—	3.7***	49.01***	6
	3	3.1***	4.9***	—	3.41	6
	4	3.0***	5.6***	−0.8	2.51	5
Females only	2	4.0***	—	4.9***	26.87***	2
	3	2.9***	5.2***	—	0.05	2
	4	2.9***	5.4***	−0.2	0.001	1
Males only	2	4.5***	—	3.2***	19.51***	2
	3	3.4***	4.4***	—	1.82	2
	4	3.2***	6.0***	−1.6	0.28	1

Significance level: *** $P < 0.001$.

ciated df for the two sexes. Thus, the overall fit of model 3 to the neuroticism data is given by $\chi^2 = 0.06 + 0.88 = 0.94$ for $2 + 2 = 4$ df ($P > 0.90$). The test of heterogeneity of V_A and E_W over sexes is given by $\chi^2_2 = 5.34 - 0.94 = 4.40$ ($P > 0.10$).

The analysis of this unique sample, of uniform age and comparatively large numbers of twin pairs, yields results that are both striking and simple. For neuroticism and extraversion, model 2 gives a very bad fit, whatever we assume for the values of parameters in males and females. That is, a model that assigns all of twin resemblance in personality to the effects of the shared environment, E_B, cannot account for the observations in the US study. In contrast, the fit of model 3 is uniformly good, suggesting that there is little need to invoke the family environment over and above the genes as a cause of the correlation between relatives. It can be argued that the goodness-of-fit test is relatively weak because the effects of the family environment are combined with those of all other sources of poor fit. In this respect, our test is conservative in requiring very strong evidence that a model fails before we begin to consider more subtle hypotheses. We believe this is an important safeguard in an area in which complex models have been advanced with little more than the support of random fluctuations in measures of family resemblance. However, a more powerful test for the effects of the family environment is obtained by comparing the chi-squares for model 4 with those for model 3. The difference for neuroticism is 0.46 and 0.90 for extraversion. For neither trait does the change in chi-square attributed to the family-environmental parameter even approach significance. The same follows if the test is constructed separately for males and females. The estimates of E_B in the tables are close to zero. In fact, the parameter estimates are slightly but not significantly negative. Such a finding, if significant, would favour a hypothesis of genetic non-additivity. The estimate of -2.2 for E_B in the case of neuroticism and -0.8 for extraversion correspond to (non-significant) dominance estimates of 4.4 and 1.6 respectively. Even if these estimates were accepted at their face value, they would imply a relatively small contribution of genetic non-addivity to variation in personality. However, the relative magnitudes of the additive and dominance variance components are a poor guide to the relative size of additive and dominance deviations (see above, p.56). On the untestable assumption that increasing and decreasing alleles are equally frequent, the best estimate of the dominance ratio (see e.g., Mather and Jinks, 1982; Jinks and Fulker, 1970) is $\hat{H}_R/\hat{D}_R$, where $\hat{H}_R = 4\hat{V}_D$ and $\hat{D}_R = 2\hat{V}_A$.

Incorporating our estimates from the NMSQT data in the formula yields estimated dominance ratios of 0.70 and 0.76 for neuroticism and extraversion respectively. These large values are consistent with large heterozygous effects at every locus affecting the traits, but assume equal gene frequencies. However, they illustrate two important points in the genetic analysis of natural populations: (a) that very large dominance deviations produce relatively small amounts of dominance variance in randomly breeding populations; (b) that even in twin studies as large as the NMSQT study large amounts of dominance can remain undetected by statistical tests of significance for traits of intermediate heritability. The latter point has been investigated systematically by computer-simulation studies (see Chapter 3).

Comparing the fit of model 4 with that of model 2 confirms the importance of genetic factors, since the improvement ($\chi^2_1 = 26.1$ for N and 45.6 for E) is due to the addition of V_A into a purely environmental model. Thus not only does model 3 fit very well, it fits very much better than model 2, gives non-significant estimates of E_B and highly significant estimates of the additive genetic component for extraversion and neuroticism.

The analysis of the US data yields results for neuroticism and extraversion that are remarkably similar to those obtained with the London sample. The causes of family resemblance are the same in males and females, and the magnitudes of the genetic and environmental parameters do not differ significantly between the sexes. The data confirm that most of the correlation between relatives for personality-test scores is due to genetic factors rather than cultural transmission. If cultural effects were overwhelmingly responsible for individual differences then estimates of E_B would be statistically significant. A similar conclusion is reached by Loehlin and Nichols on the basis of their inspection of the twin correlations: "As far as personality and interests are concerned, it would appear that the relevant environments of a pair of twins are no more alike than those of two members of the population paired at random." The results do *not* support the conclusion that environment is unimportant, since the correlations between even identical twins are only in the 0.5–0.6 range, indicating that nearly half the variation we see in personality is a reflection of environmental differences within families ("E_W").

5.3 THE SWEDISH STUDY

In 1980 Floderus-Myrhed *et al.* published the first analysis of extraversion and neuroticism data using data from an unselected sample of 12 898 like-sex twin pairs from the Swedish Twin Registry. The measures employed are termed "psychosocial instability" and "psychosocial extraversion" by the authors and are described by them as "close to the two dimensions generally termed neuroticism and extraversion" (1980, p.154). Each dimension was measured by nine items selected in a pilot study in which the EPI was given to 400 non-twins.

Scale scores were employed as variables, and a pair was dropped from the analysis if either member answered fewer than six of the nine items in each dimension. Floderus-Myrhed *et al.* (1980) report internal consistency (Cronbach's alpha) of 0.75 for N and 0.63 for E. E and N scores correlated -0.29 overall. Zygosity determination was based on replies to questions concerning similarity. Because of the unusually large samples employed in this study, the authors were able to report analyses of variance for sexes and three age-cohorts separately. The mean squares given in Tables 5.18 and 5.19 are derived from the original paper (1980, Tables II–IV). The authors' original analysis focuses mainly on discussion of the results for the data pooled across age cohorts. In our analysis we retain the separate cohorts

Table 5.18   Mean squares for neuroticism in Swedish sample.[a]

| | | | Cohort | | | | | |
| | | | 1926–1935 | | 1936–1945 | | 1946–1958 | |
Zygosity	Sex	Source	df	MS	df	MS	df	MS
MZ	M	Between pairs	509	6.87	715	6.46	1052	6.83
		Within pairs	510	2.95	716	2.82	1053	2.09
DZ	M	Between pairs	813	5.85	1103	5.69	1751	5.88
		Within pairs	814	4.21	1104	3.86	1752	3.62
MZ	F	Between pairs	610	8.76	868	8.70	1239	9.44
		Within pairs	611	3.55	869	2.83	1240	2.19
DZ	F	Between pairs	1050	6.94	1293	7.56	1797	7.11
		Within pairs	1051	4.86	1294	4.67	1798	3.89

[a] From Floderus-Myrhed *et al.* (1980, Tables II and III).

Table 5.19   Mean squares for extraversion in Swedish sample.[a]

| | | | Cohort | | | | | |
| | | | 1926–1935 | | 1936–1945 | | 1946–1958 | |
Zygosity	Sex	Source	df	MS	df	MS	df	MS
MZ	M	Between pairs	506	7.54	715	6.21	1050	6.66
		Within pairs	507	2.80	716	2.86	1051	2.15
DZ	M	Between pairs	814	5.87	1101	5.73	1742	5.04
		Within pairs	814	3.81	1102	3.93	1743	3.50
MZ	F	Between pairs	605	7.70	867	7.52	1238	7.62
		Within pairs	606	2.52	868	2.54	1239	2.04
DZ	F	Between pairs	1047	5.51	1288	5.78	1792	5.93
		Within pairs	1048	4.18	1289	4.15	1793	3.41

[a] From Floderus-Myrhed *et al.* (1980, Tables IV and V).

throughout because, as we shall see, there are significant differences between parameter estimates derived from the different age groups.

In analyzing the Swedish data by biometrical–genetic methods, we had only the published mean squares to work from. Since the data included only like-sex pairs, we followed essentially the same procedure as Young employed in his NMSQT reanalysis, except that we worked with the untransformed scale. The same three models were fitted to sexes jointly and separately. The main difference between the two analyses lies in our treatment of the age effects. In the NMSQT study all the twins were of much the same age and so were treated as a single cohort. In the Swedish study the availability of samples large enough to warrant separation of three age cohorts meant that we could test not only the consistency of parameters across sexes, but also over cohorts using a similar rationale. Thus we began by assuming identical parameter values in every cohort and then asked what improvement in fit, if any, followed if parameters were allowed to take their own value in every cohort. Two basic models were fitted to each set of mean squares, models 3 and 4 of the previous analysis of the US data. The purely environmental model (E_B, E_W) is not described in detail here because it always fitted very badly and nothing new is introduced by discussing it.

Table 5.20 summarizes the main trends revealed by the model-fitting analysis. For each variable and basic model (models 3 and 4) four tests of significance are given, corresponding to different sets of constraints on the values of parameters in different subgroups of the data.

A model specifying the same values of V_A and E_W to all six twin groups (sexes and age cohorts), gives a goodness-of-fit chi-square of 239.6 in the case of the neuroticism data. The chi-square has 22 df because two parameters are being estimated from 24 mean squares. The fit is thus exceptionally bad. Allowing for sex differences, but not age differences, in the parameters yields a chi-square of 121.4 for 20 df (two further df are lost from the residuals because two additional parameters are now included in the model). Clearly, the improvement in fit from allowing for sex-differences is highly significant since the change in chi-square is 239.6 – 121.4 = 118.2 for 2 df. Similarly, fixing the parameters at the same values in males and females, but allowing them to differ between age cohorts (a six-parameter model, when model 3 is assumed), yields $\chi^2_{182} = 131.9$, a highly significant improvement of 106.7 for 4 df over the simplest model, which assumes identical parameter values for each age cohort and sex. In the case of neuroticism, however, none of the models considered so far comes close to fitting the data. Finally, therefore, we relax all constraints on the parameter values and allow each sex/age combination to have its own values of V_A and E_W. When this is done for the

Table 5.20 Goodness-of-fit tests for Swedish data.

	Neuroticism				Extraversion			
	$V_A E_W$		$+E_B/V_D$		$V_A E_W$		$+E_B/V_D$	
Model	χ^2	df	χ^2	df	χ^2	df	χ^2	df
Parameters same in all groups	239.6	22	234.2	21	118.4	22	100.0	21
Sex differences	121.4	20	116.1	18	99.5	20	80.7	18
Age differences	131.9	18	126.8	15	55.5	18	37.4	15
Age and sex differences	12.6	12	6.6	6	32.9	12	5.0	6

neuroticism data, under model 3, the goodness-of-fit chi-square falls to 12.6 for 12 df, a non-significant value that represents a highly significant improvement over simpler models. Thus the Swedish data give very strong support for both sex and age differences in the size of genetic and environmental components of personality. The effect of adding dominance or the family environment into the model for neuroticism is judged by comparing the chi-squares under model 3 with those under model 4 (denoted $+ E_B V_D$ in Table 5.20). In no case does the addition of dominance or the family environment to the model improve the fit significantly except when the parameters are assumed to be the same in all groups. We cannot interpret the latter difference with any confidence, since the model with parameters common to all groups fits so badly whatever assumptions are made about genes and environment. The results support a simple model for neuroticism that assumes no family-environmental effects on personality or dominant genetic effects. However, there is ample evidence that genetic and environmental effects are not the same in males and females and that they differ significantly between groups of twins born in different decades.

The finding that the family environment is not required to account for the inheritance of neuroticism replicates the results from the London and US data. However, the much bigger samples in the Swedish investigation provide the first significant evidence that the genetic and environmental effects on personality are not the same for males and females.

Similar arguments may be applied to the analysis of extraversion (see Table 5.20). Once again, the data reveal significant heterogeneity of genetic and environmental components over sexes and age groups. The main difference, however, lies in the significant contribution made by dominance to the variation in extraversion scores. In every case model 4 fits the extraversion data better than model 3, whatever is assumed about the effects of sex and age. It turns out that the improvement is better explained in terms of dominance than the family environment, since estimates of E_B are negative. The only model that fits the extraversion data adequately assumes both additive and dominant genetic effects and environmental differences within families but allows for different parameter values in each sex and age cohort ($\chi^2_{26} = 5.0$). Table 5.21 gives the parameter estimates for each subgroup of twins under the most parsimonious model that fits the data for the two personality dimensions.

Both variables show a general increase in the contribution of genetic factors to personality as we pass from older to younger cohorts. It is dangerous to read too much into trends in variances of scales based on only nine items since changes in mean are likely to be associated with changes in variance. Some speculation, however, may be in order. The differences between age groups could reflect developmental changes in gene expression

Table 5.21 Parameter estimates for Swedish data.

Dataset		Neuroticism			Extraversion			
Sex	Cohort	V_A	E_W	h_n^2	V_A	V_D	E_W	h_b^2
Male	1926–1935	1.92	3.07	0.38	2.15	0.11	2.71	0.45
Male	1936–1945	1.86	2.87	0.39	1.63	0.16	2.92	0.35
Male	1946–1958	2.47	2.19	0.53	0.98	1.21	2.13	0.51
Female	1926–1935	2.39	3.59	0.40	0.32	2.13	2.48	0.50
Female	1936–1945	3.06	2.92	0.51	0.84	1.62	2.53	0.49
Female	1946–1958	3.41	2.18	0.61	2.42	0.26	2.02	0.27

or could be due to the interaction of genetic and environmental effects with secular change or simply to selective mortality. These hypotheses can only be resolved by an appropriate longitudinal study of different birth cohorts. Heath *et al.* (1985a, b) report similar trends for a large Norwegian study of educational attainment. The relative contribution of the family environment decreases in the postwar period and the genetic contribution has increased. Heath *et al.* argue that the change reflects the "socialization" of education in Norway after World War II, in which parental influence gave way to merit as the main determinant of access to higher education. These authors therefore opt for a secular account of the observed changes. Unfortunately, it is not so easy to document the environmental processes involved in the etiology of personality differences. The Swedish study shows that the absolute contribution of the within-family environment is less in the younger cohort, suggesting that some of the explanation lies with greater uniformity in the environment.

The large samples involved in the Swedish study impart to the analysis much greater power than either the London or the US studies. As a result, we have been able to detect heterogeneity in the effects of genes and environment across sexes and birth cohorts, and have found the first strong evidence of non-additive genetic effects on extraversion. Some of the implications of the latter finding, and alternative explanations, will be pursued later.

5.4 THE AUSTRALIAN STUDY

The Swedish sample was very large, but used an abbreviated instrument and the issues of scaling were not considered. Furthermore, since the study, like the NMSQT study, omitted unlike-sex pairs, certain crucial hypotheses about sex interaction of gene expression remained untested. A recent paper by Martin and Jardine (1986) presents the first account of a fourth substantial study of the main dimensions of personality that combined the EPQ in a large study of like-sex and unlike-sex twins, which was analyzed using bio-metrical genetic methods.

Between November 1980 and March 1982, questionnaires were mailed to 5967 twin pairs over 18 years of age enrolled on the Australian NH&MRC Twin Registry. After one or two reminders to non-respondents, completed questionnaires were returned by 3810 pairs, or 64% of pairs to whom questionnaires were mailed. Among other items, the questionnaire used in the Australian study included the whole EPQ. Zygosity diagnosis was based on questionnaires concerning physical similarity and how often the twins were mistaken for one another in childhood. The sex, zygosity and age distribution of the respondents is given in Table 5.22.

A number of transformations of the raw P, E, N and L scores were tried in the attempt to remove the association between twin pair means and absolute intrapair differences. In every case the angular transformation was adopted, i.e. the same as we employed in our earlier analysis of the EPQ. Significant mean differences were found between twin groups. However, these were relatively small and only detected because of the vary large sample sizes. *F*-tests were conducted to compare the total variances of the five twin groups. The few small significant differences in variance detected for the raw scores were removed completely by the angular transformation. The transformed scores were summarized by analysis of variance for each group of twins. The mean squares between pairs are corrected for significant linear regression on age. The mean squares within pairs of unlike sex were corrected for the average difference between sexes. Since this is the largest single twin study for which complete EPQ data have been published, it is helpful to examine the age correlations (Table 5.23).

For P, E and N, there is a general reduction in score with age. Scores on the Lie scale increase significantly with age. The results are highly consistent over males and females. Initially, the unlike-sex pairs were omitted from the analysis, so the model-fitting analysis is as described above for the US study. Models 2, 3 and 4 were fitted for each variable to sexes jointly and to sexes separately, thus permitting a test of heterogeneity of parameters across

Table 5.22   Age, sex and zygosity composition of the Australian sample.[a]

	MZ females	MZ males	DZ females	DZ males	DZ opposite-sex
Number of pairs	1233	567	751	352	907
Mean age (years)	35.66	34.36	35.35	32.26	32.90
Standard deviation	14.27	14.02	14.27	13.88	13.85
Age range	18–88	18–79	18–84	18–83	18–79

[a] From Martin and Jardine (1986, Table 1).

Table 5.23 Correlations of the transformed personality scores with age.[a]

	Female	Male
Extraversion	−0.16***	−0.14***
Psychoticism	−0.20***	−0.28***
Neuroticism	−0.13***	−0.14***
Lie	0.36***	0.38***

[a] Based on Martin and Jardine (1986, Table 7).
* $0.01 < P < 0.05$; ** $0.001 < P < 0.01$; *** $P < 0.001$.

sexes. Subsequently, models were fitted to the whole dataset, including pairs of unlike sex, in the attempt to test whether there is sex limitation for the expression of genetic differences in personality (cf. our approach to the London data, above).

5.4.1 Psychoticism

For psychoticism, the model that omits genetic effects ("E_W, E_B") fails very badly in both sexes (cf. Table 5.24). In contrast, the "V_A, E_W" model fits very well, although there is highly significant heterogeneity between the sexes in the parameter values. With the inclusion of unlike-sex pairs in the analysis, Martin and Jardine were able to estimate the three parameters of additive genetic variation under the sex-limitation model: V_{AM}, V_{AF} and V_{AMF}. From these parameters, they estimated the correlation between male and female genetic effects to be 1.09. This value is close to 1 and suggests that sex differences in gene expression are "scalar"; that is, the same genes affect P scores in both sexes, but the genetic variances differ for males and females. In males, Martin and Jardine found that 50% of the variation in transformed P scores was due to genetic effects, whereas only about 35% was genetic in females.

Previous analyses of the environmental component into its long- and short-term effects have had to rely on statistical arguments rather than empirical data (see e.g. Eaves and Eysenck, 1977). Martin and Jardine pretested some of their twins ($N = 96$ individuals) as part of a pilot investigation and were able to estimate the contribution of test–retest fluctuations to values of E_W obtained from single personality-test scores. They estimate that between 30% and 60% of variation ascribed to effects of the within-family environment on P scores is actually due to relatively short-term influences that contribute to the inconsistency of test scores obtained on different occasions. Employing an argument from statistical theory alone led Eaves and Eysenck

Table 5.24 Summary of model fitting to Australian psychoticism data.[a]

	$\hat{E}_W$	$\hat{E}_B$	$\hat{V}_A$	$\hat{V}_D$	df	χ^2	h^2
Female							
$E_W E_B$	37.74***	16.56***	—	—	2	14.87***	
$E_W V_A$	34.20***	—	20.14***	—	2	2.81	0.37 ± 0.02
$E_W E_B V_A$	34.65***	4.15	15.56***	—	1	1.59	
$E_W V_A V_D$	34.65***	—	28.00***	−8.29	1	1.59	
Male							
$E_W E_B$	43.36***	25.34***	—	—	2	13.16**	
$E_W V_A$	37.78***	—	30.91***	—	2	0.28	0.45 ± 0.03
$E_W E_B V_A$	38.09***	3.38	27.26***	—	1	0.05	
$E_W V_A V_D$	38.09***	—	37.41**	−6.77	1	0.05	
Female and Male							
$E_W E_B$	39.52***	19.33***	—	—	6	63.69***	
$E_W V_A$	35.32***	—	23.56***	—	6	36.35***	
$E_W E_B V_A$	35.71***	3.84	19.35***	—	5	35.12***	
$E_W V_A V_D$	35.71***	—	30.86***	−7.67	5	35.12***	
Female and Male and Opposite-sex							
$E_W E_B$	42.37***	18.67***	—	—	8	88.22***	
$E_W V_A$	35.80***	—	25.37***	—	8	48.39***	
$E_W E_B V_A$	36.39***	3.34	21.46***	—	7	46.94***	
$E_W V_A V_D$	36.39***	—	31.47***	−6.67	7	46.94***	

[a] From Martin and Jardine (1986, Table 19).
* $0.01 < P < 0.05$; ** $0.001 < P < 0.01$; *** $P < 0.001$.

(1977) to propose a substantially larger contribution of error. However, their model for test scores assumed equivalent items and, if markedly wrong, would inflate the component assigned to sampling variance.

5.4.2 Extraversion

The results obtained for the extraversion scale are given in Table 5.25. There is no evidence of significant heterogeneity over sexes, but a strong hint of dominance in females since adding dominance significantly improves the fit of the model ($\chi^2_1 = 5.25$, $P < 0.05$). This finding is replicated in males. The V_A, V_D, E_W model was fitted to all the mean squares, including those for unlike-sex pairs, and gave a very adequate fit to the data ($\chi^2_7 = 5.42$).

The estimates of all three parameters differ significantly from zero, and the relative magnitudes of the additive and dominance components of gene

Table 5.25 Summary of model-fitting to Australian extraversion data.[a]

	$\hat{E}_W$	$\hat{E}_B$	$\hat{V}_A$	$\hat{V}_D$	df	χ^2	h_n^2	h_b^2
Female								
$E_W E_B$	143.5***	96.3***	—	—	2	91.18***		
$E_W V_A$	115.2***	—	125.3***	—	2	5.26	0.52 ± 0.02	
$E_W E_B V_A$	112.4***	−37.9	165.3***	—	1	0.01		
$E_W V_A V_D$	112.4***	—	51.6	75.8*	1	0.01		
Male								
$E_W E_B$	157.9***	84.4***	—	—	2	53.99***		
$E_W V_A$	124.7***	—	119.7***	—	2	10.35***		
$E_W E_B V_A$	118.6***	−64.7	189.6***	—	1	2.26		
$E_W V_A V_D$	118.6***	—	−4.4	129.3**	1	2.26		
Female and Male								
$E_W E_B$	148.1***	92.5***	—	—	6	150.84***		
$E_W V_A$	118.1***	—	123.7***	—	6	17.33**		
$E_W E_B V_A$	114.3***	−47.3	173.9***	—	5	4.19		
$E_W V_A V_D$	114.3***	—	32.2	94.5**	5	4.19		
Female and Male and Opposite-sex								
$E_W E_B$	158.7***	82.7***	—	—	8	166.92***		
$E_W V_A$	119.7***	—	122.9***	—	8	19.59*		
$E_W E_B V_A$	114.4***	−38.2	165.6***	—	7	5.42		
$E_W V_A V_D$	114.4***	—	50.9**	76.4***	7	5.42	0.21 ± 0.09	0.53 ± 0.02

[a] From Martin and Jardine (1986, Table 17).
* $0.01 < P < 0.05$; ** $0.001 < P < 0.01$; *** $P < 0.001$.

action might be taken to imply substantial dominance deviations or comparable epistatic interactions. Together with the Swedish data, therefore, the Australian data are showing the statistical pattern expected if there are non-additive genetic effects on extraversion. The fact that the earlier studies did not detect such effects is easily explained by the comparatively small numbers of twins in the samples. However, there is some danger in the uncritical acceptance of the dominance hypothesis without considering alternatives.

Competitive social interactions based on the genotypes of siblings will also tend to reduce the correlation between DZ twins relative to that between MZ pairs (Eaves, 1976b) in a manner that bears some superficial resemblance to the effects of dominance. However, it may be shown theoretically that competition based on genotype, unlike dominance, also produces a significant reduction in the total variance of MZ twins compared with DZs, so that the effects of dominance ought not to be confused with those of competitive interaction. In order to test this hypothesis, we fitted the competition model (cf. Table 5.6) to the Australian data and to the pooled mean squares from the Swedish investigation.

In the case of the Australian sample the competition model gave a fit close to that of the additive-dominance model. For the Swedish data the additive-dominance model fitted slightly better than the competition model when applied to the pooled mean squares published by the authors (Floderus-Myrhed *et al.*, 1980). Thus it is clear that even exceptionally large samples have difficulty discriminating between the effects of dominance and those of competition. In none of the data sets have we found evidence of significant differences between the variances of MZ and DZ twins that the competition theory would predict. However, the differences are not expected to be very great, even in the presence of competition, so the test is comparatively weak. If we had a large sample of children reared without siblings then we should expect competitive effects to result in a greater variance of twins than only-children, because having a twin constitutes an additional source of environmental variation and genotype–environment covariance not shared by singletons (Eaves, 1976a). In the absence of such a sample, however, we have to be cautious about inferring dominance rather than competition. With the data available to us, the balance of evidence comes down somewhat on the side of dominance rather than competition. Power studies by Jardine (1985) explored the problem of resolving competition and dominance in more detail, and suggest that extremely large samples would be required to resolve the effects of dominance and competition with twin data alone for effects of the magnitude we see for extraversion.

The effects of dominance might also be simulated in the analysis of twin data involving unlike-sex pairs if there is sex-limited expression of purely additive genetic effects. Such effects tend to reduce the correlation of unlike-sex DZ twins, and may thus, in analyses that do not take sex limitation into account, be sufficient to reduce the overall DZ correlation enough to give spurious effects resembling those of dominance. In the case of extraversion, however, the evidence for dominance is confirmed even without recourse to the data on unlike-sex pairs in the Swedish and Australian samples.

5.4.3 Neuroticism

Martin and Jardine's analysis of the neuroticism scores is summarized in Table 5.26. These data illustrate the possibility of mistaking sex limitation for dominance in studies employing unlike-sex pairs if same-sex data are not analyzed separately. When pairs of unlike sex are omitted from the analysis there is no evidence of a significant dominance component whether the sexes are analyzed separately or pooled. The model that omits genetic factors altogether, however, fits very poorly. Thus, as long as we consider only the like-sex pairs, there is no suggestion of dominance, and the data are consistent with the simple additive-genetic/within-family environment model that has fitted all previous data so well. Furthermore, there is little evidence of sex differences in parameter values. This finding is consistent with the London EPQ data on neuroticism. When the unlike-sex pairs are included the picture changes somwhat because the estimate of V_D is highly significant when judged both by its standard error and the change in chi-square when the dominance parameter is included in the model. A more consistent explanation of the results is that the lower correlation of unlike-sex pairs results from sex differences in gene expression and environmental effects. When such a model is fitted to the Australian data (see Table 5.27) consistent estimates for the sex-limitation parameters are obtained and the model fits as well as that which explains the correlations in terms of dominance.

The data suggest that the effects of genes and environment on neuroticism are comparable in magnitude across sexes (as we found for the NMSQT sample), but that the genetic effects are not identical in males and females, i.e. different genes have different effects in the two sexes. The consistency of gene expression over sexes is estimated to be:

$$r_{AMF} = 59.4/(95.4 \times 108.0)^{1/2} = 0.58.$$

This value is close to that for the London sample, but the larger Australian sample shows that the correlation in gene effects across sexes is significantly less than unity. The Australian study therefore supplements the

Table 5.26 Summary of model-fitting to Australian neuroticism data.[a]

	$\hat{E}_W$	$\hat{E}_B$	$\hat{V}_A$	$\hat{V}_D$	df	χ^2	h^2
Female							
$E_W E_B$	125.1***	90.5***	—	—	2	51.12***	
$E_W V_A$	104.7***	—	110.5***	—	2	0.42	
$E_W E_B V_A$	104.8***	1.2	109.2***	—	1	0.42	0.51±0.02
$E_W V_A V_D$	104.8***	—	112.8***	-2.4	1	0.42	
Male							
$E_W E_B$	141.8***	76.8***	—	—	2	28.48***	
$E_W V_A$	118.9***	—	100.3***	—	2	1.72	
$E_W E_B V_A$	116.5***	-26.4	128.8***	—	1	0.27	0.46±0.03
$E_W V_A V_D$	116.5***	—	49.7	52.7	1	0.27	
Female and Male							
$E_W E_B$	130.3***	86.1***	—	—	6	86.65***	
$E_W V_A$	109.1***	—	107.4***	—	6	5.85	
$E_W E_B V_A$	108.4***	-8.2	116.2***	—	5	5.30	0.50±0.02
$E_W V_A V_D$	108.4***	—	91.5***	16.5	5	5.30	
Female and Male and Opposite-sex							
$E_W E_B$	140.9***	72.0***	—	—	8	136.90***	
$E_W V_A$	110.9***	—	102.1***	—	8	18.42*	
$E_W E_B V_A$	107.6***	-24.1	128.9***	—	7	12.26	0.27±0.09
$E_W V_A V_D$	107.6***	—	56.7**	48.1**	7	12.26	

[a] From Martin and Jardine (1986, Table 22).
* $0.01 < P < 0.05$; ** $0.001 < P < 0.01$; *** $P < 0.001$.

Table 5.27 Neuroticism: estimates ($\pm$s.e.) obtained after fitting a model allowing different genetic and environmental components of variation in males and females for transformed neuroticism data from Australian sample.[a]

$\hat{E}_{W_M}$	$\hat{E}_{W_F}$	$\hat{V}_{A_M}$	$\hat{V}_{A_F}$	$\hat{V}_{A_{MF}}$
117.4***	104.2***	95.4***	108.0***	59.4***
±6.4	±3.9	±8.0	±5.6	±13.9
		$\chi^2_5 = 5.78\ (P = 0.33)$		
$h^2_{males} = 0.45 \pm 0.03$			$h^2_{females} = 0.51 \pm 0.02$	

[a] From Martin and Jardine (1986, Table 23).
* $0.01 < P < 0.05$; ** $0.001 < P < 0.01$; *** $P < 0.001$.

interpretation of the sex effects found in the Swedish study by suggesting strongly that some genes are sex-specific in their effects on neuroticism scores.

5.4.4 The Lie scale

Table 5.28 summarizes the model-fitting results for the lie scale. The analysis of all ten mean squares shows that none of the simple models fit as long as sex-limited effects are ignored. This echoes the previous finding for the London data, in which the lie scale behaved unlike the other three personality dimensions. Analysis of the like-sex pairs only confirms that there is heterogeneity over sexes. Fitting E_W, E_B and V_A to both sexes simultaneously gives a residual chi-square of 15.34 for 5 df. Allowing the parameters to differ over sexes gives a total residual of $0.18 + 1.52 = 1.70$ for 2 df, an improvement in fit of 13.64 for 3 df. Thus sexes are highly heterogeneous. A purely environmental model cannot explain the data for females, but the E_W, V_A model fits very nicely. For males there is less to choose between the two models. If anything, the environmental model has a slight advantage. The three additional parameters required to allow for sex-limited genetic and within-family environmental effects (see Table 5.11) reduce the chi-square from 24.90 to 11.13. This highly significant gain confirms the presence of sex interactions, but does not yield an altogether satisfactory fit ($P = 0.05$).

None of the *a priori* models gives a good fit to the data on the lie scale, and we have to proceed more tentatively if we are to avoid overinterpreting the data. We adopt a conservative position, and opt for the model that allows within family environmental variance to be larger in males and the genetic component to be larger in females. The difference in gene effects seems to be largely scalar since $r_{AMF} = 0.93$. There is some gain ($\chi^2_1 = 3.46$) from adding a family-environmental parameter to males, and although the change is of marginal significance, it is sufficient to make the model fit ($\chi^2_4 = 7.67$). The

Table 5.28 Summary of model-fitting to Australian transformed and age-corrected lie data.[a]

	$\hat{E}_W$	$\hat{E}_B$	$\hat{V}_A$	$\hat{V}_D$	df	χ^2	h^2
Female							
$E_W E_B$	90.8***	65.8***	—	—	2	42.96***	
$E_W V_A$	76.8***	—	79.6***	—	2	0.55	0.51 ± 0.02
$E_W E_B V_A$	77.3***	6.7	72.5***	—	1	0.18	
$E_W V_A V_D$	77.3***	—	92.6***	−13.4	1	0.18	
Male							
$E_W E_B$	101.2***	53.2***	—	—	2	3.37	
$E_W V_A$	93.8***	—	60.0***	—	2	4.95	0.39 ± 0.03
$E_W E_B V_A$	96.7***	32.7*	24.8	—	1	1.52	
$E_W V_A V_D$	96.7***	—	122.8***	−65.3	1	1.52	
Female and Male							
$E_W E_B$	94.1***	61.8***	—	—	6	49.39***	
$E_W V_A$	82.3***	—	73.2***	—	6	18.23**	
$E_W E_B V_A$	83.5***	14.9*	57.3***	—	5	15.34**	
$E_W V_A V_D$	83.5***	—	102.1***	−29.9	5	15.34**	
Female and Male and Opposite-sex							
$E_W E_B$	98.9***	53.5***	—	—	8	82.00***	
$E_W V_A$	81.9***	—	70.0***	—	8	24.90**	
$E_W E_B V_A$	82.6***	5.0	64.4***	—	7	24.09**	
$E_W V_A V_D$	82.6***	—	79.4***	−10.0	7	24.09**	

[a] From Martin and Jardine (1986, Table 25).
* $0.01 < P < 0.05$; ** $0.001 < P < 0.01$; *** $P < 0.001$.

London data display a similar anomaly, but there it is the female data that require a shared environmental component.

The proportions of variance in each of the personality traits that may be attributed to genetic effects and the environment within families are summarized in Table 5.29. Since retest reliabilities are available on the sample, the environmental variance is partitioned into long-term effects due to different environmental experiences between individuals and short-term components due to changes in behavior within individuals. The estimated genetic effects are split into additive and dominance components of gene action for extraversion.

5.5 SUMMARY AND DISCUSSION

In assessing the results of the four large twin studies, several basic issues have

Table 5.29 Sources of variation in transformed P, E, N and L scores for Australian sample.

	P		E	N		L	
	\multicolumn{8}{c}{Percentage of variance}
Source	Males	Females	Males + Females	Males	Females	Males	Females
Environment	64	50	47	49	55	50	62
−Error	24	30	17	13	18	21	17
−Individual	40	20	30	36	37	29	45
Genes	36	50	53	51	45	50	38
−Additive	36	50	21	51	45	50	38
−Dominant	0	0	32	0	0	0	0

to be borne in mind. What are the consistencies over all samples? Of the inconsistencies, how many can be attributed to differences in resolving power? What results do we feel confident about? What results do we accept more cautiously? What questions remain?

The overwhelming and consistent pattern to emerge from these studies is that there is a significant genetic component to all the major dimensions of personality studied. These findings are based on twins reared together, and have to be tested for their consistency with other types of data. Our analysis of personality makes robust what has been suspected for many years on the basis of smaller samples with diverse measures (see Chapter 3).

Our studies, however, have gone beyond this. All of the studies that we describe are consistent in finding no trace of a shared environmental component of twin resemblance for personality, apart possibly from the scores on the Lie scale. That is, there is little evidence that shared features of the environment such as parental attitudes, education and SES play a significant part in the determination of personality. If the shared environment were very important (explaining more than 20% of the total variance, for example) then its effects on personality would have been detected easily in studies as large as those from Sweden and Australia. The consistency of this finding across samples from different populations leads us to considerable confidence in this result.

While our analyses lead us to discount the "shared" environment, we recognize that all the studies are consistent in assigning upwards of 50% of the total variation in personality test score to environmental factors within the family. In etiological terms, this means that the personality of each individual is molded by his/her unique genotype and the unique experiences that he shares with none of his family members. The fact that the MZ twin correlations are all lower than the test–retest reliabilities for the personality measures suggests that a substantial part of the environmental variation within families is due to long-term environmental differences rather than day-to-day fluctuations in behavior.

The two larger studies suggest that genetic and environmental effects may not be the same for males and females. The Swedish dataset shows heterogeneity over sexes for extraversion and neuroticism. The Australian data show no heterogeneity for extraversion but suggest that part of the gene action underlying neuroticism is sex-specific. The simplest interpretation is that there are some genes that affect neuroticism scores in females but do not do so in males and *vice versa*. There are many aspects of physique, for example chest girth, which might be expected to display the same basic genetic mechanism. Our London data are consistent with this finding, but were not large enough to detect modest heterogeneity between the sexes.

We are less confident about other findings. The Swedish and Australian data have suggested a significant non-additive component of genetic variation for extraversion. The London and US datasets are probably too small to detect a dominance component even if one were there. The resolution of dominance from sibling competition is weak with twin data and could not be achieved in the Australian dataset. In the Swedish data the dominance hypothesis was favored slightly over the competition hypothesis.

Chapter 6

Further Tests of the Model: Studies of Adoptees and Extended Families

The studies described in the previous chapters have some serious limitations. The data on which we have so far based our attempt to resolve the genetic and social components of family resemblance come primarily from twins. Although data on twins reared together provide one of the major stepping stones in building a model for human differences, any model derived from twin studies should ultimately be tested against other kinds of data. In this chapter we briefly describe attempts by ourselves, and others, to test basic assumptions about the genetic model using three main kinds of data: nuclear families and extended pedigrees; separated twins; and adoptees and their non-biological relatives. Much of the data described in this chapter were collected and analyzed by other investigators. Our analysis of these data will focus on three important issues: assortative mating; cultural inheritance; and developmental change in gene expression.

6.1 ASSORTATIVE MATING FOR THE DIMENSIONS OF PERSONALITY

In our treatment of personality in twins, we noted that the assumption of random mating was implicit in the coefficients specified for the additive genetic component, and, as long as we had only twins reared together, the excess genetic resemblance between twins arising from assortative mating was confounded with estimates of the between-families environmental component. The fact that there was no evidence of a significant E_B for P, E and N is therefore consistent with there being no assortative mating for the transmissible causes of personality. For the lie scale, the twin data gave stronger indication of a significant E_B, which suggests that either there is a

Table 6.1 Correlations between spouses for personality test scores.

Sample	N (pairs)	P	E	N	L
Spouses	445	0.156	0.059	0.128	0.276
Parents of:					
MZ_m	59	−0.137	0.143	0.144	0.399
MZ_f	50	0.164	0.228	−0.155	0.297
DZ_m	40	0.100	−0.011	−0.153	0.404
DZ_f	37	0.377	−0.209	−0.005	0.299
DZ_{mf}	76	0.227	0.009	0.027	0.355
Male singletons	85	0.161	0.022	0.015	0.328
Female singletons	97	0.328	0.170	−0.089	0.215
Heterogeneity χ_7^2		10.30	5.96	9.15	2.59
P		<0.10	<0.50	<0.20	<0.90
Pooled z [a]		0.171	0.065	0.052	0.305
σ_z^2		0.034	0.034	0.034	0.034
Pooled r		0.170	0.065	0.052	0.296

[a] Pooled zs and correlations corrected for bias (df = 865)

non-genetic component to twin resemblance, or there is marked assortative mating giving rise to additional genetic resemblance between twins, or both. In each case the evidence for the mating system is indirect. Based on the findings of the twin data alone, we expect no marked spousal correlation for P, E or N, but might expect a significant correlation between mates for the lie scale. To what extent do the results of our twin analyses coincide with the empirical findings with respect to the resemblance between spouses for personality?

Table 6.1 summarizes two large bodies of data relating to the resemblance between spouses for the EPQ scales. The first set of data comprises 445 pairs of spouses ascertained as a quota sample in London. The entire dataset comprises 568 pairs, but any individual who omitted either a single personality item or a single item relating to social attitudes was excluded from the analysis. The second dataset comprises the parents of juvenile twins and only children. The juvenile data are analyzed when we consider developmental effects in the next chapter. There are eight groups of spouses altogether.

Correlations for each personality dimension were transformed to zs and tested for homogeneity prior to pooling (see Chapter 3). The pooled correlations were corrected iteratively for the small biases that accumulate over several correlations. The correlations were reasonably homogeneous for all personality scales. The pooled correlations are based on 865 df and are known very precisely if we discount the negligible heterogeneity ($\sigma_z^2 = 0.034$). The pooled correlations for extraversion and neuroticism do not differ significantly from zero. The correlation for P is highly significant but quite small ($r = 0.170$). A correlation this small is not expected to contribute greatly to family resemblance for moderately heritable traits as long as mate selection is based on the phenotype and not on the underlying genotype. The largest correlation for any of the measures is for the lie scores ($r = 0.296$). The implications of these correlations for the correlations between other kinds of relatives including twins, will be examined later in this chapter and elsewhere in the book. For now, it is important to note that there are some significant correlations between mates, that mating is random for the major dimensions of extraversion and neuroticism, and that there is not the slightest suggestion that "opposites" attract, otherwise the correlations would be negative. The findings are remarkably consistent with the results of the model-fitting analyses of the twin data in the previous chapter. The spousal correlation is greatest for the lie scale, which is the variable for which the twin data gave the strongest evidence for a significant "E_B" component.

6.2 ADOPTIONS AND EXTENDED KINSHIPS: THE LONDON STUDY

6.2.1 Source of data

At an early stage in the London study an attempt was made to supplement the data on twins by the collection of personality measures on adopted individuals and their relatives and also on extended kinships comprising nuclear families and more remote relationships. All the individuals, like the twins, were volunteers, and so the data may be expected to suffer from similar uncertainties of ascertainment as the twin data. In the "extended-kin-ship" sample we secured 178 families. The "adoption" sample comprised 150 adult adoptees with at least one non-biological relative and a further 191 adoptees for whom no further relatives were available. These single individuals, however, were retained in the full analysis because the variance of adopted individuals provides some information about the contribution of genotype-environment covariance to individual differences arising because the environment of children is a function of the phenotype or genotype of the parents. These data were supplemented, for the purposes of the present analysis, by the data on 543 pairs of adult twins available at the time the family and adoption data were coded. These pairs are a subset of the twin pairs analyzed in the previous chapter (Section 5.1).

When we analyzed twin data by themselves it was easy to suggest a number of simple data summaries (e.g. analyses of variance) that allowed us to use relatively straightforward model-fitting methods without violating the assumptions (e.g. normality and independence of summary statistics) on which the estimation and tests of significance depend. A major problem for the analysis of highly irregular kinships is that there is no convenient data summary which can be fitted straight into a model-fitting analysis without violating some of the statistical assumptions of the analysis. Each family has its own unique structure and each individual may enter into many different biological and social relationships within its pedigree. In theory, it is possible to employ the maximum-likelihood method to estimate a general covariance matrix that incorporates all possible types of relationship in the data. This approach is possible as long as the number of unique relationships in a study is small (see e.g. McGue *et al.*, 1984). In practice, however, studies that yield a large number of different biological relationships generate a correlation matrix that is too large to be estimated by this approach at the present time.

There are two approaches to this problem. The first involves computing correlations between relatives and treating them as if they were independent for model-fitting purposes, while recognizing that the correlations between the statistics might lead to incorrect tests of significance. The second approach is to use the maximum-likelihood method to estimate parameters of a genotype–environmental model from the unreduced data. This method takes account of the irregularity of the pedigrees and the correlations between the observations, but gives no direct statistical test of whether the causal model can explain the pattern of family resemblance. However, alternative hypotheses about the causes of family resemblance may still be compared using likelihood-ratio tests. This is arguably the best approach for deciding between alternative hypotheses. We used both methods in our analysis, and, by and large, they gave fairly consistent results.

6.2.2 The correlations between relatives

We computed the correlations by identifying every possible pair from each type of biological and adoptive relative in the sample and entering the scores of each pair into the formula for product-moment correlations on the assumption that the pairs were independent. A consequence of this method is that the same individual may enter into several different kinds of relationships. For example, in a three-generation pedigree, the same individual may contribute to a grandparental correlation, a parent–offspring correlation and a sibling correlation. Furthermore, the same individual may be included several times in the *same* correlation. Thus a mother will contribute to the mother–child correlation as many times as she has children for whom data are available.

This approach of treating correlations as independent has been shown to yield relatively good estimates of the parameters compared with the true

Table 6.2 Correlations between relatives for personality dimensions (decimal points omitted).

Relationship	Number of pairs	Correlation			
		P	E	N	L
MZ$_m$	70	533	648	510	530
MZ$_f$	233	408	460	446	560
DZ$_m$	47	157	248	025	021
DZ$_{mf}$	68	067	067	108	267
DZ$_f$	125	384	179	089	523
Spouses	155	273	036	063	367
Son–father	88	310	261	−054	036
Son–mother	110	107	018	190	223
Daughter–father	148	126	179	186	150
Daughter–mother	199	051	304	100	204
Sibs m	72	083	161	−012	261
Sibs m–f	195	146	187	041	264
Sibs f	151	215	360	073	379
Grandson–grandfather	12	−218	477	334	−399
Grandson–grandmother	10	023	−294	−243	−318
Granddaughter–grandfather	12	−164	495	−038	109
Granddaughter–grandmother	23	129	−264	397	192
Nephew–uncle	57	−155	−046	042	146
Nephew–aunt	65	008	−178	−058	136
Niece–uncle	87	070	058	−082	048
Niece–aunt	105	068	435	204	201
Cousins m	18	201	−106	−194	575
Cousins m–f	56	−023	027	015	020
Cousins f	39	−029	270	209	143
Second cousins m	9	−471	543	−342	−247
Second cousins m–f	19	−157	077	−284	−117
Second cousins f	4	540	−485	−420	−910
Foster-son/foster-father	18	−395	−277	410	−146
Foster-son/foster-mother	26	051	−116	248	104
Foster-daughter/foster-father	75	099	027	159	−048
Foster-daughter/foster-mother	101	−019	004	−098	080
Foster sibs m	—	—	—	—	—
Foster sibs m–f	34	−170	−228	331	−120
Foster sibs f	24	286	052	088	321

maximum-likelihood estimates, but tends to underestimate their standard errors because observations that are not independent are assumed to be so (McGue *et al.*, 1984). However, the resulting correlations do give the reader an important "feel" for the properties of the data against which substantive claims based on more rigorous statistical methods may be judged. The scores were not age-corrected prior to correlating the relatives, but subsequent analysis by maximum-likelihood allows for effects of age on the phenotype. The raw correlations were computed separately with respect to the sexes of individuals in a given relationship and are given in Table 6.2.

In practice, for data on extended kinships, the exact set of relationships chosen for summary depends partly on the complexity of the model to be tested. Thus, in our case, we do not distinguish between "maternal" and "paternal" uncles and aunts. However, when testing for the effects of assortative mating and cultural inheritance in the analysis by maximum-likelihood, we do distinguish between "cognate" and "affine" uncles and aunts. The correlations tabulated relate to cognate uncles and aunts. In the correlational analysis, we included no relatives by marriage, apart from spouses. Neither do we separate cousins related through brothers from those related through sisters and unlike-sexed siblings. However, these correlations are expected to be different under mechanisms of maternal inheritance (see e.g. Haley *et al.*, 1981) or in the presence of sex-dependent gene expression ("sex limitation") when there is assortative mating (see e.g. Eaves and Heath, 1981a, b). Furthermore, in the presence of assortative mating, relationships by marriage, which should generate zero correlations under random mating, are no longer expected to be zero.

Inspection of the raw correlations reveals considerable sampling variation and even heterogeneity over sexes of the correlations within each biological relationship. Some of the correlations for more remote relationships are aberrant, but tend to be based on small samples, so it is difficult, without further tests, to judge whether they represent genuine anomalies or just sampling variation. Some of the correlations, especially those for nuclear families, are based on very large samples indeed, and, given that there are no pertinent sampling biases, should give quite precise estimates of the population correlations.

In Table 6.3 we give the results of pooling the correlations over sexes by transforming the observed correlations to their corresponding inverse hyperbolic tangents ("z values") to improve the approximation to normality, testing for the heterogeneity, and fitting common z values with iterative refinement of the pooled value as outlined above in the treatment of twin correlations (Chapter 3). The heterogeneity of the component correlations was tested by the chi-square statistic, which compares the observed variation in the zs with that predicted from chance alone. The pooled z values were then transformed back into the pooled correlations tabulated. With one or two exceptions, the apparent heterogeneity across sexes of the correlations is not statistically significant. To avoid over-interpreting multiple tests of significance, the chi-squares for each variable were added across all the relationships tabulated to provide an overall test of heterogeneity for each variable. Overall, there is highly significant heterogeneity with respect to sex

Table 6.3 Correlations pooled over sexes.

Relationship	N	df	P		E		N		L	
			r	χ^2	r	χ^2	r	χ^2	r	χ^2
MZ	297	1	0.438	1.35	0.507	3.77	0.459	0.33	0.551	0.10
DZ	231	2	0.257	5.43	0.160	0.99	0.081	0.20	0.367	11.30
Spouses	155	—	0.273	—	0.036	—	0.063	—	0.367	—
Parent	533	3	0.125	4.36	0.207	6.47	0.117	3.89	0.166	2.20
Sibling	409	2	0.160	0.94	0.246	3.62	0.044	0.35	0.305	1.58
Grandparent	45	3	-0.016	1.02	0.043	7.07	0.199	3.12	-0.026	3.19
Uncle/aunt	302	3	0.016	2.13	0.127	19.60*	0.042	4.69	0.135	1.13
Cousin	104	2	0.008	0.68	0.093	2.05	0.052	1.92	0.151	4.64
Second cousin	23	2	-0.214	1.25	0.178	1.77	-0.293	0.02	-0.197	1.47
Foster-parent	208	3	0.001	3.42	-0.022	1.47	0.068	6.11	0.022	1.27
Foster-sibling	52	1	0.016	2.72	-0.115	1.02	0.232	0.83	0.061	2.57
Overall heterogeneity over sexes		22		23.30		47.82*		21.46		29.53

* Significant at the 1 % level.

in the correlations for extraversion, most of which can be explained by the avuncular correlations. A glance at Table 6.2 suggests that the correlations between relatives of unlike sex may be smaller than those between like-sex individuals. Our experience in the analysis of twin data (see e.g. Chapter 5) suggests that such a finding may indicate sex differences in the expression of genetic effects on the trait.

6.2.3 Inclusion of cultural inheritance and assortative mating

The spousal correlations in this sample are quite similar to those reported for two other large samples in Table 6.1. There is not the slightest positive or negative correlation between mates for extraversion and neuroticism. There is a significant correlation for P and a larger correlation for L. Both of these correlations are slightly larger than the values tabulated earlier. Thus a general model for family resemblance in personality cannot ignore the possible effects of assortative mating, even though mating may be random for specific variables.

The second main issue is that of cultural inheritance. In the previous chapter we noted that DZ twins were more alike in their Lie scores than could be predicted from a model in which the effects of the family environment were small. For the other principal personality dimensions, however, we reached the fairly strong conclusion that the family environment played almost no direct role in the development of adult personality differences. The pooled correlations for the adopted individuals in Table 6.3 offer further support for this aspect of the model. In no case is there significant correlation between the phenotypes of foster parents and their adopted children. Most models in which children are assumed to learn from their parents would predict a significant correlation between foster parent and adopted child. The standard error of these pooled correlations is about 0.07. The sample sizes for the foster–sibling correlations are unfortunately much smaller, so, although these correlations also do not differ significantly from zero, their standard errors are quite large (approximately 0.14), with the result that their upper 95% confidence interval includes the value of the correlation for DZ twins and natural siblings.

The problem with the *ad hoc* examination of correlations is that it is difficult to integrate data from multiple sources into a single estimate of the role of biological and cultural factors and easy to select correlations which happen to support the biases of the investigator. Before concluding that cultural inheritance plays no role, we should attempt to derive a model for both biological and cultural inheritance that gives a satisfactory fit to the data and allows us to test alternative hypotheses about the causes of variation.

The effects of cultural and biological inheritance in the presence of assortative mating have been the subject of extensive theoretical investigation (see e.g. Rao *et al.*, 1974; Rice *et al.*, 1978; Heath and Eaves, 1985). We outline a simplified treatment at this point in order to make their effects on the correlations between relatives more explicit.

In order to develop an effective model, we exploit the approach of path analysis (Wright, 1921), which has become the preferred method for deriving the consequences for family resemblance of diffferent theories of causation. An outline of the main principles of path analysis is given by Li (1976). The main restrictions of the method are the assumption of linearity and additivity, which preclude any prediction of the consequences of genetic interaction and genotype × environment interaction. However, with sufficient ingenuity, it is possible to specify a wide range of models for genetic and social effects, including most of the those considered so far.

The starting point for path analysis is the "path model", which represents the hypothesized causal connections within a system. The path model is a regression equation, or a series of such equations, that describe the relationships between measurements made by the investigator (e.g. "phenotypic measures") and hypothesized latent variables ("genotype" and "environment"). Given that no causes are omitted from the model (i.e. that the requirement of "causal closure" is satisfied), the relatively simple statistical rules for evaluating the variances and covariances of linear combinations of variables may be used to obtain predicted values for the correlations between measurements (see e.g. Li, 1976).

A helpful tool in developing a path model, and in deriving theoretical expectations for covariances and correlations, is the "path diagram". The path diagram in Figure 6.1 represents one model for the cultural and biological effects of parents on their offspring.

Direct causal influences, the "paths", are represented by the single-headed arrows. In our diagram, for example, there are paths from the genotype and environments of individuals to their phenotypes. The "strength" of the causal connection is measured by the "path coefficient", which is simply the partial regression of a particular variable on a given causal variable. The path coefficients are normally represented in the diagram by writing an appropriate symbol alongside the corresponding arrow. Thus the regression of phenotype P on (latent) genetic effect G is the path coefficient h.

Genetic effects are assumed to be additive. The path from parental genotype to the genotype of offspring is established by genetic theory to be $\frac{1}{2}$, so this value is specified as a constant in the diagram and in subsequent derivations. Cultural inheritance (or the effects of children "learning" from their parents) is assumed to depend on the direct impact of the parental phenotype on the offspring's environment ("P-to-E" transmission). The contribution of cultural inheritance from parents is measured by the path b. Thus we are

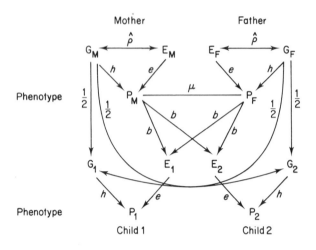

Figure 6.1   Simple path model for biological and cultural inheritance in nuclear families.

assuming that cultural inheritance is mediated through the measured phenotype of the parents. Other models can be written (see e.g. Rice *et al.*, 1980; Cloninger *et al.*, 1979) assuming alternative mechanisms of non-genetic inheritance, including transmission directly from parental environment to offspring's environment ("E-to-E" transmission). The fact that power calculations have shown the resolution of such alternative models of cultural inheritance to be difficult in practice (Heath *et al.*, 1985a, b) ranks as one of the major recent disappointments to progress in the field. As a consequence, the broad outline of our findings is likely to hold regardless of the mechanism that we assume for the environmental effects of parents on their children. A further implication of our model is that the environmental correlation between relatives is explained fully by parent-to-offspring transmission. This assumption can be relaxed, if desired. In this simplified model, we assume that mothers and fathers contribute equally to the phenotypes of their children of both sexes. We assume that there are no interactions of individual differences with age. The latter assumption can be changed in a number of ways. We consider two possibilities in this chapter and the next.

As far as assortative mating is concerned, we assume that mate selection is based on the actual measured phenotype (i.e. "test score") rather than an underlying trait. This assumption is testable, given appropriate data, and is employed here only for heuristic reasons. In a later chapter we shall consider some data that allow this assumption to be tested for social attitudes. The correlation between mates is represented by the "copath" μ (Cloninger, 1980). Use of the copath notation precludes our having to include explicit correlations between all possible parental effects in the diagram and leads to significant simplification in this case although the notation has to be used with some care because it leads to wrong expectations in more complex cases (for a generalization of the "copath" concept see Carey, 1986).

Since parents provide both genes and environment for their offspring, we should expect a correlation between genetic and environmental effects in their offspring. Indeed, when these offspring become parents, they will have genes and environments that are correlated because of biological and cultural inheritance from *their* parents. Obviously, if the cultural-inheritance parameter b is positive then the genotype–environment correlation is also positive, an effect that Jencks *et al.* (1973) describe as the "double-advantage" phenomenon in relation to genetic and environmental effects on performance because children whose parents provide them with an average genetic "advantage" may provide social advantages as well. In terms of personality, positive genotype–environment covariance would translate into extraverted parents encouraging outgoing and sociable behavior in their children. As far as neuroticism is concerned, positive genotype–environment covariance would become a "double disadvantage" for the development of high anxiety because neurotic parents would provide more stressful environments for their offspring.

If the biological and cultural inheritance of the trait continues over several generations with constant marital correlation and regression of childhood environment on parental phenotype then the phenotypic variance and the genotype–environment correlation will, under some circumstances, approach an equilibrium value. That is, for given h, b and μ, we obtain an equilibrium value of the genotype–environment correlation ρ, which can be expressed as a function of these other parameters. In the path diagram, we represent the correlation by a two-headed arrow between the genotypes of parents and their corresponding environments. The arrow is not shown in the offspring because the correlation is implied by the paths from the parents.

The rules of path analysis (see e.g. Wright, 1921; Li, 1976) are employed, with the modification suggested by Cloninger (1980) to allow for the copath between spouses, in order to derive the predicted correlations between relatives. To obtain the correlation between two variables, for example the genotype–environment correlation in offspring, the rules require that we trace every connecting pathway between the variables in question by going backwards along paths, then forwards, but not *vice versa*, without going through the same arrow twice in a single pathway. A "pathway" thus comprises several "paths". The contribution of a given pathway to a particular correlation is the product of the path coefficients beside all the connecting paths. There may be many pathways contributing to a single correlation. In tracing a given pathway, we may go along a double-headed arrow only once, in either direction, but may not go along two double-headed arrows in the same pathway because the correlation between multiple variables cannot generally be assumed to be transitive (i.e. $\rho_{AC} \neq \rho_{AB} \times \rho_{BC}$).

Phenotypic assortative mating generates correlations between the latent variables of spouses. In the traditional path-analytic treatment, these have to be derived from a separate diagram for the process of mate selection, then specified explicitly in the current diagram. The ensuing clutter is avoided by employing Cloninger's copath convention. His modification of the rules states that we may trace along any pair of two-headed arrows either side of the copath. The additional correlations between the spouses' latent variables are subsumed by this operation. The phenotypic variance and the variances of all the latent variables are assumed to be unity. Expected covariances are obtained simply by multiplying the expected correlations by the total phenotypic variance V.

Employing these conventions, we are able to derive the correlation between genotype and environment in the offspring:

$$\rho^* = b(h + e\rho)(1 + \mu).$$

However, if we assume that the population is at equilibrium under cultural and biological inheritance then the genotype–environment correlation in parents is expected to be the same as that in their children. Setting the derived correlation in the above expression to the equilibrium value $\hat{\rho}$ gives, after some rearrangement, the expression for genotype–environment correlation in terms of the other parameters of the model:

$$\hat{\rho} = \frac{(1 + \mu)bh}{1 - (1 + \mu)be}.$$

This expression is important when we come to fit the model, because it represents a constraint that must be imposed on the parameter estimates if the requirements of equilibrium are to be satisfied. It is reasonable to ask whether it is legitimate to assume the equilibrium conditions for behavioral variables since cultural expectations are likely to change with time. In theory, there is no need to assume equilibrium, so that ρ may be estimated as a free parameter. In practice, there is likely to be little information from which to estimate the genotype–environment correlation as a separate parameter. Furthermore, for moderate values of h, b and μ, much of the approach to equilibrium occurs in a single generation (assuming that $\rho = 0$ in the parents), so that the equilibrium values are not likely to be seriously misleading.

A second constraint on the parameter values is implied by the fact that the parameters are defined relative to a total variance of unity. Thus we have

$$h^2 + e^2 + 2eh\hat{\rho} = 1.$$

So far we have concentrated on the more formal aspects of the model by defining the constraints that must be satisfied by the parameter values. The same reasoning may now be applied to derive the correlations between relatives expected under the model.

From the original figure we may derive the parent–offspring correlation as:

$$r_{PO} = [be + \tfrac{1}{2}h(h + e\hat{\rho})](1 + \mu).$$

In the figure, we represent a second offspring having its own genotype and environment derived independently from the same parents. From the diagram, the correlation between siblings (also between dizygotic twins) is found to be:

$$r_{SIB} = \tfrac{1}{2}h^2[1 + (h + e\hat{\rho})^2\mu] + 2b^2e^2(1 + \mu) + 2eh\hat{\rho}.$$

The diagram for MZ twins is not shown, but can be easily drawn by allowing the genotype of one of the children in Figure 6.1 to have a second path to a new phenotype to represent the replication of the identical genotype in two individuals. A second environment for the new twin must be generated with separate paths b from the parental phenotype. If the diagram is drawn properly, the reader can verify that the correlation between MZ twins is expected to be:

$$r_{MZ} = h^2 + 2b^2e^2(1 + \mu) + 2eh\hat{\rho}.$$

Expected covariances for these and some additional relationships are summarized in Table 6.4.

Two further points need to be discussed: the correlations in adoptive families and the derivation of correlations for more remote relationships. The tabulated expectations show that the correlations involving adoptees are divided by a scaling factor. This is because, in the absence of placement, genes and environment of adoptees are expected to be uncorrelated. Thus if h and e are the paths from genotype and environment to phenotype for individuals reared by their natural parents (who are assumed to have unit total variance) then the variance of adopted individuals is only $h^2 + e^2 = 1 - 2eh\hat{\rho}$, assuming no placement effects. So, although the latent variables E and G are assumed to have unit variance in both biological and adopted

Table 6.4   Expected covariances between relatives with additive genetic inheritance, vertical cultural inheritance and phenotypic assortative mating (assuming no placement effects).

Variances	
Reared by biological relatives	1
Reared by foster parents	$h^2 + e^2$
Covariances	
MZ twins reared together	$h^2 + 2b^2e^2(1+\mu) + 2eh\rho$
MZ twins reared apart	h^2
DZ twins/siblings living together	$\frac{1}{2}h^2[1+(h+e\rho)^2\mu] + 2b^2e^2(1+\mu)$ $\qquad\qquad\qquad + 2eh\rho$
DZ twins/siblings living apart	$\frac{1}{2}h^2[1+(h+e\rho)^2\mu]$
Unrelated foster siblings	$2b^2e^2(1+\mu)$
First cousins[a]	$\frac{1}{4}h^2[\gamma+2\mu(h+e\rho)\,(h\gamma+e\rho) + \tau\mu^2(h+e\rho)^2]$ $+ \tau b^2e^2(1+\mu)^2$ $+ ehb(1+\mu)\,[h\gamma+e\rho+\tau\mu(h+e\rho)]$
Parent–offspring	$[\frac{1}{2}h(h+e\rho)+eb]\,(1+\mu)$
Foster-parent/foster-child	$eb(1+\mu)$
Avuncular (cognate)[b]	$\frac{1}{2}h[\gamma h+e\rho+\tau\mu(h+e\rho)] + \tau eb(1+\mu)$

Note: $\rho = bh(1+\mu)/[1-(1-\mu)be]$.

[a] For first cousins related through MZ twin parents ("MZ half-siblings") τ is the expected correlation of MZ twins and $\gamma=1$. For ordinary first cousins (related through sibling parents) τ is the expected sibling correlation and $\gamma=\frac{1}{2}h^2[1+(h+e\rho)^2\mu]$.

[b] The same substitutions should be made in avuncular covariances as appropriate.

children, and although we assume the regressions of phenotype on genetic and environmental effects to be the same, the covariances involving adopted individuals have to be rescaled, to reflect the differences in variance. In relationships involving one adopted individual (i.e. foster-parent–foster-child) only the foster-child variance needs to be standardized, so the covariance is divided by the standard deviation $(1-2eh\hat\rho)^{1/2}$. When both relatives are adopted, the variances of both need to be rescaled, and the covariance is simply divided by the total variance of adoptees. The reader may reduce the complex expressions to those applicable under simpler hypotheses (e.g. random mating, or no genetic effects or no cultural inheritance) simply by setting particular parameters to zero in the expectations of Table 6.4. Examination of the table helps to arrive at some generalizations about the effects of assortative mating and cultural inheritance. Both increase the sibling and parent–offspring correlations. In both cases the sibling correlation exceeds one-half the MZ twin correlation. For a wide range of parameter values, the model also predicts that the parent–offspring correlation will exceed that between siblings. This result does not hold for all types of assortative mating. For example, if assortative mating is based on the additive genetic deviation rather than the phenotype (what Fisher (1918) termed the "essential genotype") then the sibling correlation will still equal the parent–offspring correlation in the absence of cultural inheritance, but will exceed one-half the MZ correlation. The model confirms algebraically what we have already said in words—that the effects of assortative mating and cultural inheritance will be confounded if we only have data on twins.

We do not give the correlations between more remote relationships (e.g. second cousins) because they do not have to be calculated algebraically in order to be used in a computer program. Instead, we define an algorithm that enables the correlations for more remote relationships to be built up from the correlations for simpler relationships. In Figure 6.2 we present the path diagram that shows how this is done. We begin with a pair of relatives for whom the phenotypic correlation is already evaluated. In the simplest case, the pair may be siblings. The correlations between the latent genetic and environmental components of the phenotypes are assumed to be known. In the case of a sibling pair Figure 6.1 gives

$$\alpha = \beta = \hat{\rho},$$
$$\eta = 2b^2(1 + \mu),$$
$$\gamma = \tfrac{1}{2}[1 + (h + e\hat{\rho})^2\mu].$$

For other relationships, α and β will not necessarily be equal, but will be obtained from previous results. We now assume that one of the pair becomes a parent and produces the offspring shown in the figure. If the original relatives were siblings then the offspring will be a nephew/niece of the non-parental relative, and the derived relative–offspring correlation will be the avuncular correlation ϕ'. The rules of path analysis can be used to yield the expected correlation for the new relationship:

$$\phi' = h^2\gamma' + e^2\eta' + eh(\alpha' + \beta'),$$

where

$$\alpha' = \tfrac{1}{2}[\alpha + \mu(h + e\hat{\rho})(e\eta + h\alpha)],$$
$$\beta' = b(1 + \mu)(e\beta + h\gamma),$$
$$\eta' = b(1 + \mu)(e\eta + h\alpha),$$
$$\gamma' = \tfrac{1}{2}[\gamma + \mu(h + e\hat{\rho})(e\beta + h\gamma)].$$

The new values of the genetic, environmental and genotype–environmental correlations can be used as starting values for computing correlations for other relationships. The same algorithm may be used, for example, to compute the cousin correlation by starting with the avuncular correlation and allowing the "uncle" to become the parent in Figure 6.2 and the nephew to be the "relative". Some care needs to be taken in making sure that the correct substitutions are made for the genotype–environment correlations α and β, because these are not always identical and need to be correctly defined for a particular relationship.

6.2.4 Minimization subject to constraints

We fitted various forms of the model to the thirteen correlations from Table 6.3 for each variable by non-linear weighted least-squares. The model was fitted to the z-transformed correlations to improve the approximation to normality (see Rao *et al.*, 1977). Writing θ for the vector of unknown parameters (h, e, b, etc.), we seek the values of θ that minimize the weighted sum of squared residuals

$$s^2 = \sum_i w_i (z_i - Ez_i)^2,$$

subject to the side-conditions implied by the equilibrium constraint on the genotype–environment correlation and the fact that the total variance of individuals reared by their natural parents is unity. A convenient approach to minimization subject to constraints is the method of Lagrange multipliers (see e.g. Greig, 1980). If s^2 has to be minimized subject to the constraints $c_i(\theta) = 0$ then we define a new function $F(\theta, \lambda) = s^2 + \sum_i \lambda_i c_i(\theta)$. The new parameters λ are the "Lagrange multipliers". The problem of minimization then becomes an unconstrained one of minimizing F with respect to θ and λ, since any solution for which F is minimized will also be a minimum for s^2 satisfying the side-conditions on θ. That this must be the case may be seen fairly simply because the derivative of the augmented function with respect to the ith Lagrange multiplier is $c_i(\theta)$, so that any values of λ_i and θ that set this derivative to zero must *de facto* satisfy the side-condition that $c_i(\theta) = 0$. Most books on college calculus describe the method in more detail.

To attempt to write a general program for constrained optimization is foolish, given that there are a number of commercially available specially written programs that exploit the skills of numerical analysts and computer scientists in providing robust and accurate solutions to complex numerical problems. We used the Numerical Algorithms Group's (NAG) FORTRAN subroutine EO4UAF in this example to minimize s^2 subject to constraints; it automatically specifies the necessary Lagrange multipliers and constructs the augmented Lagrangian function. The user has to develop ancillary routines to supply the data and evaluate s^2 and the $c_i(\theta)$.

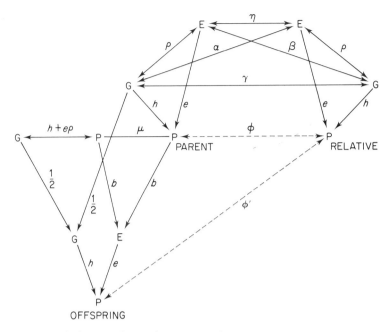

Figure 6.2 Path diagram for evaluating correlations between remote relatives.

6.2.5 Results of correlational analysis

The results of fitting selected models to the pooled correlations are summarized in Table 6.5. In addition to the parameters in the original model we include a dominance parameter d, which contributes d^2 to the total variance and the correlation of MZ twins and $\frac{1}{4}d^2$ to the correlation of siblings and DZ twins. The parameters h^2 and d^2 are analogous to the parameters V_A and V_D of the biometrical model expressed as proportions of the total variance. The contribution of the family environment is $2e^2b^2(1+\mu)$, which is analogous to E_B in the earlier treatment. The effects of genotype–environment correlation due to P-to-P cultural inheritance are also confounded with estimates of E_B in the earlier analyses of twin data.

Not every possible combination of parameters was fitted, but the tabulated results give sufficient indication of the general findings. Models that assume only cultural inheritance ($h = d = 0$) give a very bad fit in every case. The simplest situation occurs in extraversion, for which the pooled correlations are consistent with a purely additive genetic model. However, we recall that this was the trait for which there was strong evidence that the correlations were heterogeneous across sexes. The small estimated dominance

Table 6.5 Model fitting to pooled correlations.

Trait	h	μ	b	d	$e^{(a)}$	$\rho^{(a)}$	χ^2	df	P%
Psychoticism	0.62	—	—	—	0.78	—	22.9	10	1
	0.60	0.21	—	—	0.80	—	15.7	9	7
	0.69	0.24	−0.11	—	0.78	−0.09	11.2	8	19
	0.47	0.27	−0.00	0.49	0.74	−0.00	6.4	7	50
	—	0.17	0.23	—	1.00	—	75.9	9	$<10^{-6}$
Extraversion	0.68	—	—	—	0.73	—	5.2	10	88
	0.68	−0.00	—	—	0.73	—	5.2	9	82
	0.73	0.00	−0.07	—	0.72	−0.05	3.8	8	87
	0.67	0.04	−0.03	0.26	0.71	−0.02	3.6	7	83
	—	−0.11	0.34	—	1.00	—	77.9	9	$<10^{-6}$
Neuroticism	0.57	—	—	—	0.82	—	24.8	10	0.6
	0.57	0.01	—	—	0.82	—	24.8	9	0.3
	0.65	0.02	−0.09	—	0.80	−0.05	22.1	8	0.5
	0.00	0.07	0.11	0.65	0.76	0.00	10.8	7	15
	—	−0.00	0.22	—	1.00	—	64.7	9	$<10^{-6}$
Lie	0.74	—	—	—	0.68	—	37.4	10	$<10^{-6}$
	0.70	0.26	—	—	0.71	—	24.6	9	$<10^{-6}$
	0.81	0.32	−0.14	—	0.70	−0.13	19.6	8	1
	0.61	0.36	−0.02	0.47	0.65	−0.02	13.5	7	6
	—	0.13	0.34	—	1.00	—	119.6	9	$<10^{-6}$

^{a} Derived parameters.

component ($d^2 = 0.07$) might reflect the effect of sex limitation in reducing the pooled correlations for relationships in which there are unlike-sex pairs (i.e. all but MZ twins).

In the case of psychoticism the best-fitting model assumes additive genetic effects and assortative mating. However, the fit improves substantially with the addition of a dominance component that accounts for about the same proportion of the total variance as the additive genetic component. The only model that fits the correlations for neuroticism requires an enormous amount of dominance. We shall consider the implications of this result subsequently. No model fits the "lie" data very well, but a model allowing for additive and dominant genetic effects, with assortative mating just about fits at the 5% level.

The fourth model tabulated in every case is the most important because it allows for additive and dominant genetic effects, cultural inheritance and assortative mating. In each case the model fits the correlations, although the data yield a suspiciously high estimate of d for neuroticism and the model only just fits the data for the lie scale. Without exception, however, the estimate of the cultural parameter is small in comparison with the total estimated genetic effect.

6.2.6 Model fitting to unreduced data by maximum-likelihood

There are many difficulties with the above statistical analysis. The statistical method depends on the assumption of independence in the observed correlations. A number of potentially important correlations between relatives were ignored. For example, we have not included correlations for relationships by marriage apart from the spousal correlation itself. These correlations will not be zero if there is assortative mating. The effects of age on personality were ignored. The raw correlations were not corrected for the regression of personality on age. No less important is the fact that average correlations are no use if there are developmental changes in the expression of genetic and environmental effects. If different genes are expressed at different ages then no amount of conventional "age correction" will identify or remove such effects. These effects will be considered in the next chapter.

All of these considerations lead us to require a method of analysis that takes into account the uniqueness of every family in the study with respect to its biological and social structure and its age composition. The maximum-likelihood method may be applied to the raw pedigree data in a way that allows us to estimate the parameters of a particular set of models by a method that does account properly for the correlations between observations and the unique structure of every pedigree. The statistical basis of the method is described by Lange *et al.* (1976), but is extended to allow for assortative mating and cultural inheritance and applied to the analysis of personality and social-attitude data (Eaves, 1977; Eaves *et al.*, 1978). The raw data comprise a series of independent "pedigrees" of arbitrary and variable structure. For our purpose a "pedigree' may be anything from a single individual, through a pair of twins, to a complex pedigree comprising multiple generations of individuals of many different kinds of biological and social relationship.

We consider a given pedigree with p individuals. The log-likelihood of the ith pedigree is given by

$$L_i = \tfrac{1}{2}\ln|\Sigma_i| - \tfrac{1}{2}[x - E(x)]'\Sigma_i^{-1}[x - E(x)] + \text{constant}$$

(Lange *et al.*, 1976). The vector of expected values $E(x)$ may depend on covariates such as age and sex, or it may be assumed to be constant across the population. In our application we allow for the effects of sex and age by fitting a separate regression equation to each sex at the same time as fitting the genetic model. Thus for males we let

$$E(x_i) = a_m + b_m y_i,$$

where y_i is the age recorded for the ith individual. A similar function is written for females, involving a different constant, a_f, and regression coefficient, b_f. The same approach may be extended to more complex regression models and other covariates.

The expected covariance matrix Σ_i will vary in structure from one pedigree to another, but under the simpler types of model it will involve the same basic parameters in different combinations. Under the developmental model that we outlined above, the expected covariance matrix of every pedigree will be unique because every pedigree will have a unique age structure. However, we start by fitting models that assume developmental effects to be absent. Thus, given additive genetic inheritance and random environmental effects, families comprising monozygotic twins and their parents will have the following structure:

$$
\Sigma_i = \begin{bmatrix}
\text{twin 1} & \text{twin 2} & \text{mother} & \text{father} \\
V & h^2V & \frac{1}{2}h^2V & \frac{1}{2}h^2V \\
h^2V & V & \frac{1}{2}h^2V & \frac{1}{2}h^2V \\
\frac{1}{2}h^2V & \frac{1}{2}h^2V & V & 0 \\
\frac{1}{2}h^2V & \frac{1}{2}h^2V & 0 & V
\end{bmatrix}.
$$

Other families, with different relationships, will have different expected covariance matrices, but their elements will still be functions of the same parameters, h and V. The same principle applies to the other, more complex models, such as that in Figure 6.1. In some applications, as in the case of developmental changes in gene expression considered in Chapter 7, the elements of the expected covariance matrix may be unique to each pair of relatives and a function of other variables measured on the subjects.

Given that the likelihood may be expressed in this form for an individual pedigree for given parameter values, it is only a matter of tedious computation to find the parameter values that maximize the joint likelihood over all k pedigrees: $L = \Sigma_i L_i$.

The algorithm described by Lange *et al.* requires the algebraic differentiation of the likelihood function with respect to the unknown parameters. This is difficult for all but the most basic models. Fortunately, the optimization subroutines of the NAG FORTRAN library employ numerical differentiation with satisfactory results. The main disadvantage of the maximum-likelihood method is the very great increase in computer time required because the likelihood has to be computed separately for each family, and each likelihood requires computation of a determinant and an inverse matrix. The weighted-least-squares analyses for a number of models can all be conducted in under thirty seconds on a IBM 370. The maximum-likelihood approach, with our algorithm, required about eight minutes to fit one model starting with reasonable trial values.

In the extended pedigrees there are many types of relationship by marriage that have to be given non-zero expectations when there is assortative mating. As long as we restrict ourselves to models of assortative mating based on the measured phenotype, the expectations may be computed quite easily by a simple algorithm. If ϕ is the correlation for a known relationship (e.g. siblings) then the correlation by marriage (the brother/sister-in-law correlation) is $\phi\mu$, where μ is the marital correlation. The correlation between the spouses of siblings is then $\phi\mu^2$. The same approach can be used to evaluate other expected correlations between more remote relatives. When we allow for relationships by marriage, the number of unique relationships is very large and would present a considerable problem in coding for our data if every relationship were to be given a unique code at the outset. Fortunately, much of this work can be computerized. The data were coded in a three-stage process. At the original data entry stage each individual in a pedigree is simply identified by his parents, with "dummy" parents being created for individuals who are "founders". A method derived from graph theory (Maruyama and Yasuda, 1970) was then modified to identify each biological relationship and assign it a unique code. The codes for each family were stored as a matrix defining the biological relationships in the family. These codes were output with the raw data into a preprocessed datafile to simplify the substitution of expected correlations in the model-fitting program. Eaves and Eysenck (1980, pp. 247–250) illustrate this stage of coding pedigrees in more detail. In the final stage, codes were generated for relationships by marriage by combining the information on biological relationships with information on spouses.

One major drawback of the maximum-likelihood approach is the lack of any overall test of the model. Likelihood ratios can be computed to test alternative hypotheses, but there is no completely general model against which to test all the others. If a second model is a restricted form of a first more general model then twice the difference between the logarithms of the two likelihoods is approximately a chi-square having df equal to the number of parameters omitted from the reduced model. If the chi-square is not significant then we may be justified in omitting the parameters from the model. Otherwise, it must be assumed that they represent effects that make a significant contribution to individual differences.

6.2.7 Results of maximum-likelihood analysis

Tables 6.6–6.9 summarize the model-fitting results for the full dataset. A model in which there is no correction for age (model I) is generally very much worse than models in which we allow for linear regression of phenotype on

Table 6.6 Maximum-likelihood estimates: psychoticism.

Parameter	Model							
	I	II	III	IV	V	VI	VII	VIII
a_m	0.326	0.441	0.441	0.443	0.440	0.442	0.444	0.439
a_f	0.259	0.316	0.315	0.315	0.315	0.314	0.316	0.317
$b_m \times 10^2$	—	-0.306	-0.306	-0.315	-0.306	-0.314	-0.317	-0.316
$b_f \times 10^2$	—	-0.161	-0.160	-0.160	-0.161	-0.159	-0.161	-0.168
$V \times 10$	0.281	0.269	0.268	0.268	0.265	0.264	0.267	0.270
h	0.654	0.655	0.668	0.638	0.712	0.714	0.525	—
d	—	—	0.093	—	—	—	0.464	—
μ	—	—	—	0.228	—	0.243	0.270	0.256
b	—	—	—	—	-0.086	-0.098	-0.016	0.217
e [a]	0.756	0.755	0.744	0.770	0.745	0.759	0.719	1.000
ρ [a]	—	—	—	—	-0.058	-0.079	-0.010	—
L	3214.28	3271.75	3273.85	3275.58	3273.10	3277.61	3279.90	3240.45

[a] Derived parameter.

Table 6.7 Maximum-likelihood estimates: extraversion.

Parameter				Model				
	I	II	III	IV	V	VI	VII	VIII
a_m	0.865	1.029	1.031	1.030	1.030	1.030	1.031	1.037
a_f	0.863	0.911	0.912	0.911	0.911	0.911	0.911	0.909
$b_m \times 10^2$	—	−0.438	−0.443	−0.440	−0.440	−0.442	−0.445	−0.461
$b_f \times 10^2$	—	−0.137	−0.139	−0.137	−0.137	−0.136	−0.137	−0.135
$V \times 10$	0.828	0.803	0.801	0.803	0.797	0.797	0.799	0.810
h	0.711	0.698	0.655	0.695	0.729	0.731	0.693	0.810
d	—	—	0.274	—	—	—	0.214	—
μ	—	—	—	0.048	−0.049	0.056	0.061	0.061
b	—	—	—	—	—	−0.052	−0.035	0.267
e [a]	0.703	0.716	0.704	0.719	0.710	0.712	0.706	1.000
ρ [a]	—	—	—	—	−0.034	−0.039	−0.025	—
L	1911.16	1942.64	1943.03	1942.86	1943.10	1943.20	1943.51	1898.82

[a] Derived parameter.

Table 6.8 Maximum-likelihood estimates: neuroticism.

Parameter	Model							
	I	II	III	IV	V	VI	VII	VIII
a_m	0.667	0.746	0.746	0.746	0.745	0.745	0.747	0.746
a_f	0.831	0.939	0.940	0.939	0.939	0.938	0.941	0.937
$b_m \times 10^2$	—	-0.213	-0.212	-0.213	-0.209	-0.209	-0.213	-0.219
$b_f \times 10^2$	—	-0.309	-0.308	-0.309	-0.306	-0.306	-0.309	-0.219
$V \times 10$	0.750	0.731	0.728	0.731	0.725	0.725	0.729	0.736
h	0.606	0.608	0.439	0.605	0.657	0.657	0.339	—
d	—	—	0.483	—	—	—	0.549	—
μ	—	—	—	0.045	—	0.052	0.071	0.086
b	—	—	—	—	-0.071	-0.073	0.047	0.203
$e^{(a)}$	0.796	0.794	0.758	0.796	0.784	0.786	0.758	1.000
$\rho^{(a)}$	—	—	—	—	-0.044	-0.047	0.018	—
L	1992.42	2024.62	2027.52	2024.80	2025.60	2025.85	2028.18	1997.03

[a] Derived parameter.

Table 6.9 Maximum-likelihood estimates: lie scale.

Parameter	Model							
	I	II	III	IV	V	VI	VII	VIII
a_m	0.638	0.376	0.376	0.377	0.377	0.377	0.378	0.380
a_f	0.694	0.468	0.466	0.467	0.468	0.467	0.465	0.469
$b_m \times 10^2$	—	0.700	0.697	0.700	0.698	0.697	0.694	0.700
$b_f \times 10^2$	—	0.645	0.647	0.649	0.645	0.648	0.651	0.646
$V \times 10$	0.611	0.668	0.521	0.648	0.712	0.709	0.487	0.646
h	0.681	0.668	0.521	0.648	0.712	0.709	0.487	—
d	—	—	0.487	—	—	—	0.521	—
μ	—	—	—	0.269	—	0.278	0.316	0.304
b	—	—	—	—	-0.071	-0.079	0.003	0.198
$e^{(a)}$	0.732	0.744	0.701	0.761	0.737	—	0.701	1.000
$\rho^{(a)}$	—	—	—	—	-0.048	-0.067	0.002	—
L	2272.63	2523.17	2527.58	2529.01	2524.14	2530.48	2535.71	2482.74

[a] Derived parameter.

age. Although the parameter estimates do not agree exactly, there is broad agreement between the results we obtain by maximum-likelihood and the results of the much simpler weighted-least-squares method.

For every variable, we find that a model that leaves out genetic effects entirely (model VIII) and allows only for cultural inheritance and assortative mating fits very much worse than a model in which genetic effects are included (model VII). In the case of P, for example, twice the difference in log-likelihoods yields $\chi_2^2 = 78.9$ as a test of the joint contribution of additive and dominant genetic effects. Results for the other traits are comparable. Significant assortative mating is confirmed for P and L. Estimates of the cultural inheritance parameter b are uniformly small, as might be predicted from the small adoptive correlations. There is a small estimate of the dominance parameter d for extraversion, significant estimates of d for P and L, which are about the same size as h, and a very large value of d for neuroticism, which is inconsistent with the results of almost all the twin data. Taking the value of h^2 under model VII as our most conservative estimate of the contribution of additive genetic factors to personality differences, we have h^2 = 0.276, 0.480, 0.115 and 0.237 for P, E, N and L respectively. Only the value for E compares well with the estimates derived from the twin data alone in the previous chapter. If the contributions of dominance d^2 are added to yield estimates of the broad heritability then we obtain h_b^2 = 0.491, 0.526, 0.416 and 0.508 for the four traits. These results are much closer to the heritability estimates reported in the previous chapter and reflect the fact that the correlation between MZ twins is higher than the additive genetic model would predict, given the correlations for other relationships. There are many ways of looking at these findings. The contribution of additive genetic factors may indeed be small and the contribution of dominance may have been overlooked in the twin data because the test of non-additive effects is notoriously weak in twin data. Estimates of the heritability from the twin data therefore correspond much more closely to "broad heritability" because the effects of dominance will be present as a bias in the estimate of V_A even though there is no detectable dominance. This amounts to saying the twin heritabilities are right, but they confound two distinct sources of genetic variation.

A second interpretation of the same results is that the (narrow) heritabilities from the whole dataset are about right (i.e. small) but that the estimated contribution of dominance is a spurious effect of a specially high *environmental* correlation for MZ twins which our current model ignores. Since the contribution of dominance is greatest in the MZ correlation (d^2) and only ¼ d^2 in siblings and DZ twins, the excess correlation of MZ twins will be ascribed to dominance effects when it could equally be due to the special MZ twin environment. In the absence of separated MZ twins (though see below) there is no clear way of resolving these alternative hypotheses for the apparent excess resemblance of MZ twins.

A third possibility, which has great potential importance psychologically, is that there may be age differences in the expression of genetic effects. That is, when individuals are measured at the same age (as in the case of twins) the same genetic effects are expressed so that the twin correlations are high. When individuals are measured at different ages different genetic effects are being expressed because different genes control behavior at different stages of development. Thus when we measure individuals at different ages (e.g. parents and children or siblings) we find small correlations. Since much of the information about the narrow heritability comes from parent–offspring and avuncular data, estimates of h will be forced downwards so that the excess resemblance of twins will tend once more to be assigned to the effects of dominance. This possibility will be explored in detail in the next chapter.

6.3 THE EXTENDED KINSHIPS OF TWINS

An appealing research design for the analysis of genetic and cultural inheritance is the "extended twin kinship design". The strategy involves supplementing twin pairs by various informative relatives. Options include the offspring of twins (e.g. Nance and Corey, 1976), the spouses of twins (e.g. Eaves, 1980; Heath and Eaves, 1985) (see Chapter 16), the parents of twins (Young *et al.*, 1980; Jardine and Martin, 1984; Heath *et al.*, 1986), and the siblings and parents-in-law of twins (Heath, 1983). Each of these strategies has specific strengths, especially with respect to the resolution of assortative mating and cultural inheritance; these have been explored from a theoretical perspective by Heath *et al.* (1986).

Currently, the only application of a greatly extended twin design to personality data is that of Price *et al.* (1982), who publish correlations for extraversion and neuroticism, *inter alia*, for twenty-two biological and non-biological relationships derived from 138 pairs of MZ and like-sex DZ twins, and their spouses and offspring drawn from the Swedish twin registries. Only offspring aged 20 and over were included in the study in order to minimize developmental effects. Adjustments were made for age, sex and generational differences. The authors originally fitted a complex statistical model devised by Williams and Iyer (1981), which was subsequently criticized by Heath *et al.* (1984) for inconsistency with any conceivable causal model for biological and cultural inheritance in the presence of assortative mating. Price *et al.* conclude that personality traits seem to be influenced primarily by the environment, especially influences outside the home, and that, insofar as genetic effects are indicated at all, they are confounded with cultural factors. They conclude:

> Our analysis provides little support for simple additive models of genetic trans-
> mission for characteristics of normal personality. All of the traits that we have
> examined yield narrow-sense heritability estimates, h^2, which are essentially
> zero.

This finding is contradictory to that offered by most investigators, including ourselves.

Hewitt (1984) described a reanalysis of their data that suggests that their claim was premature and probably founded on "over-fitting" complex models to data in which chance departures from additive genetic inheritance were assigned to other genetic and environmental causes. In support of his contention, Hewitt outlined results of his own reanalysis in which he showed that the observed pattern of family resemblance could be described perfectly adequately in terms of the conventional additive genetic model that Price *et al.* had claimed to reject. We have repeated Hewitt's analysis and report one or two additional tests that strengthen his conclusion without altering it substantively.

The analysis employed the same model and algorithm as that already des-cribed for the analysis of the correlations from the London family and adop-tion data. A restricted set of four models was fitted that are sufficient to give further weight to Hewitt's contention that additive genetic effects alone are sufficient to explain these data. The results of fitting four crucial models to the extraversion and neuroticism correlations of Price *et al.* are given in Table 6.10.

A single parameter h is clearly sufficient to account for the observed varia-tion in the correlations from the extended twin dataset. This reproduces Hewitt's claim exactly and suggests that, after fitting h, there are no residual effects that can reasonably be ascribed to anything other than chance. How-

Table 6.10 Model fitting to correlations from extended kinships of twins.[a]

Extraversion						Neuroticism					
h	m	b	d	χ^2	df	h	m	b	d	χ^2	df
0.70	—	—	—	12.3	21	0.62	—	—	—	13.4	21
0.69	0.11	—	—	10.5	20	0.61	0.09	—	—	12.3	20
0.70	0.11	−0.02	—	10.5	19	0.68	0.10	−0.07	—	12.0	19
0.70	—	—	0.00^b	12.3	21	0.62	—	—	0.00^b	13.4	21

[a] Correlations published by Price *et al.* (1982).
[b] No real value of d^2 gives a better fit.

ever, the overall goodness-of-fit test is conservative (since all possible errors of specification are swept up in a single test), and it is instructive to test for improvement as a result of fitting some specific additional parameters to the model. No significant improvement is obtained by allowing for assortative mating, dominance or cultural inheritance because the reductions in chi-square that accompany the attempt to include dominance (d), assortative mating (μ) or cultural inheritance (b) do not even begin to approach significance for either trait.

6.4 SEPARATED TWINS

Perhaps the strongest claim that we have made so far relates to the absence of any significant effect of the family environment on personality. Although we arrive at this conclusion by fitting models to all kinds of data, we have relied heavily on family members reared by the natural parents or biologically unrelated individuals living together. We have not considered data on the resemblance of biological relatives reared separately. It is important that any model be tested as widely as possible, and one of the most compelling tests to be suggested is the correlation between separated monozygotic twins. If the postnatal environment shared by family members contributes significantly to the development of behavior, it is argued, then identical twins who have been separated at birth should be less alike than identical twins who have grown up together. There are, of course, objections to this strategy as there are to virtually all the others: the salient environment may be prenatal; placement after separation may not be random; separated twins may not be typical; the twins are measured after being reunited so "convergence" may take place. Our approach to such data is to ask not so much "Why won't it work?" as to find out whether our model gives answers consistent with other sets of data. Our predictions for separated twins, on the basis of the models in Chapter 4, are quite straightforward. Since we have found no evidence of a family-environmental effect in any of the large twin studies of extraversion and neuroticism, we should expect the correlation of separated identical twins to be no lower than that for identical twins reared together. If this indeed is the case then the model that we have outlined in Chapter 4 is capable of predicting other results. The same prediction holds for adoption data. If there is no need to invoke a shared environmental effect for the twin data, and if the model is of any predictive value, then there will be zero correlations between foster-parent and foster-child and between unrelated individuals reared in the same home.

Two studies have secured measures of extraversion and neuroticism on separated twins. An earlier, well-known, study by Shields (1962) yielded responses to an early form of the Eysenck Personality Inventory (EPI) of 38 pairs of separated MZ twins. Shields also gives results for identical and non-identical twins reared together. Shields' own analysis of the twins show that the correlation of separated MZ twins is not significantly less than that for MZ twins reared together, while the correlations for extraversion show a slight excess in favour of separated twins. The difference, however, is not significant. Shields' study is exemplary in that he provides a detailed appendix in which he gives raw scores and life histories for all the separated twins in his sample. By doing so, Shields has exposed himself to considerable criticism, directed chiefly at the results of his cognitive measures, but has also provided a unique opportunity for others to apply new methods to his data.

In their seminal paper, Jinks and Fulker (1970) employed the model-fitting methods that we have outlined (Chapters 4 and 5) in order to test hypotheses about the inheritance of extraversion and neuroticism in Shields' data. They returned to the raw data and computed the within-pair variances and variances of family means for each twin group in Shields' study. Jinks and Fulker note that the samples were larger for females than males. They note (p. 326):

> In general, we have included the male data, where they agree with the female, in order to provide replication and augmentation of them. Where there was serious disagreement between male and female data, the males were discarded. In consequence, our conclusions apply more reliably to females than to males. However, as may be seen, the agreement between sexes was satisfactory on the whole.

Jinks and Fulker base their model-fitting analysis on the within-pair variances (identical to our "mean squares within families") and the variances of pair-means, which are equal to one-half of the mean squares between pairs. For uniformity with the rest of our book, we present the observed mean squares in Table 6.11 derived from Shields' raw extraversion and neuroticism scores.

Jinks and Fulker used non-iterative weighted least squares to estimate parameters and test hypotheses. They concluded that the model with only additive gene action, random mating, and within-family environmental effects (V_A, E_W), gave an adequate fit to the neuroticism data and showed that the contribution of the family environment (E_B) was not significant. For extraversion, they obtained negative estimates of the family-environmental component E_B and the genetic variance between families G_B. They suggested that this might indicate competition affecting extraversion—a suggestion that anticipated one of the possible interpretations of our analyses of the Swedish and Australian datasets.

Jinks and Fulker put the idea of competition into words as follows (p. 333):

> The linear statistical model used in deriving the expectations of these com-

Table 6.11 Mean squares from Shield's twin study.

Twin type	Source	df	Extraversion	Neuroticism
			Mean square	
Separated MZM	Between pairs	13	22.57	30.11
	Within pairs	14	3.57	5.64
Separated MZF	Between pairs	27	28.17	25.87
	Within pairs	28	7.77	9.90
Together MZM	Between pairs	13	21.46	8.05
	Within pairs	14	6.59	7.83
Together MZF	Between pairs	28	14.48	22.16
	Within pairs	29	6.49	8.12
Separated DZF	Between pairs	6	7.87	4.35
	Within pairs	7	12.13	20.11
Together DZF	Between pairs	15	13.73	22.81
	Within pairs	16	26.55	13.10

ponents does not allow them to become negative unless the individuals within pairs are negatively correlated. This will occur with dizygotic twins if they react against each other in such a way as to develop opposite characteristics with respect to a trait. In doing this, they will be reacting on the basis of differences due to G_1 as well as E_1, whereas the same tendency in the monozygotic twins will have only E_1 effects to build upon. The negative covariance in the dizygotic pairs will, therefore, be more pronounced than in the monozygotic pairs . . . This phenomenon is termed "competition" and often takes place during the early part of the lives of many wild plants and animals.

While Jinks and Fulker do not discount a social explanation of competition, they suggest that the effect could be the developmental consequence of intrauterine competition. Their model predicts the same total variance for twins reared together and apart. This would not be the case if competition were based on postnatal social interaction. The "social-interaction" model of competition (see Eaves, 1976a, and Chapter 5) predicts a different total variance for separated MZ twins from that expected for MZ twins reared together since separated twins are removed from the environmental influence of their cotwins. Given their assumptions, Jinks and Fulker outline what they describe as a "tentative model" for the effects of competition on extraversion. They introduce a parameter CG_1 to describe "that part of the covariance between dizygotic twins due to their genetic differences (G_1)". A competitive effect corresponds to a negative CG_1 and will increase the variance within DZ pairs while reducing the covariance between them (cf. Jinks and Fulker, 1970, Table 14). Their analysis yields a negative estimate of CG_1 that is just significant, giving support to the idea that competitive effects contribute to the development of extraversion.

There are differences between the competition model outlined by Jinks and

Fulker and that developed by Eaves (1976b). This is clear in the different predictions that they make about the total variances for MZ and DZ twins. Jinks and Fulker's model assumes them to be equal, Eaves predicts that they will be different. The difference seems to lie in the part that interactions play in the development of MZ twins. Jinks and Fulker suggest that competition will play no role because MZ twins are genetically identical. They are, it appears, thus postulating different mechanisms of social interaction for MZ and DZ twins or for the social effects of genetic differences within and between families. Eaves' model assumes that the social interaction is a general process that occurs whenever two individuals are raised together. Competition will have consequences for both types of twin, but the effects will differ because of the smaller genetic correlation between DZ twins. Thus, for example, Eaves' model predicts a smaller total variance for MZ twins than DZs. The total variance of separated twins will depend on whether the competitive effects occur before or after separation, on whether the twins were fostered as singletons or with siblings, and on how far the social interaction is based on genetic effects on introversion–extraversion.

We refitted models to the mean squares derived from Shields' data. The expected mean squares are given in Table 6.12. Expectations for twins reared together have been given already (Table 5.6) but are modified now to reflect the additional information that can be derived from separated monozygotic twins. The main differences in the expectations of separated twins are:

(1) the between-families environmental component E_B contributes to σ_W^2 for separated twins rather than to σ_B^2;

(2) the effects of postnatal competition are absent in separated twins.

Both these differences follow from the assumption that effects of the

Table 6.12 Expected variances and covariances for MZ and DZ twins together and apart in the presence of prenatal or postnatal competition and cooperation (ignoring dominance and other shared environmental effects).[a]

Relationship	Statistic	E_W	V_A	V_{AS}	C_{AS}
Separated MZ	Variance	1	1	(1)	(2)
	Covariance	0	1	(1)	(2)
Separated DZ	Variance	1	1	(1)	(1)
	Covariance	0	$\frac{1}{2}$	$(\frac{1}{2})$	(2)
Together MZ	Variance	1	1	1	2
	Covariance	0	1	1	2
Together DZ	Variance	1	1	1	1
	Covariance	0	$\frac{1}{2}$	$\frac{1}{2}$	2

[a] Terms in parentheses are omitted in the model for postnatal interaction.

family environment and competition are postnatal. In addition, the expectations for competition assume that the separated twins were reared as singletons and did not experience any developmentally significant social interaction based on the genotypes of foster-siblings. All the expectations assume that fostering is random with respect to the genetic and environmental determinants of extraversion.

In addition, since we now have separated twins who were reared (we have assumed) as singletons, we must now separate the direct and indirect effects of genes in the expectations for the additive genetic component. It was stated in Chapter 4 that the effects of competition generate an additional component of environmental variation in twins reared together that is completely confounded with the genetic effects. Now, in the presence of competition as we have described it, the "true" additive genetic variance, V_A may be separated from the contribution V_{AS} of additive genetic differences to the environmental variation. Both parameters are included in the expectations of Table 6.12. Conventional genetic effects V_A contribute to the expected mean squares for all kinds of twins, but the environmental effects and genotype-environmental covariance generated by competition (V_{AS} and C_{AS}) contribute only to expectations for twins reared together.

The sample sizes are much smaller than those we have previously analyzed, so the standard errors of parameters are expected to be very much larger and the comparisons between alternative hypotheses much weaker.

In Table 6.13 we present the results of testing the conventional models which ignore competitive interactions. For neuroticism the simplest model that fits the data assumes that the family environment and dominance do not contribute, since a model that includes either or both of these effects fits no better than the V_A, E_W model. For extraversion, however, we reproduce Jinks and Fulker's finding that the simple model is not adequate. Superficially, the fit is improved very greatly by the addition of a dominance parameter to the model ($\chi_1^2 = 14.3$), but a nonsensical negative value of V_A results. Taken by itself, however, this finding is not too surprising, because we know that there is a very high negative correlation between estimates of additive and dominant genetic effects in twin studies (see Chapter 5).

Table 6.13 Model fitting to Shields' data, ignoring competitive effects.

Extraversion						Neuroticism					
V_A	E_W	E_B	V_D	χ^2	df	V_A	E_W	E_B	V_D	χ^2	df
7.70	7.34	—	—	21.5	10	7.10	8.54	—	—	8.56	10
7.76	7.40	−0.12	—	21.5	9	7.02	8.46	0.15	—	8.53	9
−24.01	6.56	—	32.19	7.2	9	−3.23	8.31	—	10.51	8.16	9
−23.78	6.62	−0.12	32.03	7.2	8	−3.62	8.16	0.28	10.75	8.10	8

Table 6.14   Competitive effects on extraversion in Shields' data.

Competition	Parameter				χ^2	df
	E_W	V_A	V_{AS}	C_{AS}		
Prenatal	6.40 ± 0.98	19.19 ± 6.96		-6.16 ± 3.31	7.5	9
Postnatal	6.45 ± 0.98	9.05 ± 2.59	13.58 ± 9.78	-8.79 ± 4.72	4.3	8

The analysis of the anomalies of the extraversion twin data are pursued further in Table 6.14, which gives the results of fitting two versions of the competition model. The first model, which ascribes all competitive interaction to effects *in utero*, fits no better than the additive-dominance model, although all three parameters are sensible and significant. This model predicts the same pattern of mean squares for twins reared together and apart, and cannot separate the direct contributions of genes to the variance from those of sibling interaction because they are all assumed to occur prior to separation.

The residual chi-square for the postnatal interaction model is somewhat smaller, although we cannot provide a valid significance test because the two competition models are not submodels of an identifiable more general model. Nevertheless, the finding is consistent with an appealing theory of the etiology of individual differences in extraversion in which genetic differences are reinforced by competitive social interactions which occur when individuals are raised in close proximity to one another. The fact that the effects are more likely to be postnatal suggests that these are "learned" responses rather than acquired as a result of competition for prenatal resources.

A study by Lytton (1977) generated data on parent–child interaction that were consistent with a model for competition between twins. Lytton measured the amount of time 32-month-old male twins and singletons spent playing with their fathers. As might be expected, the mean time individuals played with their fathers was greater for singletons than twins, presumably because fathers did not have to divide their attention between the twins. Of particular interest for the genetic analysis, however, was the finding that the variances and covariances between the twins were consistent with a model of competitive interaction based on the genotypes of the twins (see Eaves *et al.*, 1978). Our analysis of Shields' data is consistent with Jinks and Fulker's own earlier analysis, which is not surprising since we used the same data and very similar models and methods. Our results for neuroticism agree closely with theirs and such small differences as are found can be traced to our use of iterative weighted least-squares. The data yield no hint of non-additive genetic effects on neuroticism (as we should expect with such small samples) and give a very small estimate of the family environmental component. On balance, therefore, our own twin data are consistent with Shields' earlier study in suggesting that individual differences in neuroticism scores are primarily the result of additive genetic effects and environmental influences unique to the individual. Our interpretation of Shields' data on extraversion is also very close to Jinks and Fulker's, with the exception that our model for competition differs from theirs. There is no convincing evidence of non-additive genetic effects, and no support for a shared environmental effect. Estimates of V_D and E_B do not differ significantly from zero. However, there is some weak evidence for competition because of the slight differences in variance between MZ and DZ twins and the very small correlation between DZ twins.

It is indeed remarkable that Jinks and Fulker's first application of the model-fitting approach to what, by current standards, are such small samples should nevertheless have forseen the main features of the findings of the next decade of twin studies.

A more recent study in the USA by Bouchard *et al.* (1981) has begun to accumulate a series of separated twins around the world and is likely to yield a further benchmark for testing some of the important aspects of the model. As part of an extensive battery of psychometric assessments, Bouchard and his colleagues administered the Differential Personality Questionnaire, which yields two higher-order factors resembling the extraversion and neuroticism dimensions of Eysenck's theory. Although the samples are still small, the preliminary results of model-fitting analysis (Bouchard, unpublished) are quite comparable to those of Shields and consistent with our model based on twins reared together. There is little evidence of a major shared environmental effect on personality, and some evidence of a non-additive genetic component.

6.5 OTHER ADOPTION STUDIES OF EXTRAVERSION AND NEUROTICISM

Scarr *et al.* (1981) administered the EPI to a total of 120 nuclear families and 115 adoptive families as part of a study of personality resemblance in families. The full set of familial correlations for extraversion and neuroticism, giving separate values for male and female offspring are tabulated in the original paper. We have extracted summary statistics, pooled over sexes of offspring. These are given in Table 6.15.

In their paper Scarr *et al.* give the *correlation* of offspring with midparent, rather than the *regression*. We present the latter statistic to summarize the parent–offspring resemblance because it is more directly related to the heritability (in the case of children reared by their biological parents) than

Table 6.15 Summary statistics from adoptive and biological families (modified from Scarr et al., 1981).[a]

Statistic	Extraversion		Neuroticism	
	Biological	Adoptive	Biological	Adoptive
Sibling correlation	0.06	0.07	0.28	0.05
Regression of offspring on midparent	0.27	−0.00	0.35	0.07

[a] See text for assumptions.

the corresponding correlation. When the variances of males and females are equal, the correlation is equal to the regression of offspring on midparent divided by $\sqrt{2}$. The rescaling removes an apparent inconsistency in the original tables between heritability estimates based on midparental correlations and single-parent offspring correlations. When mating is random, the offspring–midparent regression ought to be twice the latter.

For both extraversion and neuroticism, the resemblance between foster-parents and adopted children is very slight, suggesting that parental extraversion and neuroticism have little cultural impact on these dimensions in their offspring. The fact that foster-siblings are also uncorrelated confirms that the personality of parents has little direct effect on personality of children, and furthermore is consistent with there being no residual effects of the environment due to shared experiences that do not depend on parental personality. The adoption data are therefore completely consistent with the other data that we have discussed so far in showing little evidence that the family environment affects personality. In addition, Scarr's data confirm the very small correlation between spouses for extraversion and neuroticism.

So far, the results of Scarr's analysis are consistent with the model that we have developed for adult personality. However, she notes the same discrepancy between the correlations for twins and those for other relationships, notably siblings and parents and offspring. Having indicated the comparatively low correlation between biological parents and children, she observes (Scarr et al., 1981, p. 114):

> . . . family studies of personality do not agree with studies of MZ and DZ twins, even when the ages of the participants are limited to those who can answer the same personality instruments. If we believe twin studies, about half the variance in the personality traits is due to genetic differences among people. If we believe the studies of ordinary siblings and parent-child pairs, the explanatory power of genetic differences shrinks to about 0.25. What are we to make of this contradiction?

Scarr et al. conclude that neither genes nor family environment have much effect on personality, but that the course of development is largely a function of individual experiences (E_W, in our model). They remark that "Individuals within families are vastly different in the personality characteristics we measured, and psychology has no theory to explain that individuality" (Scarr *et al.*, 1981, p.118). We suspect that the data on DZ twins are not so inconsistent with Scarr's sibling data, in view of the substantial non-additive component that we found in twin studies of extraversion and in the joint analysis that included the family data. It is probable that the low sibling correlation might be more of a problem than the high MZ correlation. An alternative, which we shall consider in the next chapter, is that there are developmental changes in gene expression that result in the decay of phenotypic resemblance between relatives as the age difference between them increases. Under such circumstances, even if gene action is additive, we should expect the correlation of DZ twins to exceed that for siblings (because the twins are measured at the same age) and the correlation of adult siblings to exceed that between parents and their adult children, since parents and children are measured at the most widely separated ages. Such an explanation is different from the environmental hypothesis suggested by Scarr *et al.*, but recognizes that gene expression may change with age when there is a substantial genetic component.

6.6 SUMMARY

The results of the adoption studies, involving separated twins, adopted siblings, and foster-parents with their adopted children, are all consistent with the extensive data on twins reared together. They give absolutely no indication that the environment shared by children within a family has any lasting effect on their resemblance for the two main dimensions of personality most widely studied: extraversion and neuroticism. These data, taken together with the other data that we have already described, suggest that personality theory can gain little by assuming that social learning from parents plays a major role in the creation of personality differences. For neuroticism and extraversion we find that the correlations between mates, based on large samples of spouses, are close to zero. Significant correlations are found for the psychoticism and Lie scales, but these are still too small to have gross effects on the distribution of biological and cultural differences in the population.

The results strengthen, if anything, our earlier conclusion that there is a significant environmental component in the development of personality, but that its source must be sought in the idiosyncratic "slings and arrows of outrageous fortune" that afflict every individual irrespective of the control and influence of his parents and teachers. The adoption data yield personality correlations close to zero. Separated twins are not less alike than those reared together.

The data on non-twin biological relatives, however, do not give strong support to the genetic aspect of the model derived from twin data. The results of Scarr *et al.*, and those outlined here from the London study, are all consistent with a much smaller contribution of additive genetic factors than might be predicted from the twin data alone. However, these estimates are based upon relatives of different ages in whom different genes may be acting. The analysis so far suggests that non-additive genetic factors may be important for neuroticism, though much of the twin data gave no hint of dominance. An alternative explanation of this apparent non-additivity, namely the interaction of genetic differences with age, will be pursued in the next chapter. In the case of extraversion the overall estimate of the additive genetic variance was reduced by the inclusion of other relationships. In the combined family, twin and adoption data there was no strong indication of dominance, but the data on twins alone suggest that competitive interactions between twins may be important, so a developmental process unique to twins may be needed to account for the discrepancy between the twin and family data. The data on separated twins suggest that competitive effects arise after birth rather than before it, but sample sizes are small, so that the effects would have to be very large to generate the observed effects.

Chapter 7

Personality Development and Change: A Genetic Perspective

Until now, we have regarded the effects of age on behavior as little more than a nuisance. If personality changes with time then it is assumed that the rates of change are roughly the same for everyone and that regression corrections should be sufficient to allow a genetic analysis to continue unimpeded. If changes in behavior were all explained by age regression then it is arguable that there would be no developmental psychology because development would simply be a matter of producing "more or less" of the same basic function. There would be no suggestion that there are crucial neurological or psychological mechanisms that emerge at particular stages of development. The reality of development, however, is more complex. Different behavioral themes appear to dominate different stages in the life of the individual. Different genes operate at different times in development. The purpose of this chapter is to begin the exploration, from a genetic perspective, of some of the implications of temporal change for the genetic analysis of personality. In the study of behavioral development it is commonly assumed that the role of social factors requires no proof. There have been countless correlational studies that try to relate aspects of maternal behavior to the behavior of their children. Such relationships are assumed to reflect social rather than genetic causation. Indeed, some psychologists have difficulty with the idea that genes affect behavior because they see obvious behavioral changes with age. The genotype is fixed at birth, so if behavior changes, they assume that it must be a function of the environment. This position might be summarized as "The genotype imparts fixity, the environment stimulates variability".

Such a view, however, does not come from genetics. Barring mutation, the genotype is fixed, but gene expression certainly is not. For example, some fears may be expressed and adaptive at particular ages as a function of the greatest typical threat to survival at a given age. It is conceivable that natural selection might "tune" some genes to be expressed at certain times and "switched off" at others. In contrast, cognitive development may be the continual accumulation of adaptive information under the control of the genotype. Once genes underlying information processing are "switched on", their effects over time are cumulative as further information is incorporated into the developing phenotype. Eaves *et al*. (1986b) present a mathematical model for this process, which will be employed, in a modified form, later in the chapter. Broadhurst and Jinks (1966) showed that gene expression in rat development changed markedly with age and that, even when the phenotype measured was the same at different ages, its adaptive significance, and hence its genetic architecture, could change markedly. In young rats there is genetic dominance for relative immobility. In older rats the direction of dominance changes because, it is argued, the adaptive significance of the trait changes during development. Eaves *et al*. (1986b) show how different models for the role of genes and environment in development lead to different predictions about how such heritabilities and familial correlations may change as a function of age. For example, if all the genes do is to set the initial state of the organism at birth, and development is the successive modification of the phenotype by the environment (for example through memory of environmental input), then the heritability of a trait will decline during development to an asymptotic value that is a function of the initial heritability and the efficiency of "memory". On the other hand, if the genes are continually synthesizing a product required for information processing then heritability will increase from a small value at birth to a high asymptotic value that is a function of the initial heritability and the consistency of gene expression during development.

In the light of our growing understanding of gene action and development, we recognize that personality differences measured in juveniles may have causes quite different from personality measured in adults. Different genes may be responsible for anxiety levels at different ages. The contribution of environmental factors may increase with age if behavior is permanently changed by environmental events. The influence of the social environment may change as children become more independent of their parents. The ideal strategy for analyzing such changes is the longitudinal twin study, in which the behavior of twins is assessed repeatedly with their parents on a number of occasions. It is then possible to document with some precision how the expression of genes and environment change with age and examine the process by which the genetic and environmental differences apparent in adults unfold during development. There have been three longitudinal twin studies of cognitive development (Wilson, 1972; Fischbein, 1981; Hay and O'Brien, 1981). So far, no longitudinal genetic study of personality has examined personality development from the perspective of Eysenck's personality theory.

In this chapter we describe several analyses that begin to examine some of the issues of genetic change as a function of age. First we describe the analysis of P,E,N and L in juvenile twins, simply to test whether the causes of adult and juvenile variation are the same. Then we consider the additional information that can be gained by the inclusion of the twins' parents to see if the same genes and environmental effects are expressed in juveniles and adults. We shall ask how far genetic effects in children can predict the outcome in adults. In this analysis we shall include data from a sample of juvenile singletons and their parents in the attempt to test models for the effects of sibling interaction on personality. In a third set of analyses we shall consider the causes of long- and short-term personality changes in adults by considering how twin differences change as a function of age. Finally, we shall exploit a model for developmental change in gene expression to account for any correlation between family resemblance and age differences between relatives.

7.1 ANALYSIS OF P,E,N AND L IN JUVENILE TWINS

In the London Twin Study we secured responses from juvenile twins to the Junior Eysenck Personality Questionnaire (JEPQ, see Appendix C) which, like the Adult EPQ, yields scores on psychoticism, extraversion, neuroticism and the lie scale (P,E,N and L). The twins were aged between 7 and 17 years, with a mean age of 11.06 years. EPQ data were also gathered from the parents of the twins, and similar data were obtained from 182 families comprising singleton children (85 male, 97 female) and their parents. A modified square-root transformation was used for the P scores. The scalar problems associated with E,N and L were removed by the angular transformation. Only the twin data are considered in the initial analysis. Mean squares within and between pairs were computed for each group of twins separately and corrected for the effects of age and sex as described in Chapter 5 (p.83). The mean squares are given in Table 7.1.

The method of weighted least squares was employed to fit to the juvenile personality data all the models applied to the data on adult twins in Chapter 5. Models were fitted with and without sex-limited effects. Tables 7.2 and 7.3 present the results of fitting genetic and environmental models that allowed for sex limitation.

In no case is there any real justification for including sex differences in gene expression or environmental variance. There are no significant reductions in the residual chi-square resulting from the inclusion of sex-limited effects in the model. We are thus free to consider the results obtained for each trait

Table 7.1 Mean squares for transformed P,E,N and L from London juvenile twin sample (JEPQ).[a]

				Mean squares			
Zygosity	Sex	Statistic	df	P	E	N	L
MZ	F	Between pairs	52	636	59	92	51
		Within pairs	54	272	13	37	20
MZ	M	Between pairs	63	579	53	93	62
		Within pairs	65	265	14	37	18
DZ	F	Between pairs	41	640	41	58	78
		Within pairs	43	288	28	23	25
DZ	M	Between pairs	42	523	34	63	53
		Within pairs	44	293	34	63	14
DZ	M/F	Between pairs	80	601	36	96	67
		Within pairs	81	367	30	53	26

[a] Corrected for age and sex multiplied by scale factors to remove decimals.

Table 7.2 Improvement in V_A, E_W model when allowance is made for sex differences in genetic and environmental effects in juvenile London twin data.

	Trait			
Parameter	P	E	N	L
E_{WM}	258***	16***	41***	16***
E_{WF}	259***	13***	32***	20***
V_{AM}	169***	18***	29***	22***
V_{AF}	214***	21***	25***	26***
V_{AMF}	204*	7	32**	34***
r_{AMF}	1.078	0.330	1.204	1.439
Goodness-of-fit χ_5^2	2.4	5.4	11.2	12.3
P	0.80	0.37	0.049	0.03
Improvement-of-fit χ_3^2	0.05	3.2	1.1	0.0

Note: Improvement in fit is judged by subtracting the goodness-of-fit chi-square under this model from that obtained when sex-limited effects are deleted from the model (cf. Tables 7.4–7.7). Significance levels as follows:
* $P < 0.05$; ** $P < 0.01$; *** $P < 0.001$.

Table 7.3 Improvement in E_B, E_W model when allowance is made for sex differences in environmental effects in juvenile London twin data.

Parameter	Trait			
	P	E	N	L
E_{WM}	281***	22***	50***	17***
E_{WF}	284***	20***	32***	23***
E_{BM}	147***	11***	19***	22***
E_{BF}	191***	15**	26***	22***
E_{BMF}	103*	3	16**	17***
r_{BMF}	0.613	0.242	0.720	0.774
Goodness-of-fit χ_5^2	0.8	21.9	12.3	4.63
P	0.97	0.0006	0.031	0.47
Improvement-of-fit χ_3^2	2.5	3.7	4.9	4.4

Note: Improvement in fit is judged by subtracting the goodness-of-fit chi-square under this model from that obtained when sex-limited effects are deleted from the model (cf. Tables 7.4–7.7). Significance levels as follows:
* $P<0.05$; ** $P<0.01$; *** $P<0.001$.

Table 7.4 Results of fitting models to psychoticism scores of juvenile London twins assuming no sex differences in genetic and environmental effects.

Model	Parameter estimates					Goodness-of-fit		
	E_W	E_B	V_A	V_D	C_{AS}	χ^2	df	P
1	447***	—	—	—	—	32.7	9	0.0002
2	303***	147***	—	—	—	3.3	8	0.92
3	257***	—	192***	—	—	2.9	8	0.94
4	272***	76	102	—	—	1.9	7	0.96
5	272***	—	330**	−152	—	1.9	7	0.96
6	270***	—	152**	—	19	2.4	7	0.94

Note: Significance of parameter estimates tested by comparing with theoretical error. Such standard errors only apply for models which fit the data. Significance levels as follows:
** $P<0.01$; *** $P<0.001$.

when the simpler models, which assumed no sex-limitation, were fitted to the juvenile data.

7.1.1 Psychoticism

Table 7.4 presents the analysis of the P scores. As we might expect, the model that assumes no family resemblance is soundly rejected ($\chi^2_9 = 32.7$,

$P < 0.001$). However, both the E_B, E_W model and the V_A, E_W model fit equally well. None of the three parameter models leads to a significant improvement. Clearly, the data give no hint of dominance. Model 4 gives positive but not significant estimates of the additive genetic component and the between-families environmental component. Our conclusion for juvenile P scores is therefore that there is significant family resemblance of uncertain origin. Perhaps both genetic and cultural factors are involved, but their contribution is beyond resolution with these small samples.

7.1.2 Extraversion

Results for extraversion are somewhat clearer (Table 7.5). The genotype–environmental hypothesis (model 3) gives an adequate fit to the observed mean squares, which is far better than either of the simple environmental models (1 and 2). Model 4 yields a significant negative estimate of the between-families environmental component, so is rejected in favor of model 5, which includes dominance. The dominance model represents a significant improvement over the additive model ($\chi^2_1 = 7.5$, $P < 0.01$), but gives a negative estimate of the additive genetic contribution. The competition/cooperation model gives a significant negative estimate of the genotype–environment covariance but fits slightly worse than the dominance model. Although these results are based on small numbers, they are remarkably consistent with our findings for extraversion in adult twins (Chapter 5), for whom there was strong evidence of a dominance or competition parameter in the very large Swedish and Australian studies.

Table 7.5 Results of fitting models to extraversion scores of juvenile London twins assuming no sex differences in genetic and environmental effects.

Model	Parameter estimates					Goodness-of-fit		
	E_W	E_B	V_A	V_D	C_{AS}	χ^2	df	P
1	34***	—	—	—	—	48.0	9	<0.0001
2	23***	11***	—	—	—	25.6	8	0.0012
3	15***	—	19***	—	—	8.6	8	0.38
4	13***	−14**	35***	—	—	1.1	7	0.99
5	13***	—	−7	28**	—	1.1	7	0.99
6	13***	—	26***	—	−4*	2.7	7	0.91

Note: Significance of parameter estimates tested by comparing with theoretical error. Such standard errors only apply for models that fit the data. Significance levels as follows: * $P < 0.05$; ** $P < 0.01$; *** $P < 0.001$.

Table 7.6 Results of fitting models to neuroticism scores of juvenile London twins assuming no sex differences in genetic and environmental effects.

Model	Parameter estimates					Goodness-of-fit		
	E_W	E_B	V_A	V_D	C_{AS}	χ^2	df	P
1	63***	—	—	—	—	43.1	9	<0.0001
2	43***	20***	—	—	—	17.2	8	0.028
3	36***	—	28***	—	—	12.3	8	0.14
4	37***	4	23*	—	—	12.5	7	0.085
5	37***	—	34*	−7	—	12.5	7	0.085
6	37***	—	25***	—	1	12.7	7	0.081

Note: Significance of parameter estimates tested by comparing with theoretical error. Such standard errors only apply for models that fit the data. Significance levels as follows: * $P<0.05$; ** $P<0.01$; *** $P<0.001$.

7.1.3 Neuroticism

Table 7.6 presents the model-fitting results for neuroticism in the juvenile twins. Both of the environmental models fail. None of the models gives an excellent fit, but the two-parameter genotype–environmental model (model 3) fits as well as any. Adding dominance, the family environment or sibling interactions yields no improvement. Model 4 is significantly better than model 2, and has a statistically significant value for the additive genetic component. The estimate of the family environmental component in model 4 is not significant, and this model fits no better than model 3, which omits the family environment. Thus the model that we arrive at for juvenile neuroticism is the same as that which has emerged consistently from our analysis of the adult data. Family resemblance appears to have no environmental component, and reflects genes of mainly additive effect. There is a large environmental component due to differences within families.

7.1.4 The lie scale

The results for the lie scores are quite different from the main personality dimensions and intrinsically interesting because they yield the first clear example we have encountered of a trait for which individual differences are caused entirely by the environment (Table 7.7). The second environmental model (E_B,E_W) fits the data very well indeed, while the two-parameter genotype–environmental model (V_A,E_W) does not. Adding a genetic parameter (model 4) to the environmental model yields no improvement in fit ($\chi^2_1 = 0.9$, $P>0.5$) and gives a non-significant estimate of the additive genetic

Table 7.7 Results of fitting models to lie scores of juvenile London twins assuming no sex differences in genetic and environmental effects.

Model	Parameter estimates					Goodness-of-fit		
	E_W	E_B	V_A	V_D	C_{AS}	χ^2	df	P
1	41***	—	—	—	—	78.4	9	<0.0001
2	21***	21***	—	—	—	9.0	8	0.34
3	17***	—	24***	—	—	16.1	8	0.04
4	20***	17***	5	—	—	8.1	7	0.32
5	20***	—	57***	−34***	—	8.1	7	0.32
6	19***	—	16***	—	5**	10.7	7	0.15

Note: Significance of parameter estimates tested by comparing with theoretical error. Such standard errors only apply for models that fit the data. Significance levels as follows: * $P<0.05$; ** $P<0.01$; ***$P<0.001$.

parameter. A model that tries to explain twin resemblance in terms of cooperative interactions based on genotype C_{AS} involves more parameters and fits worse than the simple environmental model. Thus we conclude that the similarity between juvenile twins for the lie scale of the EPQ is due solely to the social environment. There are also significant intrapair differences, which must be due to individual differences in experience within the family. The twin data alone are insufficient to determine whether the shared environment of twins is due to the influence of the parental phenotype, to social interaction between the twins based on environmental (rather than genetic) differences, or simply due to shared experiences that do not depend on the parents. The results for the lie scale among juveniles do not correspond exactly to those in the adult samples from London and Australia because the adults gave some indication of sex differences in the determination of lie scores. However, both populations showed evidence that the family environment was important in at least one sex.

7.2 THE CAUSES OF DEVELOPMENTAL STABILITY AND CHANGE

Separate analyses of adult and juvenile data allow us to compare the causes of variation in the two age groups. Our studies show that the determination of extraversion and neuroticism in children is much the same as in adults. There is evidence of genetic factors for both dimensions, and there is a suggestion that non-additive genetic effects may affect extraversion. There is no clear evidence that the family environment plays a significant role. The adult psychoticism data gave little support to a contribution of the family environment in addition to additive genetic factors. The results for juveniles are more ambiguous. Resolution of genetic and social components of family resemblance in P scores was not possible in this age group. There is evidence of shared environmental factors in the determination of adult and juvenile lie scores, with the effect being especially marked in juveniles.

The separate analyses of adults and juveniles, however, do not permit us to assess the extent to which the same genes and social factors contribute to variation at both ages. The fact that genetic factors affect the same personality trait at two ages does not mean the same genes are involved on both occasions. For example, different genes may be involved in juvenile and adult anxiety because different environments are salient at different ages. Alternatively, there may be some genes whose effects are responsible for emotionality at both stages and some genes that determine age-specific sensitivity to particular features of the environment.

We have already suggested that longitudinal data are best for determining the long-term overlap between genetic and environmental effects on behavior. We do not have such data yet. However, we can begin to test hypotheses about the long-term consistency of gene expression because we have data on the parents of juvenile twins in addition to the juvenile and adult twins. The basic principle of our analysis is simple. We use the adult twins to give estimates of the contribution of genetic and environmental factors to adult personality. The juvenile twins give estimates of genetic and environmental effects on juvenile personality. The extent to which the same genes and cultural effects are involved at both ages can be estimated from the parent–offspring resemblance. For example, if the parent–offspring covariance is zero for a personality trait then we should conclude that different genes affect juvenile and adult behavior whatever heritability estimates are generated by the adult and juvenile twins by themselves. In theory, it would be possible to have high heritabilities in both adults and juveniles and yet have a zero parent–offspring covariance because the genes affecting adult and juvenile behavior are quite distinct. Even though the trait is given the same name, and measured by similar items, the scale could be measuring fundamentally different aspects of the adaptive response. Alternatively, if the same genes affect behavior in adults and juveniles then we should find a parent–offspring correlation that can be predicted perfectly from knowledge of heritability in adult and juvenile twins.

The above intuitive argument is simple. In practice, the model may be more complex because of cultural transmission from parent to child (which may mimic genetic inheritance under some circumstances) and because of non-additive genetic effects, including dominance, which tend to reduce the correlation between parents and children. However, Young *et al.* (1980) employed this basic notion to develop models for the stability and change of personality as measured by the adult and juvenile forms of the Eysenck Personality Questionnaire.

7.2.1 The model

The elements of the genetic model are the same as those for a phenotype in which there is no developmental change (see Chapter 5). We consider a single locus A/a with three genotypes AA, Aa, aa. Following Mather and Jinks (1982), we may specify the additive and dominance effects of the locus on a continuous phenotype. The only difference is that we specify the effects of the locus on adults and juveniles separately:

genotype	AA	Aa	aa
effect on adults	$m_a + d_{aA}$	$m_a + h_{aA}$	$m_a - d_{aA}$
effect on juveniles	$m_j + d_{aJ}$	$m_j + h_{aJ}$	$m_j - d_{aJ}$

The basic algebraic methods of Mather and Jinks may now be used to define additive and dominance variance components for traits measured in adults and juveniles. Thus, in the absence of epistasis and assortative mating, the contributions of several loci to the genetic variance are additive and independent such that the additive and dominance components of variance at a given age may be expressed in terms of gene frequencies and the effects of the contributing genes at a given age. Thus we may define additive and dominance components V_{AA} and V_{DA} for adults and corresponding components for juveniles: V_{AJ} and V_{DJ} respectively. If parents and offspring were measured as adults then their genetic covariance is expected to be $\frac{1}{2}V_{AA}$. However, if children are measured as juveniles, a new parameter V_{AAJ} has to be introduced to represent the covariance between adult and juvenile gene effects. The new parameter is similar to that used to account for sex-limited gene expression in unlike-sex twin pairs, since the latter represents the covariance between gene effects in males and females. The expectation of the additive genetic covariance between adults and juveniles may also be expressed in terms of gene effects and gene frequencies as

$$V_{AAJ} = 2\sum_a u_a v_a [d_{aA} + (v_a - u_a)h_{aA}][d_{aJ} + (v_a - u_a)h_{aJ}].$$

Similarly, we may define a dominance-covariance term

$$V_{DAJ} = 4\sum_a u_a^2 v_a^2 h_{aA} h_{aJ}.$$

The dominance component will not appear in expectations of the parent–offspring covariance, but would be needed, for example, to account for the non-additive genetic component of the phenotypic correlation between measures of the same individuals as children and adults.

The important aspect of these expectations is that they involve products of the form $d_{aA}d_{aJ}$ etc. For a particular gene, the term will be positive if the gene has an effect in the same direction in adults and juveniles, but zero if it affects adult behavior (say) but has no effect on children, or *vice versa*. It is mathematically possible, but difficult to imagine in practice, that a gene that increases trait expression at one age may reduce trait expression at another. The corresponding product term would then be negative for that locus.

7.2.2 Data summary

Since many of the families have a number of different relationships and involve subjects measured at different stages of development, we compute the covariance matrices between relatives for each subset of the data. For the juvenile twins, we shall have five 4×4 covariance matrices for each

Table 7.8 Covariances (upper triangles), variances (diagonals) and correlations (lower triangles) between relatives for psychoticism scales of the EPQ and JEPQ.

Group	df	Juvenile twin/singleton families				df	Adult twins	
		Mother	Father	Child 1	Child 2		Twin 1	Twin 2
MZ$_m$	58	0.402	−0.062	−0.010	0.010	69	0.557	0.303
		−0.137	0.503	−0.032	−0.036		0.543	0.558
		−0.027	−0.074	0.362	0.157			
		−0.023	0.074	0.376	0.482			
MZ$_f$	49	0.447	0.072	0.057	0.093	232	0.398	0.170
		0.164	0.431	0.060	−0.031		0.422	0.408
		0.132	0.140	0.420	0.182			
		0.200	−0.068	0.401	0.488			
DZ$_m$	39	0.529	0.047	0.034	−0.059	46	0.397	0.047
		0.100	0.412	−0.037	0.012		0.115	0.429
		0.068	−0.085	0.473	0.115			
		−0.138	0.032	0.286	0.343			
DZ$_f$	36	0.358	0.161	0.102	0.026	124	0.486	0.183
		0.377	0.509	0.111	0.051		0.344	0.548
		0.245	0.224	0.488	0.176			
		0.064	0.109	0.380	0.440			
DZ$_{mf}$	75	0.584	0.136	−0.019	0.128	67	0.523	0.100
		0.227	0.613	0.074	0.139		0.192	0.515
		−0.036	0.138	0.469	0.117			
		0.238	0.252	0.242	0.499			
Singleton$_m$	84	0.432	0.071	0.065				
		0.161	0.458	−0.013				
		0.159	−0.031	0.388				
Singleton$_f$	96	0.440	0.162	0.055				
		0.328	0.554	0.076				
		0.130	0.160	0.405				

Table 7.9 Covariances (upper triangles), variances (diagonals) and correlations (lower triangles) between relatives for extraversion scales of the EPQ and JEPQ.

Group	df	Mother	Father	Child 1	Child 2	df	Twin 1	Twin 2
MZ$_m$	58	0.070	0.010	0.010	0.016	69	0.075	0.046
		0.143	0.071	0.011	0.003		0.639	0.070
		0.212	0.228	0.033	0.020			
		0.311	0.063	0.562	0.038			
MZ$_f$	49	0.071	0.014	0.010	0.012	232	0.070	0.032
		0.228	0.056	0.013	0.006		0.472	0.063
		0.213	0.291	0.033	0.024			
		0.217	0.112	0.624	0.043			
DZ$_m$	39	0.078	−0.001	0.003	0.019	46	0.063	0.010
		−0.011	0.085	0.004	0.010		0.162	0.058
		0.051	0.076	0.034	0.000			
		0.364	0.177	0.009	0.036			
DZ$_f$	36	0.088	−0.014	0.003	0.005	124	0.080	0.013
		−0.209	0.048	0.005	−0.010		0.169	0.077
		0.053	0.141	0.030	0.006			
		0.091	−0.251	0.177	0.033			
DZ$_{mf}$	75	0.062	0.001	0.009	0.008	67	0.062	0.009
		0.009	0.079	0.002	−0.003		0.159	0.052
		0.243	0.044	0.024	0.003			
		0.154	0.057	0.091	0.043			
Singleton$_m$	84	0.086	0.002	0.000				
		0.022	0.002	0.007				
		0.001	0.133	0.046				
Singleton$_f$	96	0.072	0.014	0.008				
		0.170	0.092	0.007				
		0.134	0.104	0.045				

personality trait because each family comprises two parents and two twin children. There are two 3 × 3 matrices for the male and female singletons and their parents and there are five 2 × 2 matrices derived from the data on each group of adult twins. A few juvenile twin pairs were omitted from the analysis because no parental data were available. The covariance matrices given in Tables 7.8–7.11 have been corrected for age. The average effect of the sex difference is automatically removed for unlike-sex pairs since the variances and covariances are computed around the means for males and females. Similarly, there is no need to correct the parents' scores for sex differences. Age correction was conducted by computing the residuals for each score after extracting the linear regression on age. Separate regressions were computed for males and females in both age groups.

Table 7.10 Covariances (upper triangles), variances (diagonals) and correlations (lower triangles), between relatives for neuroticism scales of the EPQ and JEPQ.

Group	df	Juvenile twin/singleton families				df	Adult twins	
		Mother	Father	Child 1	Child 2		Twin 1	Twin 2
MZ$_m$	58	0.078	0.009	0.010	0.011	69	0.070	0.037
		0.144	0.054	0.017	0.013		0.511	0.076
		0.153	0.291	0.059	0.030			
		0.149	0.211	0.456	0.072			
MZ$_f$	49	0.070	−0.011	0.032	0.023	232	0.065	0.028
		−0.155	0.070	0.018	0.012		0.425	0.066
		0.460	0.257	0.069	0.029			
		0.342	0.178	0.436	0.065			
DZ$_m$	39	0.079	−0.009	0.028	0.004	46	0.054	0.001
		−0.153	0.042	−0.005	−0.003		0.021	0.059
		0.381	−0.085	0.069	0.000			
		0.059	0.057	0.007	0.058			
DZ$_f$	36	0.059	0.000	−0.004	0.022	124	0.063	0.004
		−0.005	0.068	−0.007	−0.005		0.066	0.065
		−0.082	−0.138	0.037	0.016			
		0.427	−0.081	0.403	0.045			
DZ$_{mf}$	75	0.041	0.001	0.009	0.008	67	0.069	0.011
		0.027	0.057	0.022	0.001		0.167	0.060
		0.156	0.329	0.077	0.023			
		0.146	0.010	0.293	0.077			
Singleton$_m$	84	0.086	0.001	0.019				
		0.015	0.071	0.006				
		0.263	0.089	0.058				
Singleton$_f$	96	0.058	−0.005	0.022				
		−0.089	0.060	0.002				
		0.359	0.040	0.063				

Variances are given along the leading diagonals of the matrices and covariances in the upper triangle. For convenience the correlations between relatives are given in the lower triangle since there are scalar differences between the variances of adult and juvenile scores. For each trait there are 77 unique observed variances and covariances in the data summary: 5×10 from juvenile twin families; 2×6 from the singleton families and 5×3 from the adult twin groups. Our goal is to find a parsimonious model for this large number of statistics.

It is helpful to consider some of the basic features of the data in advance of describing the model fitting. There is clearly marked correlation between spouses for the lie scores. The pooled, weighted correlation between spouses for L is 0.305 (Chapter 6). This correlation has substantial significance for interpretation of the analysis of the twin data presented in Chapter 5.

Table 7.11   Covariances (upper triangles), variance (diagonals) and correlations (lower triangles) between relatives for the lie scales of the EPQ and JEPQ.

Group	df	Juvenile twin/singleton families				df	Adult twins	
		Mother	Father	Child 1	Child 2		Twin 1	Twin 2
MZ_m	58	0.035	0.016	0.012	0.011	69	0.055	0.027
		0.399	0.044	0.012	0.013		0.513	0.051
		0.347	0.311	0.034	0.023			
		0.285	0.294	0.588	0.045			
MZ_f	49	0.037	0.009	0.010	0.003	232	0.044	0.022
		0.271	0.032	0.010	0.009		0.516	0.042
		0.297	0.300	0.033	0.017			
		0.090	0.239	0.460	0.040			
DZ_m	39	0.025	0.014	0.007	0.003	46	0.025	−0.001
		0.404	0.046	0.005	−0.003		−0.041	0.035
		0.209	0.123	0.038	0.021			
		0.123	−0.081	0.647	0.029			
DZ_f	36	0.047	0.011	0.029	0.005	124	0.047	0.020
		0.299	0.028	0.016	0.006		0.429	0.045
		0.525	0.366	0.067	0.031			
		0.104	0.164	0.549	0.048			
DZ_{mf}	75	0.035	0.012	0.009	0.013	67	0.051	0.008
		0.355	0.034	0.009	0.009		0.170	0.039
		0.229	0.233	0.048	0.022			
		0.304	0.209	0.442	0.050			
$Singleton_m$	84	0.032	0.012	0.002				
		0.328	0.045	0.002				
		0.083	0.048	0.028				
$Singleton_f$	96	0.036	0.008	0.008				
		0.215	0.039	0.004				
		0.251	0.112	0.028				

Assortative mating tends to increase the similarity of MZ and DZ twins, even in the absence of cultural inheritance, by creating additional genetic variation between families as a result of correlations between loci. Under random mating, it will be recalled that the genetic correlation between siblings is ½. If there is assortative mating then the genetic correlation increases to $\frac{1}{2}(1 + A)$, where A is the correlation between additive genetic deviations of spouses (Fisher, 1918). All of the models considered in Chapter 5 assume random mating. As a result, any excess DZ correlation resulting from the genetic consequences of assortative mating may be mistakenly ascribed to the effects of cultural inheritance if our genetic analysis is restricted to twin data alone. Thus some of the variation ascribed to E_B in the analysis of the lie scores might actually be genetic and attributable to assortative mating. (We shall return to this problem when we consider assortative mating for social attitudes.)

A slightly smaller correlation between spouses is found for the P scores of parents ($r = 0.171$). This degree of assortative mating will have less effect on the resemblance between twins in the next generation than we expect for the lie scale. The correlations between mates for extraversion and neuroticism are 0.065 and 0.052 respectively, and provide little evidence for mate selection based on personality. The parent–offspring correlations for personality are variable but Young *et al.* (1980) conclude that they are homogeneous for all the four personality dimensions and provide little support for models of sex-linked or sex-limited inheritance.

The total variances of adult twins do not differ in any systematic way from the parents of juvenile twins and singletons. Model fitting to the twins separately has already provided little evidence of heterogeneity in the variances of the twin groups within an age cohort in the London data. A notable feature to emerge from the new compilation, however, is the difference in total variance between juvenile twins and singletons for two traits: extraversion and the lie scale. This leads us to suspect that a simple genetic model will not be able to explain family resemblance for these traits but that we shall need to invoke sibling interaction to explain some of the results for juveniles. We recall that analysis of the twin data gave some evidence of dominance or competition for extraversion and E_B or cooperation for the lie scores.

7.2.3 Method for model fitting

The method of weighted least squares, while it is very straightforward for fitting linear models to independent mean squares, proves very cumbersome for fitting models to covariance matrices because the variances and covariances are not independent within a matrix derived from a given group of relatives. This occurs because the same individuals contribute information to many different observed statistics. Although it is possible to modify the WLS procedure to take these correlations into account, it turns out to be no more difficult to obtain estimates by the method of maximum likelihood using an iterative procedure to compute maximum-likelihood estimates and numerical methods for differentiating the likelihood function with respect to given parameter values. If the raw data comprise m observed covariance matrices $\mathbf{S}_i$, each having N_i df, then the log-likelihood for the entire set of covariance matrices is

$$\log L = -\tfrac{1}{2}N_i[\ln|\Sigma_i| + \mathrm{tr}(\mathbf{S}_i\Sigma_i)^{-1}] + \text{constant},$$

where Σ_i is the matrix of expected variances and covariances corresponding to S_i. For a given set of values for p parameters, the expected covariance matrices may be evaluated and thus the likelihood obtained for any set of parameter values. Standard numerical methods similar to those described previously (Chapters 4 and 6) may then be employed to obtain those parameter values for which L is maximized.

The likelihood may be obtained under alternative hypotheses and likelihood-ratio tests constructed in the attempt to discriminate between alternative models for family resemblance. If L_0 is the log-likelihood obtained under some general hypothesis involving p parameters, and L_1 the log-likelihood obtained when k of these are set to zero, then the difference $2(L_0-L_1)$ is a chi-square for k df. If this chi-square is not significant then it may be assumed for the time being that deletion of the additional parameters has no adverse effect on the fit of the model, so that the simpler hypothesis is to be preferred. If each expected statistic is replaced by the corresponding observed value in the expression for L_0 then the chi-square is a test of goodness-of-fit of the reduced model that yields L_1, with df equal to the number of statistics (77 in our example) minus the number of parameters estimated from the data.

7.2.4 Results for the genotype–environmental model

We first report the results obtained when the V_A, E_W model is fitted to all four personality factors. The model involves three genetic parameters: the additive genetic variances in adults and juveniles (V_{AA} and V_{AJ}) and the covariance between additive genetic effects in adults and juveniles (V_{AAJ}). Two environmental parameters were specified corresponding to the within-family environmental variances of adults and juveniles: E_{WA} and E_{WJ} respectively. Since we have no longitudinal data on the same individuals, we cannot detect long term effects of the within-family environment, and there is thus no need for a within-family environmental covariance term E_{WAJ} in our example.

The contributions of the five paramaters to the variances and covariances of parents, adult twins, juvenile twins and singletons are given in Table 7.12. An alternative model in which all family resemblance is due to shared environmental factors is also specified in the table. This model defines adult and juvenile between-family environmental components (E_{BA} and E_{BJ}) and a covariance parameter representing the environmental covariance between adults and juveniles (E_{BAJ}).

The maximum-likelihood estimates of the five parameters of the genotype–environmental model, their standard errors estimated from the square

Table 7.12 Expected variances and covariances of a simple genotype–environmental model.

Statistic	Expectation
Variances	
Juveniles	$V_{AJ} + E_{WJ} + E_{BJ}$
Adults	$V_{AA} + E_{WA} + E_{BA}$
Covariances	
Juvenile MZ twins	$V_{AJ} + E_{BJ}$
Juvenile DZ twins	$\frac{1}{2}V_{AJ} + E_{BJ}$
Adult MZ twins	$V_{AA} + E_{BA}$
Adult DZ twins	$\frac{1}{2}V_{AA} + E_{BA}$
Parent–offspring	$\frac{1}{4}V_{AAJ} + E_{BAJ}$
Spouses	0

Table 7.13 The relationship between adult and juvenile personality.

Parameter	Extraversion	Neuroticism	Psychoticism	Lie scale
E_{WA}	0.035 ± 0.003	0.038 ± 0.003	0.24 ± 0.02	0.021 ± 0.001
V_{AA}	0.037 ± 0.003	0.027 ± 0.003	0.23 ± 0.02	0.020 ± 0.002
E_{WJ}	0.017 ± 0.002	0.036 ± 0.004	0.25 ± 0.03	0.016 ± 0.002
V_{AJ}	0.021 ± 0.003	0.028 ± 0.005	0.18 ± 0.03	0.022 ± 0.002
V_{AAJ}	0.012 ± 0.003	0.023 ± 0.004	0.07 ± 0.03	0.013 ± 0.002
χ^2	73.80	87.20	70.11	145.47
P_{72}	0.4	0.1	0.5	0.001

roots of the diagonal elements of the matrix of information realized and the chi-square testing the goodness-of-fit of the model are given in Table 7.13 for all four variables. The information matrix was approximated by the numerical estimates of the second derivatives of the log-likelihood at the maximum-likelihood parameter values.

From the table we deduce that the model gives a very bad fit to the data from the lie scale. We already expect this because the separate analyses of the twin data have suggested a strong family-environment component to responses on this scale of the EPQ and JEPQ, and we have already shown a highly significant correlation between spouses for L scores. The overall fit for the other three dimensions is quite good. In the case of extraversion and psychoticism the chi-squares are actually less than the expected value (equal to the df) if all the residuals were due to chance alone. However, when the residuals have many df there is still the possibility that specific important aspects of the data might be obscured by large numbers of chance fluctuations, so some consideration is warranted of what other major changes in the model could nevertheless lead to a significant improvement over the first model (see below).

Taking the tabulated parameters at their face values, however, allows us to make a preliminary estimate of the degree of continuity in gene expression between adults and juveniles, and the heritability of adult and juvenile personality differences for P,E and N. The lie scale is excluded from these computations since the genetic model does not fit the data.

Given that the model is correct, the heritability of the trait in juveniles is

$$h^2_J = V_{AJ}/(V_{AJ} + E_{WJ}),$$

and in adults the heritability is

$$h^2_A = V_{AA}/(V_{AA} + E_{WA}).$$

Of particular interest from a developmental perspective, however, is the cross-correlation between adult and juvenile genetic effects, r_{AAJ}. If this correlation is unity then we suppose that the same genetic effects contribute to individual differences in both adults and juveniles. That is, there is considerable long-term stability in gene expression. In our case, a high correlation would imply that measures made on juveniles are good predictors of the genetic component of adult personality. The heritabilities and adult–juvenile genetic correlations are given in Table 7.14.

The heritabilities are comparable between adults and juveniles. However, the dimensions differ in the extent to which the same genes are expressed in adults and juveniles.

7.2.4.1 Neuroticism

In the case of neuroticism the adult–juvenile genetic correlation is 0.84 ± 0.14, which does not differ significantly from unity. This result is consistent with the same genes contributing to adult and juvenile measures of neuroticism. The model that replaces genetic effects by shared environmental effects cannot explain the neuroticism data ($\chi^2_{72} = 105.7$, $P = 0.006$).

Table 7.14 Contributions of genetic factors to adult and juvenile personality.

Parameter	Extraversion	Neuroticism	Psychoticism	Lie scale
Adult heritability	0.51	0.42	0.49	0.49
Juvenile heritability	0.55	0.44	0.42	0.58
Adult–juvenile genetic correlation	0.43	0.84	0.34	0.62

7.2.4.2 Extraversion

The extraversion data yield a lower genetic correlation between adults and juveniles (r_{AAJ} = 0.44 $\pm$ 0.11), which is significantly less than unity. However, it will be recalled that the twin data alone gave evidence of a significant effect of dominance or competition at both ages. As a result, the estimate of the additive genetic variance in adults and juveniles will be inflated if such effects are ignored (as they are in this model). Therefore, while the adult–juvenile covariance term is a valid estimate of the additive component of genetic covariance between extraversion in children and adults, the correlation will be biased downwards because the denominator of the equation for r_{AAJ} will be inflated by biases due to dominance effects.

We consider the effect that allowing for sibling interaction would have on the model for extraversion. We postulate that, at each locus, the indirect effect on the phenotype of a sibling is a constant multiple b of the direct effect. Thus the gene effects are as follows for a single locus:

genotype	AA	Aa	aa
direct effect	$m + d_a$	$m + h_a$	$m - d_a$
indirect effect	$m + bd_a$	$m + bh_a$	$m - bd_a$

If b is positive then there is cooperation. Competition is represented by negative values of b. If b is constant for all loci then the additional environmental variance and genotype–environment covariance resulting from sibling effects may be represented in terms of b and the additive genetic variance, in the absence of dominance. Thus the additional environmental variance resulting from sibling effects in twins will be $V_{AS} = b^2 V_A$. The genotype–environment covariance can be expressed as $C_{AS} = bV_A$. The contribution of genotype–environmental covariance to the total variance for a trait will depend on the genetic relationship between the interacting siblings. The extra environmental variance and the genotype–environmental covariance do not appear in the expected variances and covariances involving singletons. In order to derive expected variances and covariances under models of social interaction, we have to be much more explicit in defining the family structure. For example, if we were to allow for sibling interactions in adults then we should have to specify whether the parents of twins were without siblings. In our application, however, we have assumed that the social interaction is unique to juveniles. Alternative hypotheses could be tested if we had more details of the family structure, including data on number and age distribution of siblings.

The "twin-interaction" model is specified in Table 7.15. Note that different expectations for the parent–offspring covariance apply to twin and singleton

Table 7.15 Expected statistics in twin–parent families when juvenile competition is based on genetic differences.

Statistic	Expectation
Variance of MZ twins	$V_{AJ}(1+b)^2 + E_{WJ}$
Variance of DZ twins	$V'_{AJ}(1+b+b^2) + E_{WJ}$
Variance of singletons	$V'_{AJ} + E_{WJ}$
Covariance of MZ twins	$V_{AJ}(1+b)^2$
Covariance of DZ twins	$\frac{1}{2}V_{AJ}(1+4b+b^2)$
Twin–parent covariance	$\frac{1}{2}V_{AAJ}(1+b)$
Singleton–parent covariance	$\frac{1}{2}V_{AAJ}$

families. The singletons only share direct genetic effects with their singleton parents. However, each twin is correlated with his parents for two reasons: first he has direct genetic effects in common (hence the contribution from V_A); secondly, the genetic deviation of a parent is correlated with the environmental effect created for a child by the child's cotwin. Since the parent's genetic deviation and the child's environment are created by the same genes and because the twin who provides the environment is a child of the same parent as the twin who receives it, we have a contribution of genotype–environment covariance to the covariance of twins and their singleton parents. The model given here assumes that sibling interaction is a transient phenomenon applicable only to juveniles. Other expectations would apply if the parents and adult twins were also subject to sibling interaction.

When the model is fitted to the extraversion data there is indeed a significant reduction in chi-square. The parameter estimates and chi-square are given in Table 7.16.

Table 7.16 Parameter estimates when the effects of competition are specified in the model for juvenile extraversion.

Parameter	Estimate ± s.e.
E_{WJ}	0.015 ± 0.002
V_{AJ}	0.027 ± 0.004
V_{AAJ}	0.013 ± 0.003
b	-0.17 ± 0.06
χ^2	67.2
df	71
$P\%$	>50

7.2.4.3 Psychoticism

The simple model that assumes additive gene action and within-family environmental effects gives a satisfactory fit to the psychoticism data (see Table 7.13). However, the correlation between additive genetic effects in adults and juveniles is very small ($r_{AAJ} = 0.07$), suggesting either that a lot of dominance has been missed in the analysis of the twin data or that there is little in common genetically between measures of psychoticism in parents and children. The large Australian twin sample (cf. Table 5.26) gives no evidence whatsoever of dominance, and there is little evidence of dominance in the British juvenile data (cf. Table 7.4), so we believe that the most acceptable model involves a very large interaction between age and additive genetic effects. That is, virtually none of the genetic effects creating variation in measures of juvenile "psychoticism" persist into adulthood. Similarly, knowledge of the adult family history of psychoticism is likely to have very little value in attempting to predict the behavior of juveniles. Quite clearly, these results could relate only to the measures of "psychoticism" as defined by the adult and juvenile forms of the EPQ. At the very least, however, they illustrate an important aspect of the genetic study of development, namely that specific genes may be expressed at specific times in development and that a more imaginative use of the model-fitting strategy can elucidate such effects and distinguish them from other types of genetic non-additivity. At best, the method shows that a propensity to antisocial behavior in juveniles is distinct, biologically and developmentally, from propensity to antisocial adult behavior because different genes are responsible for these aspects of behavior at different times.

The results summarized in Table 7.13 assume random mating. This assumption is false for psychoticism scores since there is a small but significant correlation between mates. If this correlation indicates mutual selection of spouses on the basis of psychoticism then the resemblance of mates will introduce an additional source of genetic covariation between family members, for traits partly under genetic control. Similarly, traits subject to vertical cultural transmission will also display increased family resemblance as a result of positive assortative mating.

Allowing for the effects of assortative mating when parents and offspring are measured at the same stage of development is fairly simple because most of the theoretical expectations have already been tabulated by Fisher (1918) for the genetic model and Rice *et al.* (1978) for the general model involving genetic and cultural inheritance (see also Chapter 6). In the presence of developmental change in gene expression, however, allowance for assortative mating is slightly more difficult because not all the genes expressed at the time of mate selection are expressed in juveniles.

In allowing for the consequences of assortative mating for psychoticism, therefore, we argue that the genes expressed in juveniles can be divided into two kinds. Some of the genes are expressed in juveniles *and* in adults at the time of mate selection. These are the genes that are affected by assortative mating. Although they are the same in both generations, the scale of their effects may differ between adults and juveniles. The remaining genes are assumed to be expressed in juveniles only, and are thus not affected by assortative mating. More subtle models can be advanced, but this is the greatest degree of complexity resolvable with the present survey design. An analogous model can be devised for the effects of assortative mating on vertical non-genetic transmission when some of the cultural effects on children are transient in their impact.

Table 7.17 gives modified expectations for the "V_A, E_W" model and the "E_B, E_W" model when allowance is made for the effects of assortative mating. Both models include a contribution of the environment within families for adults and juveniles, E_{WA} and E_{WJ} respectively. The genetic model defines a parameter V_{AS} to represent the additive genetic effects *specific* to juveniles.

Table 7.17 Expectations under two models modified to allow for some effects of assortative mating.[a]

	Expectation	
Statistic	Genetic inheritance	Cultural inheritance
Adults		
Variance V	$\left(1 + \dfrac{A}{1-A}\right)V_{AA} + E_{WA}$	$E_{BA} + E_{WA}$
MZ covariance	$V - E_{WA}$	E_{BA}
DZ covariance	$\left(\frac{1}{2} + \dfrac{A}{1-A}\right)V_{AA}$	E_{BA}
Spouse covariance C_m	$\dfrac{A(1-A)V^2}{V_{AA}}$	μV
Juveniles		
Variance V_J	$\left(1 + \dfrac{A}{1-A}\right)V_{AJ} + V_{AS} + E_{WJ}$	$E_{BJ} + 2z^2(1+\mu)V + E_{WJ}$
MZ covariance	$V_J - E_{WJ}$	$V_J - E_{WJ}$
DZ covariance	$\left(\frac{1}{2} + \dfrac{A}{1-A}\right)V_{AJ} + \frac{1}{2}V_{AS}$	$V_J - E_{WJ}$
Parent–child		
Covariance	$\dfrac{1}{2}\dfrac{1+\mu}{1-A}V_{AAJ}$	$z(1+\mu)V$

[a] Parameters are defined in the text.

The coefficients of V_{AS} are those appropriate to random mating since these genes are assumed not to be expressed at the time of mate selection. The additive effects of the second set of genes contribute to V_{AA} in adults and V_{AJ} in juveniles. Now, if we assume that the effect of each locus on V_{AA} has a proportional contribution to V_{AJ} then we may define $V_{AAJ} = (V_{AA}V_{AJ})^{1/2}$ to represent the covariance between the adult and juvenile effects of genes expressed in adults. Thus we have

$$V_{AJ} = V^2_{AAJ}/V_{AA}.$$

Under assortative mating, in the absence of cultural transmission, Fisher (1918) showed that the additive genetic component of variance in the population was augmented because positive assortative mating produced positive correlations between the genes affecting the phenotype on which mate selection was based. Thus, when the total genetic variance is computed, not only do the variances of the individual loci have to be included, but also all the possible covariances between them. As the number of loci, k, gets larger, the number of covariances, $k(k-1)$, increases more rapidly, so that the overall effect of assortative mating on the genetic variance may be substantial for a polygenic trait (see e.g. Crow and Kimura, 1970, Chapter 4). Fisher showed that the correlation between loci is a function of the correlation between additive genetic deviations of spouses, A. Under assortative mating, the total additive genetic variance increases over several generations from V_A to an equilibrium value of

$$V^*_A = \frac{1}{1 - A}V_A,$$

where V_A is the additive genetic variance in a randomly mating population with the same gene frequencies. The narrow heritability under assortative mating is thus

$$h^{*2} = V^*_A/V_P,$$

where V_P is the total phenotypic variance.

The value of A depends on the mechanism of assortative mating. Fisher considers three models for assortment (see e.g. Heath and Eaves, 1985), of which the first, phenotype assortative mating, is that most commonly assumed in data analysis. The model of phenotypic assortative mating assumes that mate selection is based primarily on the measured phenotype rather than on a latent variable. If the correlation between mates in the population is μ then it follows that the additive genetic correlation between mates is $A = h^{*2} \mu$, in the absence of cultural transmission.

The increased genetic variance arising from assortative mating is equal to $V^*_A - V_A = AV_A/(1-A)$. In siblings and twins the additional component of variance contributes entirely to the genetic variance *between* families.

The expectations in Table 7.17 assume phenotypic assortative mating. The genetic model has three genetic parameters: V_{AA}, V_{AAJ} and V_{AS}, as described above. The fourth genetic parameter V_{AJ} is expressed as a function of V_{AA} and V_{AAJ}. There are two environmental parameters E_{WJ} and E_{WA}. One additional parameter A is used to represent the effects of assortative mating. The population marital correlation μ is expressed in terms of A and the additive genetic variance in adults, i.e. $\mu = A/h^{*2}$. When $A = 0$ the model reduces to a reparameterization of the original developmental model (Table 7.12), which assumes random mating, additive genetic effects and within-family environmental effects.

For the purposes of comparison with the genetic model, we examine the consequences of assortative mating for the case in which any family resemblance is environmental (see Table 7.17). It is assumed that the total variation in adults is the sum of a within-families environmental effect E_{WA} and a between-families environmental effect E_{BA}. The phenotypic correlation between mates is μ. It is again assumed that variation in juveniles is purely environmental. The within-family environmental component is E_{WJ}. The environmental variance between families, however, is assumed to arise from two sources. Part of the environment is due to transient effects, specific to juveniles. These effects are represented in the model by E_{BJ}. The remaining part of the family environment of juveniles is assumed to result from the direct effect of the parental phenotype on the shared environment of offspring. If the total variance of parents is V_P and the regression of parental phenotype on the parental component of juvenile family environment is z then the between-families environmental component of juveniles that can be attributed to the direct impact of the parental phenotype is $2z^2 (1 + \mu)V_P$.

Table 7.18 gives the results of fitting the two models to the psychoticism data on adults and juveniles. Both models give a good fit to the data,

Table 7.18 Estimates of parameters for alternative genetic and environmental models fitted to psychoticism.

		Genetic model		Environmental model
	Parameter	Estimate	Parameter	Estimate
Adult model	E_{WA}	0.25 ± 0.02	E_{WA}	0.30 ± 0.02
	V_{AA}	0.21 ± 0.02	E_{BA}	0.18 ± 0.02
	A	0.09 ± 0.002	μ	0.18 ± 0.04
Juvenile model	E_{WJ}	0.25 ± 0.03	E_{WJ}	0.30 ± 0.02
	V_{AAJ}	0.06 ± 0.02	E_{BJ}	0.13 ± 0.03
	$V_{AS}D_{RS}$	0.17 ± 0.04	z	0.07 ± 0.03

although the genetic model gives a lower chi-square than the environmental model. A test of goodness of fit is therefore unable to resolve the two models convincingly.

The earlier analysis of juvenile twins revealed ambiguity in the tests of the genetic and environmental model such that it was impossible to choose reliably between genetic and social models for the resemblance of juvenile twins. No such ambiguity exists for the adult twin data, especially when the results for the London sample are considered in the light of the very large Australian sample. In adults there is very little evidence of a between-family environmental component.

The addition of a shared environmental component to the expectation for adult twins and parents does not improve the fit of the genetic model significantly ($\chi^2_1 = 0.01$, $P > 0.9$). However, adding a genetic component to the expectations for adults leads to a marked improvement in the environmental model ($\chi^2_1 = 10.4$, $P < 0.01$). Variation in adult scores is therefore compatible with a genetic model for family resemblance with random environmental effects within families. The pattern of inheritance for juveniles is ambiguous, and social and genetic models have to be given equal weight at this stage. Given the genetic model for adult psychoticism, and allowing for the effects of assortative mating, we may estimate the heritability of adult psychoticism scores by substituting our parameter estimates in the expression for h^{*2} above. We obtain $h^{*2} = 0.48 + 0.04$. If we grant a genetic model for juvenile variation then we have

$$h^2_J = V_{GJ}/(V_{GJ} + E_{WJ}),$$

where $V_{GJ} = V_{AJ} + V_{AS}$. Substituting the parameter estimates from Table 7.18 gives $h^2_J = 0.42 \pm 0.07$. It is interesting to note that only about 10% of the genetic variation in juveniles, i.e. $(V_{GJ} - V_{AS})/V_{GJ}$, can be assigned to genes also expressed in adults.

If we adopt an environmental model for juveniles then the proportion of the total variation due to specific within-family environmental effects is E_{WJ}/V_{PJ}. Substituting our parameter estimates in this expression shows that about two-thirds of juvenile variation in P scores is due to these highly idiosyncratic environmental effects, which include errors of measurement. Of the remaining one-third that can be attributed to shared environmental effects, only about 4%, i.e.

$$\frac{2z^2(1 + \mu)V_P}{E_{BJ} + 2z^2(1 + \mu)V_P}$$

can be explained by reference to the cultural impact of the parental phenotype. Thus, whatever we assume about the causes of variation in the juvenile measure of psychoticism, it is quite clear that the causes are specific to juveniles and show no direct correlation with psychoticism scores of adults. The adult and juvenile P scales are quite distinct and may even be caused by different mechanisms. Thus it is unlikely that juvenile measures of psychoticism will be good predictors of adult psychotic behavior, since the two are genetically and culturally distinct.

7.2.4.4 The lie scale

The lie scale provides the exception in this investigation, in that the simple five-parameter genotype–environmental model (Table 7.13) fails utterly to account for the observed variances and covariances (χ^2_{77} = 145.47, $p < 0.001$). In an earlier section a highly significant marital correlation was described for the lie scale. However, although leading to a significant improvement in fit (χ^2_1 = 44.36, $P < 0.001$), allowing the matrial correlation to be greater than zero still does not yield an adequate model (χ^2_{71} = 101.10, $P < 0.01$). The equivalent six-parameter environmental model that allows for both specific environmental influences (E_W) and environmental influences shared by siblings (E_B) is no better (χ^2_{71} = 102.23, $P < 0.01$). A combination of these two models still did not fit (χ^2_{69} = 91.66, P = 0.035), but showed that for the purpose of explaining the adult variation the contribution of additive genetic effects (V_A) was far more significant than those effects of the family environment (E_B). On the other hand, the reverse seemed to be the case for the juvenile variation, where common environmental influences, or influences simulating these, appeared to play a somewhat greater role.

In preliminary investigations of the juvenile scales (see above) the possibility that sibling cooperation might be influencing juvenile lie-scale scores was considered. Sibling cooperation implies that the phenotype of each member of a sibship has an increasing effect upon the phenotype of the other members. The specification of cooperation, and the converse effect (which we call competition) against a *genetical* background is considered in Chapter 5. We now develop a model in which the primary sources of variation are environmental, instead of genetic, and include the environmental influences of the parental phenotype. Upon those sources are superimposed the effects of juvenile cooperation.

Leaving aside the sources of adult variation for the moment, and proceeding in a manner similar to that used for psychoticism, let the parents have phenotype P, with variance V_P, and further let this have an effect environmentally upon the phenotype of the offspring, with regression coefficient z. Also let there be environmental sources (E_C) of juvenile twin covariation that are not correlated with the adult lie-scale scores, represented by parameter E_{BJ}, and a source (E_S) of environmental variation specific to

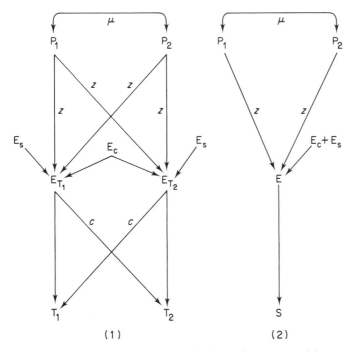

Figure 7.1   Regression model for juvenile lie-scale scores: (1) twin family; (2) singleton family.

individuals, represented by the parameter E_{WJ}. Thus far, the model is the same as one specified for juvenile psychoticism. However, now let the sum of the environmental influences upon one twin (E_T) have not only a direct effect upon the twin's own phenotype but also an indirect effect upon that of his cotwin with regression coefficient c. There are of course, by definition, no sibling influences upon the phenotype of singletons. The model is represented diagramatically in Figure 7.1. It may be shown by standard rules of regression theory that the appropriate expectations for the juvenile variation and covariation are those of Table 7.19.

The model so far has been independent of the sources of variation in the adult lie-scale scores. In the first instance let us consider a model specifying only environmental variation specific to individuals, E_{WJ}, additive gene action, V_{AJ}, and the genetic correlation between spouses, A. This model for the adult variation and covariation is the basic genotype–environmental model with assortment previously considered for adult psychoticism (Table 7.19). The combination of this model for the adult scores and the full model of Table 7.19 for the juvenile scores was fitted to the mean-squares and

Table 7.19 The contribution of mutual reinforcement ("cooperation") to twin resemblance in juveniles.

Statistic	Expectation
Variances	
Singletons	$E_{WJ} + E_{BJ} + 2z^2(1+\mu)V$
Twins V_J	$(1+c^2)(E_{WJ}+E_{BJ}) + 2cE_{BJ} + 2z^2(1+c)^2(1+\mu)V$
Covariances	
Twins	$V_J - (1-c)^2E_{WJ}$
Singleton–parent	$z(1+\mu)V$
Twin–parent	$z(1+c)(1+\mu)V$

Table 7.20 Estimates of juvenile parameters for "reinforcement" models.

Parameters	Estimate ± s.e.	
	Full model	Reduced model
E_{WJ}	0.031 ± 0.006	0.035 ± 0.002
E_{BJ}	0.003 ± 0.005	—
z	0.14 ± 0.02	0.14 ± 0.02
C	0.20 ± 0.08	0.24 ± 0.03
χ^2	87.8	88.1
df	70	71
$P\%$	7	8

mean-products matrices for the lie scale. The estimates of parameters of the full juvenile model are given in Table 7.20, while the estimates for the adult model were

$$\hat{E}_{WA} = 0.021 \pm 0.001,$$
$$\hat{V}_{AA} = 0.017 \pm 0.002,$$
$$\hat{A} = 0.16 \pm 0.02.$$

The model provided an adequate, although perhaps not good, explanation of the observed matrices ($\chi^2_{70} = 87.78$, $P = 0.07$). However, the estimate of E_{BJ} is obviously not significant and was discarded from the model, the remaining paramaters being re-estimated (Table 7.20). The estimates of parameters contributing to adult variation and covariation did not alter. The fit of the resultant model ($X^2_{71} = 88.08$, $P = 0.08$) was no worse for dropping out the superfluous parameter. Allowing for environmental sources of twin covariation in the adults, by fitting an E_{BA} parameter, did not produce a significant improvement in fit, and the resulting parameter estimate was small and not significant.

The possibility that this apparent sibling cooperation could be genetically based was also considered. However, a suitable model had to account for three important aspects of the data: (i) additive gene action in the adults; (ii) a non-zero marital correlation; (iii) a correlation between juvenile and adult genetic expression in the range $0 < r_{GAJ} < 1$. The model specified contained elements of two previous models considered with respect to extraversion and psychoticism. The sibling effects were specified by means of the parameter b employed in the investigation of extraversion. This model was combined with the basic genotype–environmental model with assortment previously considered for adult psychoticism (Table 7.17) and with the associated model for juveniles in which the additive genetical variation was separated into that also having expression in adulthood and that genetical variation specific to adolesence. The resulting model contained seven parameters: E_{WA} and E_{WJ} representing environmental variation specific to individuals in adults and juveniles respectively; V_{AA}, additive genetic variation in adults; V_{AAJ}, the covariance of gene effects in adults and juveniles; V_{AS}, the genetical variation specific to juveniles; A, the additive genetic correlation between spouses; and b, to represent the effects of sibling cooperation. The expectations for the juvenile variation and covariation are presented in Table 7.21, while the expectations of the adult statistics have already been pre-

Table 7.21 Cooperation based on genetic differences: expected statistics involving juveniles.

Statistic	Expectation
Variances	
Singletons[a]	$\dfrac{1}{1-A} V_{AJ} + V_{AS} + E_{WJ}$
MZ twins V_{MZ}	$(1+b)^2 \left(\dfrac{1}{1-A} V_{AJ} + V_{AS} \right) + E_{WJ}$
DZ twins V_{DZ}	$V_{MZ} + \dfrac{1}{2}\left(\dfrac{1+A}{1-A} V_{AJ} + V_{AS} \right)$
Covariances	
MZ twins	$V_{MZ} - E_{WJ}$
DZ twins	$V_{DZ} - E_{WJ} - \dfrac{1}{2}(1+b)^2 \left(\dfrac{1+A}{1-A} V_{AJ} + V_{AS} \right)$
Twin–parent	$\dfrac{1}{2} \dfrac{1+\mu}{1-A} V_{AAJ}(1+b)$
Singleton–parent	$\dfrac{1}{2} \dfrac{1+\mu}{1-A} V_{AAJ}$

[a] This expectation was printed incorrectly in Young *et al.* (1980).

Table 7.22   Effects of cooperation on juvenile "lie" scores.[a]

Parameter	Estimate	
	Full model	Reduced model
E_{WA}	0.0210 ± 0.0010	0.0230 ± 0.0070
V_{AA}	0.0165 ± 0.0015	0.0155 ± 0.0020
A	0.16 ± 0.02	0.15 ± 0.02
E_{WJ}	0.0210 ± 0.0020	2.30 ± 0.0020
V_{AJ}	0.0065 ± 0.0025	0.0070 ± 0.0020
V_{AS}	0.0035 ± 0.0030	—
b	1.02 ± 0.75	1.53 ± 0.67

[a] Parameters are defined in the text.

sented (Table 7.17). The fit of the model was quite good ($X^2_{70} = 84.55$, $P = 0.11$). However, two of the estimates were not significant — the sibling-effects parameter b and the parameter V_{AS} estimating the juvenile specific genetical variation. When fitting the environmental sibling-effects model, we found no evidence for shared twin environments apart from those caused by their parents. Therefore V_{AS} was dropped from the present model. This action resulted in a significant estimate of b (Table 7.22), but also failure of the model ($X^2_{71} = 93.38$, $P = 0.04$). Thus, while the evidence is by no means conclusive, the environmentally based explanation is an acceptable provisional hypothesis for the variation and covariation in adult and juvenile lie-scale scores.

Several summary statistics might be derived from the environmental sibling-effects model. Thus it was found:

(i) that the expected variance of twins is approximately 10% greater than that of singletons, i.e.

$$\frac{(1 + c^2)E_{WJ} + 2z^2(1 + c)^2(1 + \mu)V_P}{E_{WJ} + 2z^2(1 + \mu)V_P} = 1.09 \pm 0.02$$

(ii) that of the expected variance of twins, 92%, i.e.

$$\frac{(1 + c^2)E_{WJ}}{(1 + c^2)E_{WJ} + 2z^2(1 + c)^2(1 + \mu)V_P} = 0.92 \pm 0.03,$$

may be ultimately traced to environmental influences specific to individuals, and similarly for singletons, 94%, i.e.

$$\frac{E_{WJ}}{E_{WJ} + 2z^2(1 + \mu)V_P} = 0.94 \pm 0.02,$$

may be traced to the same source;

(iii) that the covariance of twins and their parents is approximately 25% greater than that of singletons and their parents, owing to "cooperative" effects, i.e.

$$\frac{z(1 + c)(1 + \mu)V_P}{z(1 + \mu)V_P} = 1.24 \pm 0.03;$$

(iv) finally, that of the covariance of twins only about 15% may be traced to the shared influences of their parents, i.e.

$$\frac{2z^2(1 + c)^2(1 + \mu)V_P}{2cE_{WJ} + 2z^2(1 + c)^2(1 + \mu)V_P} = 0.16 \pm 0.05,$$

the remainder being the result of each twin's specific environmental circumstances having an indirect "cooperative" effect upon his cotwin.

Nevertheless, the most important finding is the fact that all primary sources of juvenile twin covariation (E_{BJ}) are ultimately traceable to the environmental influences of the parental "lie" behavior.

The sources of variation in the adult lie-scale scores seem to be clearer than those for the juvenile scores. As in the case of psychoticism, the most important effects appear to be those due to specific environmental influences, contributing to E_{WA} and additive gene action V_{AA} ($h^2_A = 0.48 \pm 0.04$). We also recognize a fairly large spousal correlation ($\mu = 0.35 \pm 0.04$), which has a significant impact on the environmental resemblance of children reared in the same family.

So far, this chapter has examined the interaction of gene expression with age differences, and formulated some basic models for the influence of social interactions upon personality development. Although the idea that gene expression and interaction may be modified with age is largely accepted in the animal literature (see e.g. Broadhurst and Jinks, 1966), its rigorous analysis in man is still far from realization. The present experimental design has been able to show that a simple model that assumes consistency of gene action between adults and juveniles measured on similar personality scales cannot account for the degree of similarity between relatives. It is therefore quite likely that the traits measured in adults and juveniles differ in their causes of variation, at least in part, so that genetic effects manifest in juveniles are not expressed in the same individuals as adults and *vice versa*. The fact that somewhat different scales were used in the assessment of adults and juveniles does not alter the general conclusion, since the scales were designed expressly to measure those aspects of behavior that were factorially consistent in adults and juveniles. Other explanations of the data for dimensions of extraversion and psychoticism would invoke a substantial amount of genetical non-additivity to account for the findings in adult twins and the data in the chapter, but the inability of parental data to predict the findings for offspring is equally likely to be attributable to the interaction of genetic differences with age.

7.3 DEVELOPMENTAL CHANGES IN GENE EXPRESSION IN ADULTS

The analyses described so far are concerned with the major developmental changes associated with the differences between childhood and adulthood. However, it is equally possible that the genetic control of personality is not constant throughout adult life. Ideally, such changes are best studied by the collection of longitudinal data on twins. However, preliminary tests may be conducted that address some kinds of developmental change in a cross-sectional study of adult personality.

The Swedish twin study described in Chapter 5 was large enough to reveal age changes in the magnitudes of genetic parameters by fitting genetic models to different subgroups defined by different birth cohorts. The results indicated that the genetic variance for extraversion and neuroticism was greater for younger twins than older twins and showed that the environmental variance was generally lower in younger than older twins for both these variables, as measured by the shorter form of the EPQ used in this sample.

7.3.1 Age trends in intrapair twin differences

An alternative, simple approach is to correlate the absolute intrapair differences for MZ and DZ twins for personality scores with age. This approach was first suggested by Eaves and Eysenck (1976a) and applied to neuroticism scores derived from an early 80-item personality inventory (the "PI"), which was a precursor of the EPQ. They showed that the intrapair difference for MZ twins (pooled across sexes) did not increase with age, but that the differences for DZ twins (especially females) increased significantly with age. Similar relationships were not found for E and P scores. Martin and Jardine (1986) report results for the same analysis applied to their Australian twin sample using the EPQ (Table 7.23).

Table 7.23 Correlations of absolute within-pair differences in the transformed personality scores with age.

		MZF	MZM	DZF	DZM	DZO
Extraversion	Angle	0.03	0.03	0.07	−0.04	0.01
Psychoticism	Angle	−0.02	−0.04	−0.03	−0.01	−0.07**
Neuroticism	Angle	0.02	0.01	0.12**	0.02	0.01
Lie	Angle	−0.03	−0.01	0.09*	0.03	0.05

Significance levels as follows:
* $P < 0.05$; ** $P < 0.01$.

The results show that intrapair differences of MZ twins are uncorrelated with age for all of the personality dimensions. There is therefore no evidence that the effects of unique environmental experiences (E_W) increase with the accumulation of more experience as a function of age. However, for females at least, there is a highly significant relationship between intrapair differences in neuroticism for DZ twins. There are two main interpretations of this result: first, there may be new genetic effects expressed with increasing age. New genes are "turned on" while genes expressed earlier have persistent effects, so that the genetic variance increases with age. If the environmental variance remains constant then only the DZ twins show a significant relationship between age and absolute intrapair differences. The alternative explanation, which has greater appeal from the psychological standpoint, asserts that individuals who are genetically predisposed to neurotic behavior create environments that reinforce this behavior. There is thus a "genotype–environment" correlation in which the individual's genotype creates the environment, which then serves to augment existing genetic differences. In this way, a positive correlation between age and intrapair differences would be created for DZ twins but not for MZ twins. As long as we are restricted to cross-sectional twin data, these two hypotheses cannot be resolved. Eaves *et al.* (1986b), however, have shown that longitudinal twin data, or data on relatives measured at different ages, can be exploited to resolve these alternative hypotheses. If the first model is correct then the genetic correlations between different ages will decay as functions of age difference (see below). Under the second model, genetic variances will increase (as under the first model also), but the matrix of genetic correlations between measures made at different times should be of unit rank.

A similar trend, though less marked compared with that found for neuroticism, occurs for the lie scores in females. The absolute intrapair differences for DZ twins increase significantly with age. No such trends are found for extraversion and psychoticism in like-sex twins, suggesting that there are no developmental trends in gene expression during adult life for these two traits. However, the significant negative correlation for psychoticism scores in the unlike-sex pairs may indicate a significant reduction in the sex difference with increasing age, when the effects of average family environment and genotype are controlled by comparing siblings of the same age but different sexes.

7.3.2 Extended kinships: a developmental model

A more flexible model for developmental change in gene expression was presented by Eaves *et al.* (1986b). Indeed, such a model was already used

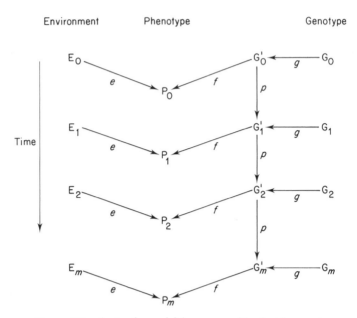

Environment   Phenotype   Genotype

Time

Figure 7.2   A simple model for personality development.

unwittingly without any clear theoretical basis in a paper by Eaves (1980). The basic idea behind the model is that the phenotype at any time is a function of the phenotype at earlier stages and that new genetic effects may be expressed at different stages of development. We consider one special case of the more general formulation which Eaves *et al.* (1986b) applied to longitudinal cognitive data in twins.

Figure 7.2 gives the simplified version of the model that we have tried with our data. The phenotype of a given individual at times 0, 1, 2,. . .,*t* is assumed to be a function of additive genetic effects and unique environmental experiences. At each occasion, there are new genetic effects G_m and new specific environmental effects E_m. For simplicity, we assume that the variances of the genetic and environmental effects are constant and standardized to unity throughout development. At any time *m* the regression of measured personality on E_m is *e*. The regression of personality on G_m is *h*. Whereas the model assumes that all environmental effects are occasion-specific (in this case), it assumes that genetic effects operate through an intervening gene product G'_m, which persists over time to some extent and influences behavior on a later occasion. Thus the genetic effects on the phenotype at a given time are the result of genes that are expressed *de novo* on that occasion together with the effects of genes expressed on all previous occasions, in so far as these persist in time. The path from G_m to G'_m is g, and the path from the product to the phenotype is f, such that $h = fg$. The effect of the previous level of the latent gene product on the present level is p. At time zero the heritability of the trait is h^2. As the trait develops, however, genetic effects will accumulate if $p > 0$, leading to an increase in genetic and phenotypic variances. The simple model assumes that the paths h, e and p are constant throughout development.

Imagine that the same people are measured repeatedly. It turns out (see Eaves *et al.*, 1986b) that the phenotypic variance increases, and, when $0 < p < 1$, reaches an equilibrium "adult" value of

$$V_{A0} = \frac{h^2}{1 - p^2} + e^2$$

when only genetic effects persist in time and environmental effects are occasion-specific. Of particular interest, however, is the fact that the genetic component of the covariance between the phenotypes of the same individuals measured on two occasions u and t decays as an exponential function of the time interval between measurements. At equilibrium, it follows that

$$V_{At,u} = h^2 \frac{e^{-\alpha \Delta_t}}{1 - p^2}$$

where $\Delta_t = u - t$, and the genetic covariance between individuals separated by Δ_t in time is

$$V_{A\Delta_t} = V_{A0} e^{-\alpha \Delta_t}.$$

Now suppose that we have a pair of adult relatives measured at different ages u and t. The additive component of their genetic covariance will simply be $V_{A\Delta t}$ multiplied by the appropriate genetic coefficient (1 for MZs, 0.5 for siblings, etc.). In this way, we may develop expectations for every pair of relatives regardless of age. The model predicts that relatives of the same degree will correlate less when they are more different in age because the phenotype is, to some extent, affected by different genes. In the case of twins, however, who are always measured at the same age, the same genes are being expressed. As a result, dizygotic twins will correlate more highly than siblings, and siblings will correlate more highly than parents and offspring, simply because the effects of development result in greater differences in gene expression between relatives who are further apart in age. These effects might contribute to spurious support for non-additive genetic effects involving dominance when twin data are supplemented by non-twin relatives.

The model was used to extend the analysis of the dataset comprising the London twins, nuclear families, extended kinships and adoptive families

Table 7.24 Parameters of developmental model for adult extraversion and neuroticism scores.[a]

	$\hat{V}$	$\hat{h}$	$\hat{d}$	α	$\hat{e}$ [b]	$2L+C$	df	P%
Extraversion	0.0803	0.698	—	—	0.716	0	—	—
	0.0801	0.655	0.274	—	0.704	0.78	1	
	0.0802	0.706	—	0.0500	0.708	0.78	1	
	0.0801	0.679	0.203	0.0279	0.705	0.84	2	
Neuroticism	0.0731	0.608	—	—	0.794	0	—	—
	0.0728	0.439	0.483	—	0.758	5.80	1	
	0.0729	0.640	—	0.2888	0.768	4.86	1	
	0.0728	0.489		0.1229	0.432	6.10	2	

[a] Constants and age regressions computed but omitted for simplicity cf. Tables 6.7 and 6.8.
[b] Obtained from $e = (1 - h^2 - d^2)^{1/2}$.

described in the previous chapter and summarized in Table 6.2. We do not yet have an adequate form of the model in the presence of assortative mating and cultural inheritance, so our analysis is currently restricted to those traits for which there is little evidence of either effect, i.e. neuroticism and extraversion.

The models were identical with the models with h, d and e in the previous chapter, with the addition of the decay parameter α to account for the exponential decay in the covariance between relatives as a function of increasing age difference predicted from the particular form of the model used here. The decay parameter was included with and without dominance in the analysis of both personality traits because we wanted to be sure that all possible ambiguity was removed from the choice between these two effects. The results are tabulated (Table 7.24) together with those for the two genetic models in which developmental effects are not included (Tables 6.7 and 6.8). Different constants and age regressions for the two sexes were included in the model, but are not tabulated since they do not differ from those already given in the previous chapter. The log-likelihoods have been multiplied by two and expressed relative to the smallest value tabulated for each variable, i.e. that associated with the simple additive genetic model. The figures tabulated are thus approximately chi-squares for testing each of the other models against this simpler alternative.

For extraversion, the model that includes both dominance and developmental effects gives no improvement over the simple additive model ($X^2_2 = 0.84$), so there is little need to consider either form of non-additivity. The joint model for neuroticism, however, is significantly better than the additive model ($X^2_2 = 6.10$, $P<0.05$), so there is some evidence for non-additivity due to either dominance, developmental changes or both. When the full model is compared with a developmental model without dominance the difference in chi-square is 1.24 for 1 df, which is not significant. Thus dominance can be omitted without serious consequences for the fit of the model. However, the reduction in chi-square that accompanies the removal of the developmental parameter from the full model is even smaller — only 0.03 — so that leaving out dominance is slightly more serious than leaving out the developmental effects. It is appealing to consider what we *would* conclude about the contribution of developmental effects under this model were we to take the decay parameters seriously. When we allow for dominance (the most conservative approach) we find α equal to 0.0028 for extraversion and 0.0123 for neuroticism. For extraversion this result would imply that the genetic covariance between adults measured with a ten-year age gap between them would have a genetic covariance that is only $100e^{-10\alpha}\%$ of that expected for the same individuals measured at the same age. For extraversion this computation yields a genetic correlation of 97% of its base-line value with a 10 year age gap. For neuroticism the same figure is 88%. Thus the data suggest a high degree of genetic continuity in the two aspects of the phenotype over time.

These results are not clear-cut, and are disappointing for the light they shed on the power of even quite complex designs to resolve non-additive effects of major importance. We might have expected this to be so because dominance only contributes to the correlation in collateral relationships (MZ twins, DZ twins and siblings), which are expected to involve individuals closer together in age than relationships that only contribute to estimates of the additive genetic component (parents and offspring, uncles and nieces, etc.). A lot of dominance may look like genotype × age interaction and *vice versa*. On balance, it looks as if measurements of adult personality made at different ages reflect the effects of the same genes throughout adult life, although the neuroticism data are consistent with a 12% decline in the genetic component of the covariance between relatives for every 10 years difference in their ages.

7.4 STABILITY AND CHANGE IN PERSONALITY

The main conclusion from this chapter is that there is substantial consistency in genetic effects on neuroticism between adults and juveniles and that, by and large, gene expression does not change much with age in adult life, although here the results of the large Swedish study and our investigations are in conflict. For the other traits, we find that the magnitude of genetic and environmental components differs between adults and juveniles, and that, even when a conscientious effort is made to devise instruments that are measuring the same trait in the two groups, different genes may be expressed in adults and juveniles. Indeed, not only may different genes be expressed in the two generations, but, as with the lie scale, genetic effects may be expressed in one generation and only environmental effects in another. Such conclusions raise the possibility, surprising at first but less so on reflection, that genetic differences in parents may partly be responsible for the family environment shared by their children.

The above analysis remains indecisive in many respects, but it has demonstrated how the model-fitting approach may illuminate the discussion of causal mechanisms in man by bringing to our attention many developmental issues that can become the points of departure for a new and more flexible understanding of the significance of human variation.

The three major dimensions of the model of personality appear, with the one possible exception of psychoticism in juveniles, to be characterized by variation resulting from specific life experiences and the additive effects of genetic differences. The effects of test unreliability contribute to our estimate of the importance of the individual environmental experiences. The covariance of genetic effects as expressed in juveniles and adults is reasonably high for extraversion and neuroticism, especially for the latter, for which there is a surprising degree of intergenerational consistency. However, in the case of psychoticism the covariance of parents and offspring is low, whether P itself be genetic or environmental. These results suggest that the prediction of adult temperament in childhood may be quite successful along the dimension of neuroticism–stability. However, the success of such a prediction is not expected to be so great if interest is in or related to the extraversion–introversion dimensions, and even less for the psychoticism dimension.

The findings for the lie scale represent a striking departure from those expected on the basis of a simple genotype–environmental model, and suggest that social interactions between parents and offspring, reinforced by the interaction of siblings in the case of twin pairs, may play an important role in the manifestation of "social desirability" in juveniles, as measured by the lie scale of the JEPQ. It is remarkable that the similarity of twins depends in no way on any environmental factors apart from those assessed by the lie scores of their immediate relatives. The possible detection of social-interaction effects in the juvenile scores and the presence of a relatively high phenotypic correlation between spouses ($\mu = 0.34$) for the lie scale suggests that whether the junior version of the scale is measuring actual deceit, lack of insight, or genuine variation in "approved behavior" (Eysenck *et al.*, 1971), the trait may well repay further examination as a paradigm of a trait for which social interactions, rather than genetical differences, are paramount determinants of individual variability.

The inclusion of singleton families in the study has highlighted the increased power of this design for detecting certain types of social interaction. This increase in power is witnessed by the detection of apparent transitory sibling effects in the variation and covariation of juvenile lie-scale, and possibly extraversion, scores. The consequences of sibling effects could not be adequately specified in the parental generation since the characteristics of the families in which the parents were reared were unknown. The collection of such information and the data to estimate the variance of adult singleton scores would be desirable. It would be interesting to test our models against data collected on, say, unrelated individuals reared together or siblings raised in sibships of sizes greater than two. Carey (1986) has recently developed an elegant model of sibling interactions that generalizes to sibships of any size. Factors in addition to sibship size such as birth interval, which collectively may be called "family density" effects (cf. Zajonc and Markus, 1975) might also prove to be important.

The final analyses of the chapter concerned changes in adult behavior. There was evidence that genetic effects on neuroticism increase with age. This finding raises the question of whether or not the same genes affect personality traits throughout adult life or whether different genes take precedence at different ages. Analysis of the change in family resemblance as a function of age differences between relatives found little to support the idea that different genes were expressed at different ages in adults, although the effects were more marked for neuroticism than extraversion. We therefore conclude that any apparent change in adult gene expression is more likely to be a function of reinforcement augmenting earlier inherited personality differences.

Chapter 8

The Genetic Analysis of Individual EPQ Items

Our analysis of the higher-order factors showed that measures of the principal dimensions of personality were scarcely affected by the shared experiences of family members. In so far as twins' personality scores are correlated, the resemblance is primarily genetic. Although the environment plays a substantial role in the genesis of individual differences, it seems that the most important environmental effects serve to differentiate members of the same family. That is, the main sources of environmental variation in personality are of an "accidental" rather than "cultural" kind.

Upon reflection, it may seem less surprising that genetic factors play such a large role in the major dimensions. Scores on personality traits are obtained by summing over a large number of specific items of behavior in order to better represent what is common to many individual items. If the underlying common processes are genetically determined then we expect the correlation between items to be genetic. Thus, when the item responses are accumulated into a test score, the contribution of genetic factors will be augmented. In contrast, we might argue, a major part of the information acquired from the environment and stored by the organism is of a highly specific kind and likely to depend far more on the idiosyncratic experiences of the individual.

Thus, while genetic factors might affect the *processes* by which information is stored and used, the specific *content* of information might be influenced more markedly by the environment. Translating this idea into a model for test responses, the implication is that genetic factors may contribute to item *covariation*, but environmental factors contribute to variation *specific* to individual items. If this is the case then composite test scores will have significantly greater heritability than individual items, since the variance of the test score includes the genetic component of the individual items *and* the covariances between them, but includes only the environmental components specific to individual items, because there is no environmental communality between them. An early multivariate study of neuroticism by Eysenck and Prell (1951) was based on the theory that covariance between tests reflected the same common genetic factor. They argued that scores on a neuroticism factor derived from combining multiple indices of the same inherited trait should be more heritable than scores on any component test.

A simple algebraic example illustrates this concept. Suppose that we combine N items of unit variance into a composite test score by unweighted summation. Let us suppose further that the heritability of each item is a constant h^2. The environmental component of each item is thus $e^2 = 1 - h^2$. Now let the genes be the only source of communality between the items, so that the interitem correlations are all equal to the heritability h^2. The total variance of the test scores will be

$$Ne^2 + N^2h^2 = Ne^2 + N(N-1)h^2$$

The genetic component of variance will be N^2h^2 and the environmental component will be just Ne^2. Thus, while the heritable variance in the test scores increases as the square of the number of items, the environmental variance increases only linearly with the number of items. Alternatively, the ratio of the genetic to environmental variance in a composite test score will increase linearly as the number of items increases when the items are equivalent.

Naturally, this relationship depends on very strong assumptions about the causes of item covariance. If, for example, the covariances were environmental and the genetic effects item-specific then the ratio of genetic variance to environmental variance would be inversely proportional to N. However, the simple model serves to make the point that combining items into a composite test score augments the contribution of genetic and environmental factors common to items (which is exactly why composite scores are derived) at the expense of factors specific to particular items.

We dwell on this point because we have failed, so far, to detect any significant contribution of the shared environment to family resemblance in test scores. If, however, the family environment primarily determined the specific profile of item responses (for example, made some people like parties and enjoy reading) then we might expect the test score to obscure a process that is nevertheless an important component of information transfer between family members.

It is against this theoretical background that we examine the genetic analysis of individual items. We expect errors of measurement, and hence estimates of E_W, to be larger for individual items, but it is conceivable that the contribution of the family environment, E_B, might be greater relative to the genetic component if the family environment affects the transmission of specific information between family members. Analysis of individual test items might therefore provide better evidence for non-genetic components of transmission than analysis of composite scores. This argument might be applied with even greater force to items in the social-attitudes domain (see Chapters 12–16).

8.1 THE DATA

The most informative data for our purpose are those of the Australian Twin Study (see e.g. Martin and Jardine, 1986) because these comprise a large single set of EPQ data, gathered on 3810 pairs of twins. The structure of the sample was described in Chapter 5, and more details may be found in the paper by Martin and Jardine and in Jardine (1985).

The items of the EPQ (see Appendix B) are identified for convenience by the scales on which they load. A factor analysis showed that the items of the four scales, P, E, N and L, all load on separate, single factors consistently in both sexes. Correlations between the factors were small in an oblique solution. The items are numbered in the order they appear on the EPQ as follows: psychoticism P1–P25; extraversion E1–E21; neuroticism N1–N23; lie L1–L21. The factor on which each item loads is identified in Appendix B.

There were significant differences in endorsement frequency between MZ and DZ twins for the following items in females: E13, E18, P3, P5, P19, L14 and L18. Zygosity differences in endorsement frequency were found in males for items E8, E17, E20, P22, N7, L3, L7 and L18.

Table 8.1 gives the frequency of "keyed" responses for each scale by sex. There are many very highly significant sex differences in endorsement frequency for items in all four personality scales.

Conventional genetic analyses of dichotomous items have been comparatively primitive. Just as analysis of continuous data in twins has focused on testing the significance of the difference between MZ and DZ twins, so analysis of dichotomous items has concentrated on comparing the so-called "concordance rates" of MZ and DZ twins. That is, given that one twin answers "Yes" to a particular item, it is asked whether the conditional probability that the cotwin will also answer "Yes" is higher in MZ and DZ twins. This crude approach suffers from all the weaknesses that we have already identified in describing the results of earlier twin studies. No genetic mechanism is hypothesized; therefore none can be tested. No environmental parameters are explicitly formulated; therefore none can be estimated.

Our approach is different, and more informative. For the analysis of continuous traits, we began with a series of models relating phenotypic variation to differences in underlying genetic and environmental sources of variation. The mean squares between and within pairs, or the twin correlations, may then be expressed as functions of these unknown parameters.

Table 8.1 Percentage of individuals endorsing an EPQ item in the direction of naming, broken down by sex. Asterisks denote significant differences between male and female endorsement frequencies.

Item	Extraversion Females	Males	Psychoticism Females	Males	Neuroticism Females	Males	Lie Females	Males
1	52.6	49.2**	19.0	13.6***	58.1	50.1***	85.5	79.5***
2	58.8	49.1***	8.5	23.0***	51.1	26.9***	58.1	51.6***
3	70.7	68.5*	30.4	40.9***	80.8	71.9***	76.9	67.6***
4	67.2	74.1***	1.0	4.1***	23.2	22.1	69.6	67.6
5	84.3	82.4*	21.9	26.9***	74.9	58.1***	40.4	35.8***
6	45.3	50.4***	4.5	8.4***	48.8	37.1***	36.0	21.0***
7	62.0	58.2**	1.1	1.9***	37.8	34.2**	61.4	58.2**
8	77.5	79.3	3.3	7.9***	31.9	24.0***	36.6	33.6**
9	79.8	79.5	2.5	8.5***	60.0	46.5***	38.1	27.6***
10	50.4	51.9	3.7	8.7***	49.0	35.5***	38.8	27.6***
11	49.4	49.1	3.7	6.6***	24.8	18.0***	15.4	14.1
12	55.1	54.1	12.6	23.1***	41.8	43.9	41.0	40.5
13	29.4	36.9***	5.3	9.7***	21.6	15.5***	48.7	49.7
14	64.2	79.1***	8.6	13.0***	55.9	42.5***	57.8	42.3***
15	56.8	84.2**	9.3	14.5***	26.6	22.1***	68.8	51.6***
16	45.0	52.4***	2.2	1.3***	50.3	37.0***	54.7	33.0***
17	54.9	71.4***	5.7	10.4***	28.6	20.4***	17.8	18.6
18	50.6	54.7***	32.0	41.0***	64.7	47.5***	58.6	56.3
19	40.0	47.8***	6.2	11.0***	34.3	27.0***	64.8	71.4***
20	57.7	56.3	24.7	29.0***	32.5	25.9***	15.9	14.5
21	63.0	59.3**	7.0	9.3***	76.0	60.2***	74.1	72.2
22			10.9	27.5***	77.6	69.1***		
23			1.9	8.7***	81.7	74.4***		
24			8.4	15.2***				
25			2.9	14.4***				

Maximum-likelihood parameter estimates may be obtained and the goodness of fit of the model tested by some appropriate statistical criterion. Our approach to the analysis of the individual items is identical. Only the initial data summary differs. Whereas continuous data may be summarized by variances and covariances, or comparable statistics such as mean squares and correlations, our analysis of individual items begins with contingency tables in which the pattern of "Yes/No" responses is tabulated for first and second twins of each sex/zygosity grouping. An illustrative set of contingency tables is given in Table 8.2 for the extraversion item "Do you have many different hobbies?" (item E1). For like-sex twins the order of first and second twin is arbitrary, though it may represent birth order if desired. For unlike-sex twins, the columns of the table represent the male responses and the rows represent the responses of females. For example, in the unlike-sex

Table 8.2 Contingency tables for item E1: "Do you have many different hobbies?".

MZF			Twin 2	
			No	Yes
	Twin 1	No	385	195
		Yes	213	440

MZM			Twin 2	
			No	Yes
	Twin 1	No	186	109
		Yes	100	172

DZF			Twin 2	
			No	Yes
	Twin 1	No	212	146
		Yes	154	239

DZM			Twin 2	
			No	Yes
	Twin 1	No	107	65
		Yes	76	104

DZO			Male	
			No	Yes
	Female	No	226	181
		Yes	233	267

twins both members of 226 twin pairs said they had no hobbies, both members of 267 pairs said they had many different hobbies and in 181 pairs the male acknowledged he had many hobbies but the female did not.

8.2 MODELS

Two basic types of model might be proposed in trying to explain the patterns of twin concordance in such tables. The first type of model assumes that a discontinuous phenotype is best approximated by a discontinuous model. For example, certain psychiatric disorders might be better explained in terms of a single gene of large effect with or without reduced penetrance (see the analysis of depression symptoms in Chapter 11). Alternatively, the environmental transmission of a dichotomous trait (such as being Catholic or not, or having some infection or not) may be better explained by the transmission of a discrete "particle" of environment in the manner of a bacterium or virus. The mathematical properties of such processes have been investigated in great detail by Cavalli-Sforza and Feldman (1981).

The second type of model assumes that the discontinuity in the phenotype is a more or less arbitrary result of classifying people by kind rather than degree (as in the case of psychiatric diagnoses) or by forcing people of varying shades of personality to choose between two options in responding to a given item. The dichotomy is therefore not fundamental, but is merely a reflection of how we choose to describe human differences. People are assumed to differ quantitatively in their level of some latent trait, and their probability of endorsing a given item is a monotonic function of their trait level.

These two kinds of models are not mutually exclusive, and may be explored on the same data (see Chapter 11). There is considerable interest among human geneticists in the so-called "mixed" model for inheritance in which transmission is due both to a single gene of large effect and to quantitative background variation which may or may not be genetic (see e.g. Lalouel *et al.*, 1983).

In analyzing personality items, we concentrate on the continuous-liability version of the model. There are two reasons for doing this. First, our original theory assumes that differences in personality are continuous. Secondly, it can be shown empirically that such a model is better able to account for the relationship of objective personality tests and psychiatric criteria (see e.g. Eysenck, 1952). Unfortunately, the resolution of continuous and discontinuous models is impossible for dichotomous items in twins and very difficult even with multicategory items for moderate heritabilities (see Reich *et al.*, 1978, and Chapter 11).

Several writers have described the basic continuous model that we employ (e.g. Falconer, 1963, 1965; Gottesman and Shields, 1968; Smith, 1971; Smith *et al.*, 1972; Eaves, 1980; Eaves *et al.*, 1978). Most of the time, however, its use has been restricted to qualitative disease states in kinships ascertained through an affected proband. Here we are concerned with normal differences in personality in an unselected sample. Most of the earlier treatments of the model (often termed the "threshold model") do not employ maximum-likelihood methods for estimation and give no tests of hypotheses. Therefore the conclusions are questionable for the same reasons that we questioned the results of earlier twin studies of continuous variables.

The simplest form of the threshold model, first made explicit by Falconer (1963) but implicit in early studies by Pearson (1900), assumes that a con-

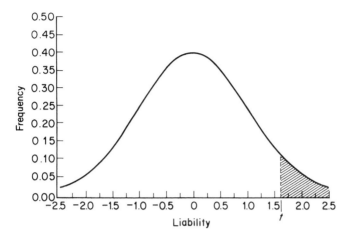

Figure 8.1 The multifactorial model (*t* is the threshold for becoming affected).

tinuous, normally distributed, scale of liability underlies a given dichotomy at the phenotypic level. Without loss of generality, it is assumed that the liability scale has zero mean and unit variance. Thus the main goal of genetic analysis is to determine the model of transmission of liability rather than the actual dichotomy itself. Clearly, the theoretical correspondence between this model and that advanced over a generation ago by Eysenck for the relationship between personality and psychiatric disorders is very striking. Falconer argues that there exists a threshold, *t*, on the liability scale above which all individuals are affected by a disorder and below which all individuals are normal. The basic features of the model are represented in Figure 8.1.

The model has certain unrealistic features. For example, it is assumed that the regression of probability of endorsement on liability is a step function that switches from 0 to 1 at the threshold *t*. A more realistic model would postulate a logistic or cumulative normal item regression (see e.g. Birnbaum, 1968) in which the probability of endorsement increased gradually as a function of liability (see Figure 8.2).

In practice, it turns out that the classical threshold model described by Falconer is sufficient to account for the pattern of association between pairs of relatives for a single dichotomous item; the more general latent-trait model is not identified with only one item.

The genetic analysis of the latent dimension assumes that the distribution of liability in pairs of twins is bivariate-normal (see Figure 8.3), with zero mean-vector and correlation ρ between the liabilities of twin pairs. If both twins of a pair are above the threshold then both will endorse the item. If

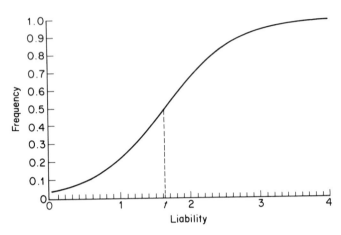

Figure 8.2 Logistic model for probability of endorsement as function of liability (*t* is liability at which probability of being affected is 50%).

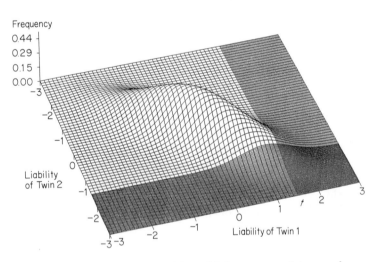

Figure 8.3 Bivariate normal threshold model for pattern of item endorsement in twin pairs. (The figure is drawn for a correlation of 0.8 in liability. Close hatching indicates parts of the distribution comprising affected individuals, i.e. having liabilities greater than the threshold, *t*.)

both are below the threshold then both will not endorse the item. If one twin is above the threshold and the other below then the first will endorse the item and the other will not.

The probability that both twins will endorse a given item is thus

$$P_{11} = \int_t^\infty \int_t^\infty \phi(x,y,\rho) \, dydx,$$

where $\phi(x, y, \rho)$ is the bivariate-normal probability function. Similarly, the probability that the first twin will endorse the item and the second will not is

$$P_{10} = \int_t^\infty \int_{-\infty}^t \phi(x,y,\rho) \, dydx.$$

Similar expressions follow for the other categories of twin responses, P_{01} and P_{00}.

Given a sample of N twin pairs, the expected number of pairs concordant for endorsement is NP_{11}. The expected numbers for the other cells of the contingency table may be computed similarly, as long as the appropriate probabilities P_{ij} can be computed for given t and ρ.

8.3 ESTIMATION

The bivariate normal probability integral can be evaluated numerically for given values of t and ρ, so that expected proportion can be derived. Given the observed cell numbers N_{ij} and the expected proportions P_{ij}, the log-likelihood of obtaining a given contingency table is

$$L = C + \Sigma_i \Sigma_j N_{ij} \ln P_{ij}.$$

If there are k groups of twins, each with log-likelihood L_i, then the overall likelihood is simply

$$L = \sum_{i=1}^k L_i$$

for given values of the expected twin correlation and threshold. Since the likelihood may be computed for known values of the thresholds and twin correlations for liability, we can find the parameter values for which the likelihood is maximized. A FORTRAN program was written for this purpose employing the Numerical Algorithms Group's optimization subroutine E04JBF to obtain maximum-likelihood estimates.

We know already that there are significant sex differences in endorsement frequency for many of the items. In our model we therefore usually postulate separate thresholds t_m and t_f for males and females. Then, for example, the proportion of unlike-sex dizygotic twins expected to be concordant for endorsement is

$$P_{11} \int_{t_m}^{\infty} \int_{t_f}^{\infty} (x,y,\rho_{dzu})dydx,$$

where ρ_{dzu} is the expected correlation for unlike-sex dizygotic twins.

Having recognized that the statistic of main concern is the twin correlation in liability, it is now a simple step to see that the expected correlations in liability may be parametrized in terms of genetic and environmental effects, just as we did for the mean squares in the continuous case (see Chapter 5). Expected correlations are given for a full model in Table 8.3. The model includes additive and dominant genetic effects and environmental effects between families. The effects may be the same in males and females, or heterogeneous across sexes, as before (cf. Chapter 5). The environmental effects within families (E_W) only contribute to intrapair differences, not correlations, and so do not appear explicitly in the model. The contribution of E_W is implied, however, since the total variance in liability is fixed at unity. For the case where parameters are homogeneous over sexes, therefore, the contribution of the within-family environment, in the absence of dominance, is simply

$$E_W = 1 - V_A - E_B.$$

When separate genetic and environmental effects have to be defined for each sex, we have

$$E_{WM} = 1 - V_{AM} - E_{BM} \qquad \text{for males,}$$
$$E_{WF} = 1 - V_{AF} - E_{BF} \qquad \text{for females.}$$

For a given model, the expected numbers in every cell of the contingency tables can be computed for the maximum-likelihood parameter values. If we

Table 8.3 Expected correlations in liability for twins.

	E_B	E_{BM}	E_{BF}	E_{BFM}	V_A	V_{AM}	V_{AF}	V_{AMF}	V_D
MZF	1	0	1	0	1	0	1	0	1
MZM	1	1	0	0	1	1	0	0	1
DZF	1	0	1	0	$\frac{1}{2}$	0	$\frac{1}{2}$	0	$\frac{1}{4}$
DZM	1	1	0	0	$\frac{1}{2}$	$\frac{1}{2}$	0	0	$\frac{1}{4}$
DZU	1	0	0	1	$\frac{1}{2}$	0	0	$\frac{1}{2}$	$\frac{1}{4}$

denote the expected number in the jth row of the kth column of the ith table by E_{ijk}, and the corresponding observed number by N_{ijk}, a goodness-of-fit test of the model is given by

$$\chi^2 = \sum_{ijk} \frac{(N_{ijk} - E_{ijk})^2}{E_{ijk}}.$$

If there are k tables, each with m rows and columns, then the chi-square has $k(m^2 - 1) - p$ df, where p is the number of parameters estimated from the data. For the threshold model, the parameters include genetic and environmental parameters (e.g. V_A and E_B) *and* any thresholds estimated from the data. Estimates of E_W do not count as independent parameters since the total variance is constrained to be unity, and the estimates of the within-family environment are obtained by subtraction.

An alternative test, which is probably more appropriate for testing different models for correlations, is to compare the likelihood under any of the reduced models for the correlations (e.g. the "V_A" model) with that under a full model in which each group of twins is represented by a unique correlation. If the full model is a significant improvement on a simpler model then it must be assumed that the simpler model cannot explain the observed correlations in liability for the twins. The logical procedure used in deciding between models with this approach is identical with that we described for the analysis of the scores on the personality dimensions in Chapters 5 and 6.

8.4 EXAMPLE

We illustrate the analytical procedure with the extraversion item E1 ("Do you have many hobbies?"), for which the full data are given in Table 8.2.

Since there were significant sex differences in endorsement frequency for many items, all the models allowed for sex differences in thresholds. In this way, all the tests of significance were directed mainly to testing hypotheses about the correlations in liability. In summarizing the models, we omit direct reference to E_W since it is implied in all models.

In Table 8.6 we give chi-squares for testing the fit of various models to the data on item E1 under the column labelled E1. By the conservative criteria of parsimony and goodness of fit, we find that the "best" model for twin resemblance is the V_A model, since E_B alone does not fit and all other models are more complex. If we consider the "improvement-in-fit" criterion then we see that marginal improvement ($X^2_2 = 5.98$, $P \approx 5\%$) follows from allowing for sex limitation of gene expression ("V_{AM}, V_{AF}, V_{AMF} model"). On balance, therefore, the model that assumes that gene effects are additive and the same in both sexes seems good enough for this item. In Table 8.7 the parameter estimates are given for the selected model. In this case separate thresholds are given for males and females (t_m and t_f) because the sex differences in endorsement frequency result in heterogeneous thresholds. Indeed, such differences are found for most of the items (cf. Table 8.1). Additive genetic effects (V_A) account for an estimated 47.2% of the variance in liability for this item.

8.5 RESULTS

The results for the remaining items are summarized in tabular form. Separate tables are given for items loading on each of the main factors.

8.5.1 Psychoticism

The likelihood-ratio tests of specific models against the full model are given in Table 8.4 for the P items. Parameter estimates are given in Table 8.5 for the best-fitting model only.

Out of the 25 P-items on the EPQ, a simple genetic model is most appropriate for 20. Item P16 ("Is [or was] your mother a good woman?") is unique in having a genetic component and a significant shared environmental effect whose effects depend on sex (i.e. the V_A, E_{BM}, E_{BF}, E_{BMF} model). It is, perhaps, not too surprising that twins should agree about the virtues of one of the most salient features of their shared environment but that sons and daughters should be influenced by different factors in making their judgement. Item P9 ("Do you enjoy practical jokes that can sometimes really hurt people?") was the only one for which no genetic effect was required to account for twin similarity. A shared environmental effect was necessary, in addition to a genetic effect, for items P3 ("Do you lock your house up carefully at night?"), P5 ("Do you believe insurance schemes are a good idea?") and P22 ("Do you sometimes like teasing animals?"). In these cases the family environment contributed between 20% and 40% to variation in liability.

8.5.2 Extraversion

The model-fitting results for the extraversion items are summarized in Tables 8.6 and 8.7. No model can adequately explain the data for items E3 and E11. All the chi-squares were highly significant for E11. There are sex

Table 8.4 Likelihood-ratio tests for analysis of P items.[a]

| | | Item | | | | | | | | | | | | |
Model	df	P1	P2	P3	P4	P5	P6	P7	P8	P9	P10	P11	P12	P13
E_B	4	15.38**	7.12*	10.28*	3.30	12.28*	26.58***	4.40	2.96	4.18	12.82*	6.60	10.50*	18.38**
V_A	4	<u>8.54</u>	<u>5.88</u>	8.32	<u>1.30</u>	12.30*	<u>8.94</u>	<u>4.30</u>	<u>1.48</u>	5.18	5.96	<u>3.10</u>	<u>3.30</u>	8.86
E_B, V_A	3	8.48*	5.10	<u>2.76</u>	1.28	<u>6.32</u>	8.94*	3.94	1.40	3.66	5.96	2.70	3.20	5.76
V_A, V_D	3	8.48*	5.10	2.76	1.28	6.32	8.94*	3.94	1.40	3.66	5.96	2.70	3.20	5.76
V_{AM}, V_{AF}, V_{AMF}	2	4.40	4.38	5.66	0.78	10.08**	6.50*	0.04	1.16	4.26	4.76	2.54	1.30	5.22
$-L$ (full model)		3427.94	2876.55	4741.40	725.28	4058.77	1622.56	549.75	1453.86	1353.10	1557.53	1400.80	3281.29	1859.32

| | | Item | | | | | | | | | | | |
Model	df	P14	P15	P16	P17	P18	P19	P20	P21	P22	P23	P24	P25
E_B	4	11.00*	11.64*	14.80**	6.28	11.34*	10.48*	44.76***	12.16*	11.60*	5.36	6.60	8.20
V_A	4	<u>3.44</u>	<u>4.46</u>	13.98**	<u>4.44</u>	<u>8.60</u>	<u>6.98</u>	<u>4.12</u>	<u>1.62</u>	8.90	<u>1.86</u>	<u>2.42</u>	<u>0.68</u>
E_B, V_A	3	0.30	4.32	9.66*	4.06	6.72	6.92	2.16	0.58	<u>4.58</u>	1.80	1.16	0.66
V_A, V_D	3	0.30	4.32	12.72***	4.06	6.72	6.92	2.16	0.58	4.58	1.80	1.16	0.66
V_{AM}, V_{AF}, V_{AMF}	2	2.36	2.02	8.26*	3.04	7.48*	2.74	2.14	0.18	8.72*	1.36	2.30	0.40
$E_{BM}, E_{BF}, E_{BMF}, V_{AF}$	1	—	—	<u>2.7</u>	—	—	—	—	—	—	—	—	—
$-L$ (full model)		2475.28	2599.86	633.15	1959.10	4856.85	2071.51	4263.41	2059.49	3199.56	1266.56	2521.30	1734.01

[a] Where twice the difference in log-likelihood of the models under comparison has χ^2 distribution with df equal to the differences in the number of parameters in the two models, values for best-fitting model are underlined. Significance levels as follows:
* $P<0.05$; ** $P<0.01$; *** $P<0.001$.

Table 8.5 Parameters of best-fitting models for P items.

Item	$\hat{t}_M$	$\hat{t}_F$	$\hat{E}_W$	$\hat{E}_{WM}$	$\hat{E}_{WF}$	$\hat{E}_B$	$\hat{E}_{BM}$	$\hat{E}_{BF}$	$\hat{E}_{BMF}$	$\hat{V}_A$	$\hat{V}_{AM}$	$\hat{V}_{AF}$	$\hat{V}_D$
P1	−1.094	−0.881	0.683	—	—	—	—	—	—	0.317	—	—	—
P2	−0.739	−1.369	0.666	—	—	—	—	—	—	0.334	—	—	—
P3	−0.224	−0.515	0.553	—	—	0.181	—	—	—	0.266	—	—	—
P4	−1.735	−2.331	0.409	—	—	—	—	—	—	0.591	—	—	—
P5	−0.606	−0.780	0.535	—	—	0.206	—	—	—	0.259	—	—	—
P6	1.365	1.691	0.295	—	—	—	—	—	—	0.705	—	—	—
P7	2.055	2.297	0.491	—	—	—	—	—	—	0.509	—	—	—
P8	1.409	1.840	0.634	—	—	—	—	—	—	0.366	—	—	—
P9	1.370	1.958	0.643	—	—	0.357	—	—	—	—	—	—	—
P10	−1.357	−1.783	0.476	—	—	—	—	—	—	0.524	—	—	—
P11	1.509	1.791	0.459	—	—	—	—	—	—	0.541	—	—	—
P12	0.731	1.152	0.604	—	—	—	—	—	—	0.396	—	—	—
P13	1.301	1.614	0.615	—	—	—	—	—	—	0.385	—	—	—
P14	−1.123	−1.367	0.724	—	—	—	—	—	—	0.276	—	—	—
P15	−1.058	−1.320	0.540	—	—	—	—	—	—	0.460	—	—	—
P16	−2.220	−2.020	—	0.455	0.108	—	0.545	0.484	0.242	—	—	0.408	—
P17	1.261	1.580	0.610	—	—	—	—	—	—	0.390	—	—	—
P18	0.218	0.470	0.650	—	—	—	—	—	—	0.350	—	—	—
P19	−1.223	−1.533	0.673	—	—	—	—	—	—	0.327	—	—	—
P20	0.558	0.677	0.467	—	—	—	—	—	—	0.533	—	—	—
P21	1.320	1.475	0.562	—	—	—	—	—	—	0.438	—	—	—
P22	0.583	1.240	0.479	—	—	0.202	—	—	—	0.319	—	—	—
P23	1.353	2.065	0.567	—	—	—	—	—	—	0.433	—	—	—
P24	1.021	1.384	0.502	—	—	—	—	—	—	0.498	—	—	—
P25	−1.061	−1.901	0.492	—	—	—	—	—	—	0.508	—	—	—

Table 8.6 Likelihood-ratio tests for analysis of extraversion items.[a]

Model	df	E1	E2	E3	E4	E5	E6	E7	E8	E9	E10	E11
							Item					
E_B	4	35.30***	79.94***	82.48***	71.80***	12.60*	42.54***	16.02**	23.16***	39.72***	62.70***	68.92***
V_A	4	8.56	17.72**	18.54***	13.00*	1.34	4.92	3.90	1.78	6.36	17.46**	29.04***
E_B, V_A	3	8.56	6.32	4.90	3.20	0.66	3.78	0.94	1.24	4.44	7.70	20.34***
V_A, V_D	3	8.56	6.32	4.90	3.20	0.66	3.78	0.94	1.24	4.44	7.70	20.34***
V_{AM}, V_{AF}, V_{AMF}	2	2.58	16.26***	17.02***	7.06*	2.00	3.16	2.00	1.24	4.44	5.84	10.72**
$-L$ (full model)		5157.80	5072.30	4530.65	4525.03	3317.34	5140.85	4973.77	3929.27	3755.81	5182.54	5188.12

Model	df	E12	E13	E14	E15	E16	E17	E18	E19	E20	E21
						Item					
E_B	4	65.58***	53.90***	31.66***	24.68***	13.02*	23.72***	30.94***	30.38***	32.04***	39.76***
V_A	4	8.64	3.22	1.34	3.04	1.24	4.00	9.58*	1.54	2.06	4.58
E_B, V_A	3	2.32	1.22	1.12	0.74	1.22	3.24	8.70*	1.52	1.10	1.24
V_A, V_D	3	2.32	1.22	1.12	0.74	1.22	3.24	8.70*	1.52	1.10	1.24
V_{AM}, V_{AF}, V_{AMF}	2	5.66	0.20	0.28	2.78	0.18	0.04	1.28	0.08	1.92	3.64
$-L$ (full model)		5110.26	4610.96	4437.48	3047.71	5201.47	4935.83	5198.00	5049.94	5106.77	4976.13

[a] Values for best-fitting model are underlined. Significance levels as follows: * $P<0.05$; ** $P<0.01$; *** $P<0.001$.

Table 8.7 Parameter estimates and position of the threshold for best fitting models for extraversion items.[a]

Item	t_m	t_f	E_1	E_{1M}	E_{1F}	V_A	V_{AM}	V_{AF}	V_{AMF}	V_D
E1	0.023	−0.068	0.528	—	—	0.472	—	—	—	—
E2	0.026	−0.225	0.456	—	—	0.037	—	—	—	0.507
E4	−0.651	−0.447	0.434	—	—	0.055	—	—	—	0.511
E5	−0.928	−1.008	0.488	—	—	0.512	—	—	—	—
E6	0.005	−0.114	0.517	—	—	0.483	—	—	—	—
E7	−0.205	−0.306	0.492	—	—	0.508	—	—	—	—
E8	0.777		0.550	—	—	0.450	—	—	—	—
E9	−0.830		0.495	—	—	0.505	—	—	—	—
E10	−0.022		—	0.616	0.517	—	0.384	0.483	0.122	—
E12	0.120		0.456	—	—	0.169	—	—	—	0.375
E13	0.330	0.546	0.437	—	—	0.563	—	—	—	—
E14	−0.803	−0.370	0.428	—	—	0.572	—	—	—	—
E15	−1.000	−1.115	0.535	—	—	0.465	—	—	—	—
E16	−0.059	0.126	0.671	—	—	0.329	—	—	—	—
E17	−0.558	−0.123	0.627	—	—	0.373	—	—	—	—
E18	−0.117	−0.013	—	0.766	0.568	—	0.234	0.432	0.359	—
E19	0.051	0.255	0.492	—	—	0.508	—	—	—	—
E20	−0.181		0.560	—	—	0.440	—	—	—	—
E21	−0.236	−0.330	0.564	—	—	0.436	—	—	—	—

[a] t_m is the threshold for males and t_f is the threshold for females.

differences in gene expression for two items. In the case of E10 ("Happy-go-lucky") the correlation between gene effects in males and females is very low. For E18 ("Take on too many activities") the correlation is very high, indicating that the sex difference in gene expression is a matter of scale for this item. Three items show statistically significant evidence in favor of a dominance component: E2, E4 and E11. For the remaining 14 items, the V_A model gave the best fit. There is no evidence of a significant E_B component for any of the extraversion items. Thus, for the component items of the extraversion scale, the individual results echo those of the factor very closely. There is no evidence that the family environment contributes to twin resemblance for the individual items. There is strong evidence for a genetic component, and some support for a non-additive genetic component. The non-additive component for extraversion cannot be confused with sex-limitation of the additive component since Jardine (1985) has shown that the dominance parameter is still required by the data even when the unlike-sex pairs are omitted from the item analysis.

8.5.3 Neuroticism

Table 8.8 summarizes the likelihood-ratio statistics obtained when the threshold models are fitted to the neuroticism items. Altogether, a model that assumes only additive gene action fits 18 of the items very well, with three of these showing evidence for sex-limited gene expression. The remaining four items provided evidence for a significant dominance component, which might reflect reduction of the correlation between unlike pairs due to sex limitation. In view of this possibility, we repeated the analysis, omitting the male–female twin pairs. The results of this analysis are summarized in Tables 8.9 and 8.10.

The reanalysis now yields estimates of the additive and environmental components that are consistent over sexes and removes any hint of dominance in all but one item. Once again, there is absolutely no evidence that the family environment plays any more role at the item level than it did at the level of factor scores.

8.5.4 The lie scale

The results for the lie scale are summarized in Tables 8.11 and 8.12. No model fits the data on item L18 ("Do you always practice what you preach?"). The simple additive genetic model fits 14 of the remaining items comfortably. A component due to the family environment is necessary for items L11 and L13 and sufficient to explain twin resemblance for item L7 ("Do you sometimes talk about things you know nothing about?"). Sex differences in gene expression are implicated for items L2 ("Greediness"), L8 ("Did what was asked as a child") and L14 ("Have you ever cheated at a game?"). However, the correlations between gene expression in males and females were estimated as 0.95, 0.90 and 0.82 for these three items respectively, indicating that the sex difference is primarily due to differences in scale or heritability rather than a qualitative sex difference in gene expression. That is, genetic differences within males and females are mostly caused by the same genes. There is little evidence that differences in lie scores are influenced by different genes in the two sexes. It will be recalled that analysis of the total lie scale scores in the Australian sample gave evidence of a significant family-environment effect in males. There is some evidence of shared environmental effects at the level of individual items also.

Table 8.8 Likelihood-ratio tests for analysis of neuroticism items.[a]

							Item					
Model	df	N1	N2	N3	N4	N5	N6	N7	N8	N9	N10	N11	N12
E_B	4	34.20***	27.04***	10.54*	25.56***	23.36***	35.16***	26.96***	60.84***	48.32***	34.94***	37.80***	17.06**
V_A	4	<u>7.22</u>	7.28	2.20	6.20	<u>6.24</u>	<u>5.04</u>	<u>2.70</u>	13.98**	10.56*	8.92	6.32	<u>0.60</u>
E_B, V_A	3	5.00	6.74	1.94	4.70	5.72	3.96	2.34	0.28	5.26	7.40	1.78	0.56
V_A, V_D	3	5.00	6.74	1.94	4.70	5.72	3.96	2.34	0.28	<u>5.26</u>	7.40	<u>1.78</u>	0.56
V_{AM}, V_{AF}, V_{AMF}	2	4.52	<u>1.10</u>	2.12	0.08	3.76	0.44	2.20	6.72*	2.20	<u>0.22</u>	0.92	0.22
$-L$ (full model)		5141.69	4908.74	3958.09	4038.46	4553.63	5092.42	4909.27	4473.87	5086.59	5085.20	3949.37	5126.18

							Item					
Model	df	N13	N14	N15	N16	N17	N18	N19	N20	N21	N22	N23
E_B	4	12.70*	12.04*	12.18*	33.60***	53.88***	44.14***	39.72***	26.48***	11.38*	10.42*	26.30***
V_A	4	<u>1.82</u>	<u>3.82</u>	2.90	5.58	11.98	10.02*	8.00	7.32	<u>0.60</u>	<u>6.26</u>	<u>4.90</u>
E_B, V_A	3	1.80	3.16	1.90	5.04	1.74	4.16	3.86	6.46	0.56	4.58	2.88
V_A, V_D	3	1.80	3.16	1.90	5.04	<u>1.74</u>	<u>4.16</u>	<u>3.86</u>	6.46	0.56	4.58	2.88
V_{AM}, V_{AF}, V_{AMF}	2	<u>0.70</u>	2.04	2.76	4.08	3.26	7.08*	2.18	<u>1.26</u>	0.46	3.74	2.60
$-L$ (full model)		3685.35	5150.75	4195.55	5034.96	4223.84	4985.60	4656.78	4582.62	4489.47	4227.92	3824.51

[a] Values for best-fitting model are underlined. Significance levels as follows: * $P < 0.05$; ** $P < 0.01$; *** $P < 0.001$.

Table 8.9 Likelihood-ratio tests for analysis of neuroticism items, omitting unlike-sex pairs.[a]

Model	df						Item						
		N1	N2	N3	N4	N5	N6	N7	N8	N9	N10	N11	N12
E_B	3	25.96***	6.50	8.16*	12.92**	16.34***	20.12***	18.24***	41.34***	32.00***	23.78***	21.32***	9.62*
V_A	3	6.38	1.12	2.14	3.02	5.28	2.84	2.28	6.78	6.64	7.44	1.68	0.26
E_B, V_A	2	4.38	0.08	1.96	3.00	5.22	2.84	2.24	0.04	4.82	7.20*	0.76	0.16
V_A, V_D	2	4.38	0.08	1.96	3.00	5.22	2.84	2.24	0.04	4.82	7.20*	0.76	0.16
$-L$ (full model)		3904.07	3749.17	3012.58	3035.64	3455.16	3856.32	3714.17	3411.91	3848.78	3855.10	2992.32	3889.74

Model	df						Item					
		N13	N14	N15	N16	N17	N18	N19	N20	N21	N22	N23
E_B	3	4.76	8.30*	6.28	20.30***	30.96***	34.76***	21.44***	7.44	10.04*	9.28*	18.20***
V_A	3.	0.70	3.88	2.90	5.48	3.74	8.38*	2.72	1.36	0.46	5.12	3.48
E_B, V_A	2	0.20	3.06	1.10	4.06	0.72	4.08	2.16	0.76	0.12	4.44	2.60
V_A, V_D	2	0.20	3.06	1.10	4.06	0.72	4.08	2.16	0.76	0.12	4.44	2.60
$-L$ (full model)		2794.91	3913.47	3185.50	3814.11	3151.25	3779.59	3550.32	3479.40	3394.28	3230.37	2892.28

[a] Values for best-fitting model are underlined. Significance levels as follows: * $P < 0.05$; ** $P < 0.01$; *** $P < 0.001$.

Table 8.10 Parameters of best-fitting models for neuroticism items, omitting unlike-sex pairs.

Item	$\hat{t}_m$	$\hat{t}_f$	$\hat{E}_W$	$\hat{E}_B$	$\hat{V}_A$	$\hat{V}_D$
1	0.041	−0.205	0.605		0.395	
2	0.617	−0.020	0.596	—	0.404	—
3	−0.565	−0.857	0.608	—	0.392	—
4	0.787	0.742	0.591	—	0.409	—
5	−0.167	−0.657	0.618	—	0.382	—
6	0.364	0.018	0.550	—	0.450	—
7	0.419	0.324	0.564	—	0.436	—
8	0.714	0.461	0.526	—	0.474	—
9	0.090	−0.258	0.560	—	0.440	—
10	0.402	0.024	0.592	—	0.408	—
11	0.949	0.678	0.528	—	0.472	—
12	0.184	0.201	0.616	—	0.384	—
13	1.025	0.796	0.601	—	0.399	—
14	0.214	−0.143	0.639	—	0.361	—
15	0.781	0.627	0.554	—	0.446	—
16	0.347	0.007	0.456	—	0.544	—
17	0.862	0.599	0.518	—	0.482	—
18	0.070	−0.369	0.566	—	0.014	0.420
19	0.619	0.397	0.562	—	0.438	—
20	0.676	0.445	0.598	—	0.402	—
21	−0.217	−0.706	0.667	—	0.333	—
22	−0.470	−0.745	0.612	—	0.388	—
23	−0.618	−0.911	0.577	—	0.423	—

8.6 RELATIONSHIPS BETWEEN ITEMS AND FACTORS

We conclude our analysis of the individual items by asking whether there are any consistent relationships between the results of the genetic analysis of the individual items and those of the factor analysis and genetic analysis of the test scores. Figures 8.4–8.7 present the heritabilities of the individual items and estimates of their standard errors for items grouped according to the personality dimensions on which they load most significantly. Within each set of items, a test was constructed for the heterogeneity of heritability estimates by fitting a model, that assumed constant heritability over all items, to the individual heritability estimates weighted by their standard errors. The chi-squares testing heretogeneity were as follows: psychoticism $X^2_{23} = 99.66$, $P<0.001$; extraversion $X^2_{20} = 86.70$, $P<0.001$; neuroticism $X^2_{22} = 45.68$, $P<0.01$ and lie scale $X^2_{19} = 101.12$, $P<0.001$. The estimates of heritability are therefore significantly heterogeneous within factors.

The heterogeneity is partly a reflection of differences in sampling error in

Table 8.11 Likelihood-ratio test of specific models as compared with the full model for lie items.[a]

Model	df	L1	L2	L3	L4	L5	L6	L7	L8
E_B	4	4.76	31.06***	18.08***	20.12***	43.20***	21.28***	6.16	44.26***
V_A	4	1.58	9.70*	1.24	7.86	2.90	2.88	7.32	8.80
E_B, V_A	3	1.04	9.32*	1.16	7.84*	2.36	2.54	3.68	7.00
V_A, V_D	3	1.04	9.32*	1.16	7.84*	2.36	2.54	3.68	7.00
V_{AM}, V_{AF}, V_{AMF}	2	0.74	1.82	0.18	3.16	2.36	0.38	5.96	1.14
L^b		3379.99	5093.81	4300.44	4669.07	4940.12	4491.50	5061.62	4848.82

Model	df	L10	L11	L12	L13	L14	L15	L16	L17
E_B	4	13.78**	12.86*	25.22***	22.80***	33.20***	25.74***	24.36	4.68
V_A	4	1.50	11.82*	6.96	6.62	13.90**	4.02	2.42	1.98
E_B, V_A	3	1.50	7.32	5.32	2.62	13.24**	4.02	1.02	0.60
V_A, V_D	3	1.50	7.32	5.32	2.62	13.24	4.00	1.02	0.60
V_{AM}, V_{AF}, V_{AMF}	2	0.32	7.72	1.62	4.94	5.92	1.88	0.10	0.76
L^b		4816.74	3114.85	5012.31	5089.62	5066.08	4825.95	4949.65	3552.65

Model	df	L19	L20	L21
E_B	4	25.34***	14.50**	6.94
V_A	4	5.78	2.04	3.10
E_B, V_A	3	3.62	1.58	1.96
V_A, V_D	3	3.62	1.58	1.96
V_{AM}, V_{AF}, V_{AMF}	2	3.82	0.10	2.44
L^b		4653.39	3228.77	4360.30

[a] Values for best-fitting model are underlined. Significance levels as follows: * $P<0.05$; ** $P<0.01$; *** $P<0.001$.

[b] L = log-likelihood under full model.

Table 8.12 Parameters of best-fitting models for lie items.

Item	$\hat{t}_M$	$\hat{t}_F$	$\hat{E}_W$	$\hat{E}_{WM}$	$\hat{E}_{WF}$	$\hat{E}_B$	$\hat{E}_{BM}$	$\hat{E}_{BF}$	$\hat{E}_{BMF}$	$\hat{V}_A$	$\hat{V}_{AM}$	$\hat{V}_{AF}$	$\hat{V}_{AMF}$	$\hat{V}_D$
1	0.827	1.055	0.656	—	—	—	—	—	—	0.344	—	—	—	—
2	0.041	0.201	—	0.632	0.458	—	—	—	—	—	0.368	0.542	0.423	—
3	−0.451	−0.734	0.585	—	—	—	—	—	—	0.415	—	—	—	—
4	0.492	0.513	0.650	—	—	—	—	—	—	0.350	—	—	—	—
5	0.363	0.246	0.473	—	—	—	—	—	—	0.527	—	—	—	—
6	−0.804	−0.359	0.510	—	—	—	—	—	—	0.490	—	—	—	—
7	0.204	0.291	0.737	—	—	0.263	—	—	—	—	—	—	—	—
8	0.424	0.341	—	0.650	0.471	—	—	—	—	—	0.350	0.529	0.388	—
9	−0.596	−0.304	0.592	—	—	—	—	—	—	0.408	—	—	—	—
10	−0.595	−0.285	0.646	—	—	—	—	—	—	0.354	—	—	—	—
11	−1.043	−1.024	0.495	—	—	0.218	—	—	—	0.287	—	—	—	—
12	−0.234	−0.233	0.482	—	—	—	—	—	—	0.518	—	—	—	—
13	0.024		0.438	—	—	0.164	—	—	—	0.394	—	—	—	—
14	−0.194	0.189	—	0.619	0.456	—	—	—	—	—	0.381	0.544	0.371	—
15	0.034	0.492	0.536	—	—	—	—	—	—	0.464	—	—	—	—
16	−0.439	0.114	0.459	—	—	—	—	—	—	0.541	—	—	—	—
17	−0.913	−0.925	0.611	—	—	—	—	—	—	0.389	—	—	—	—
18	−0.563	−0.380	0.438	—	—	—	—	—	—	0.562	—	—	—	—
19	−1.021	−0.997	0.605	—	—	—	—	—	—	0.395	—	—	—	—
20	−0.625	−0.647	0.645	—	—	—	—	—	—	0.355	—	—	—	—

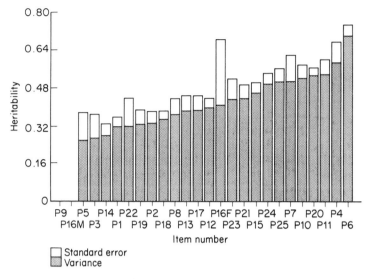

Figure 8.4 Heritabilities of psychoticism items.

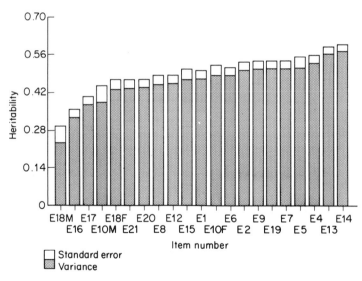

Figure 8.5 Heritabilities of extraversion items.

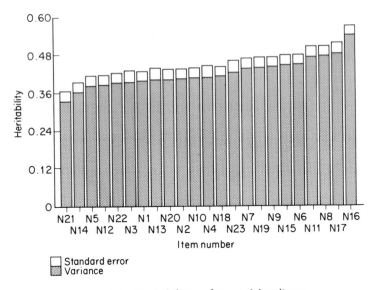

Figure 8.6   Heritabilities of neuroticism items.

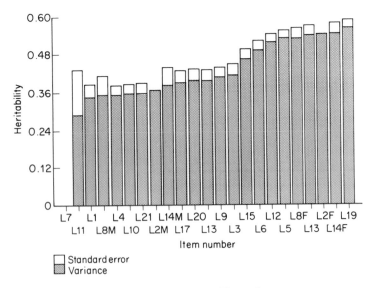

Figure 8.7   Heritabilities of lie-scale items.

Table 8.13 Correlations between item parameters and salient loadings for P, E, N and L.[a]

		Extraversion[b]	Psychoticism[b]	Neuroticism[c]	Lie[b]
$\hat{V}_A$	Males	0.44*	−0.40*	0.04	0.02
	Females	0.48*	−0.26	0.08	0.10
$\hat{V}_D$	Males	0.41***	−0.24*	0.10	−0.19
	Females	0.43***	−0.10	0.10	−0.16

$\hat{V}_A$ estimated from the E_W, V_A model; $\hat{V}_D$ estimated from the E_W, V_A, V_D model.
[a] Significance levels as follows: * $P < 0.05$; *** $P < 0.001$.
[b] Opposite-sex pairs included.
[c] Opposite-sex pairs omitted.

the individual items. However, some of the heterogeneity may reflect the relationship between the individual items and the underlying factors. Thus if the factor is substantially genetic then the heritability of the item will be a function of the loading of the item on the factor. Table 8.13 gives the correlations between the item loadings in an orthogonal factor analysis of the item intercorrelation matrix and the heritabilities of the component items, estimated from fitting the V_A model to each item. Pairs of unlike sex were omitted from the model-fitting analysis of the neuroticism items. The correlations between heritability and factor loading are only significant for the extraversion scale in both sexes, and for P items among males.

More important, perhaps, is the relationship between estimates of dominance variance obtained from the V_A, V_D model and the factor loadings because of the suggestion that the extraversion factor is characterized by genetic non-additivity. Indeed, there is a highly significant correlation between the estimated contribution of dominance among the E items and the loading of the E items on the extraversion factor, suggesting that the dominance we observe at the factor-score level is also expressed at the item level for those items which load more heavily on the extraversion factor. From the discussion in Chapter 5, we recall that the effects of sibling interaction are almost inseparable from those of dominance in twin data. If we accept the "competition" interpretation of the apparent genetic non-additivity for extraversion then it follows that the individual extraversion items display the effects of competition in proportion to their loadings on the extraversion factor.

The relationship between the estimates of additive and dominance variance is pursued further for each factor separately in Figures 8.8–8.11. The horizontal axis on each figure represents the estimated magnitude of the additive genetic component for each item, and the vertical axis represents the

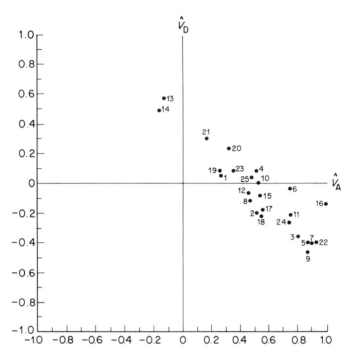

Figure 8.8 Relationship between estimated additive and dominance components ($\hat{V}_A$ and $\hat{V}_D$) for psychoticism items of the EPQ.

estimate of the dominance component. Each item is plotted on the figure corresponding to its main factor loading using estimates of V_A and V_D obtained by fitting the additive-dominance model to the contingency tables. Estimates are plotted whether or not the model actually fits the data for a given item. Negative estimates of the dominance component are plotted as such. Thus items that fall *below* the horizontal axis are those that are most likely to have significant family-environmental components.

The vector ($+ \hat{V}_A$, $+ \hat{V}_D$), which is inclined at 45° above the horizonal axis, represents the broad heritability, the expected correlation of identical twins. The vertical axis has dominance and the family environment as opposite poles and summarizes the difference between the expected MZ correlation and twice the expected DZ correlation.

The most obvious feature of all four figures is the large negative correlation between estimates of additive and dominance variance, as has already been remarked upon in our analysis of the factor scores (Chapter 5). A more important trend is revealed, however, by comparing the V_A, V_D plots for different factors. For neuroticism (Figure 8.10), 13 items out of 23 have posi-

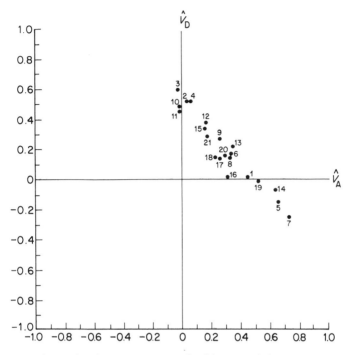

Figure 8.9 Relationship between estimated additive and dominance components ($\hat{V}_A$ and $\hat{V}_D$) for extraversion items of the EPQ.

tive estimates of V_D, even when unlike-sex pairs are omitted from the analysis. This number is close to that predicted if the effects of dominance do not differ from zero, on average, over all items, and reflects the general conclusion of our analyses of the factor scores presented in Chapter 5. In contrast, 17 out of 21 extraversion items show positive estimates of the dominance component, which is significantly different from the 1 : 1 expectation if the genetic effects were purely additive. Thus the results of the item analysis of extraversion is also consistent with that for the factor scores. Slightly over half (15/25) of the P items have negative estimates of the dominance component, which is scarcely sufficient to lend support to a specific or general contribution of the family environment. For the lie scale, the majority of the items (15/21) show negative dominance estimates, which is consistent which the results of the twin analysis for the Australian data, which showed evidence of an E_B effect for males (Chapter 5).

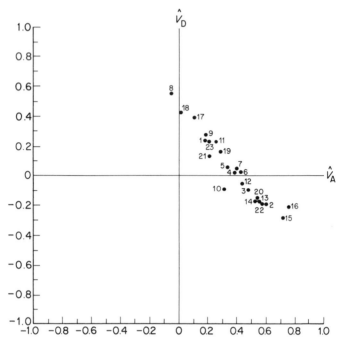

Figure 8.10 Relationship between estimated additive and dominance components ($\hat{V}_A$ and $\hat{V}_D$) for neuroticism items of the EPQ.

8.7 DISCUSSION AND SUMMARY

There are two main conclusions from the item analysis: (1) the individual items provide no greater support for the effects of family environment on personality than we gained from analysis of the factor scores — estimates of E_B for the items are mostly small; (2) the pattern of genetic and non-genetic effects on the individual items is very similar to that of the factors on which they load.

Variation in responses to individual items generally has a significant genetic component. Generally, at the level of individual items, there is little more support for an environmental hypothesis of family resemblance than there was for the major dimensions derived by adding the item responses. Our original speculation — that genes are responsible for the overall personality predisposition and the family environment creates its specific manifestation — turns out to be wrong for these items.

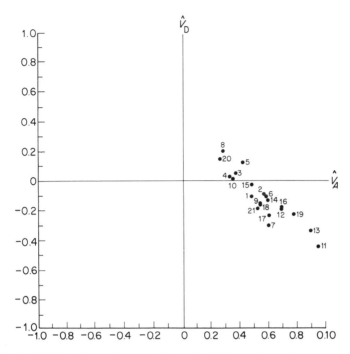

Figure 8.11 Relationship between estimated additive and dominance components ($\hat{V}_A$ and $\hat{V}_D$) for lie-scale items of the EPQ.

Although E and N items show little difference in overall heritability, our finding that the apparent contribution of non-additive genetic factors is greater for extraversion items than the neuroticism items is important for theories of personality for two reasons: (1) it suggests that the dimensions of Eysenck's personality theory are biologically distinct and may not be rotated arbitrarily to a new position without losing this distinction; (2) it may indicate that the two personality traits have a different adaptive role in the life of the organism.

Gray has argued that the fundamental consistent patterns of individual differences in personality are better understood in terms of sensitivity to reward and sensitivity to punishment rather than extraversion and neuroticism (see e.g. Gray, 1970, 1973). Gray's model corresponds to a 45° rotation of Eysenck's E and N dimensions. Our genetic analysis suggests that extraversion and neuroticism, as measured by questionnaire, are subject to different patterns of genetic architecture. Genetic effects of neuroticism are mainly additive. There is evidence of non-additivity for extraversion. Rotating the axes as Gray's theory suggests would result in psychometric scales that combine genetic elements of both extraversion and neuroticism and obscure what is, genetically speaking, a significant difference between the major dimensions. The basic principle remains unaltered if we argue that what we have called "dominance" is really sibling competition, since Gray's model still has to explain why the extraversion dimension is more subject to the effects of social interaction than neuroticism.

Chapter 9

The Specificity of Gene Effects: Implications for the Interaction of Persons and Situations

9.1 PERSONALITY TRAITS AND PERSONS × SITUATIONS INTERACTION

Most of the studies that we have described so far have focused on personality "traits" — consistent patterns of individual differences that are characterized by a relatively high degree of temporal stability and contribute to relatively large correlations between items in a behavioral inventory. That is, the measures studied so far have high retest reliability and relatively high internal consistency.

It has been argued that concentrating on traits, or common factors, as the main object of psychological interest is to miss important components of behavioral variation attributable to the situations in which behavior is studied and the interaction between persons and situations (P × S). There is a substantial body of empirical data to show that, in any attempt to measure personality through inventories, or to seek consistency of individual differences over situations, a significant component of variance may be ascribed to P × S interaction (see e.g. Endler and Hunt, 1976; Argyle and Cook, 1976). The psychological and biological implication is expressed most clearly by Mischel (see e.g. Mischel, 1977), who has doubted the value of any concept of personality as a persevering and consistent set of behavior patterns:

> Although traditional trait ratings may serve as summaries in everyday language of the gist of our impressions of each other, they do not capture the interaction between persons and conditions as the ongoing behavior is generated; and they certainly do not illustrate the causes of behavior.
>
> (Mischel, 1977, p. 335)

The significance of such interactions from the standpoint of personality

theory may have been overstated. Finding a significant interaction component by analysis of variance, or finding that a single common factor does
not explain all the covariance among a set of items or situations, cannot by
itself be taken as evidence that the trait model fails. There may equally be
"traits" of interaction, which can be represented parsimoniously in terms of
individual differences in one or two components that describe the sensitivity
of individuals to environmental changes. From a genetic perspective, the
existence of genotype × environment interactions is well documented for
experimental organisms, but it can often be shown that the interaction can be
explained in terms of relatively simple genetic parameters accounting for the
response of the genotype to changes in the environment (see e.g. Mather and
Jinks, 1982).

Indeed, whether one measures a "personality trait" by adding up the average performance in a number of situations or by examining parameters of
P × S interaction is often a matter of the paradigm being used for personality
measurement. In questionnaire studies extraversion emerges as a main effect
in the analysis of variance of subjects and items (or in the factor analysis of
the interitem correlation matrix), but in studies of conditioning, the extraversion dimension emerges as a parameter in the interaction of subjects and
trials (see e.g. Eysenck, 1976b) and as an interaction of subjects and drug
doses in psychopharmacological investigations (see e.g. Eysenck, 1960).

Thus the ability to show that subjects and treatments interact in a psychological experiment does not necessarily undermine the concept of personality. Before the trait concept can be discarded, it must be shown that the
interaction does not itself reflect consistent patterns of individual differences
in responsiveness to situations when these are properly ordered. Similarly,
the demonstration that subjects and situations may interact does not
necessarily favor a social rather than a biological model of behavior at the
level of P × S interaction. Organisms have evolved their behavioral patterns
in order that they might respond more efficiently to salient features of the
environment. The most effective schedules of reinforcement, for example,
are likely to be those that come closest to simulating those occurring in the
environment in which natural selection has "tuned" the genetic control of
behavior. Many common fears are entirely realistic in the light of the most
serious threats to survival. The geneticists' view that sensitivity to the
environment is under genetic control and may itself be important for adaptation finds an echo in Mischel's own thinking:

> In sum, a recognition of the continuity and coherence of perceived personality
> attributes must coexist with the finding of "specificity" at the behavior level.
> The latter may be viewed as reflecting man's discriminative facility, not merely
> the biases of faulty measurement . . . discriminative facility is highly adap-

tive, a reduced sensitivity to changing consequences may be characteristic of an organism adapting poorly.

(Mischel, 1977, p. 335)

Even at the level of persons × situations interaction, therefore, we expect the issue of social and genetic determination to be fundamentally important. A biologically based model of adaptation sees the environmental modification of behavior as an aspect of the phenotype that may itself be under genetic control and subject to evolution by natural selection. Indeed, investigations in plants and animals have already documented the relationship between the environment in which selection occurs and the genetic control of sensitivity to the environment. Furthermore, in man, geneticists have begun to examine the genetic control of parameters describing the homeostatic response. Murphy and Williams illustrate how the genetic control of homeostasis (or "sensitivity to the environment") can be represented in terms of relatively few genetic parameters (see e.g. Murphy and Williams, 1984). Sensitivity to the environment may thus be a "trait" and "genetic" in every practical sense, but its effects not emerge as a main effect in a simple analysis of variance or correlational study.

In this chapter we describe a basic attempt to explore the interaction of subjects and situations using repeated responses of twins to eleven items of a neuroticism scale obtained at two-year intervals. The study, as we conducted and analyzed it, does not address all the psychological issues raised by persons × situations interactions, nor does it address all the important theoretical issues about the adaptive significance of appropriate sensitivity to the environment. However, it is still almost unique in presenting a genetic analysis and interpretation of data that involve significant interactions of persons and situations. It thus lays the groundwork for a more thoughtful design of such studies with better selections of situational effects than those that we included in our study.

When we consider several behavioral responses from a large sample of subjects on more than one occasion, we may recognize, in addition to the main effects of persons, occasions and items (tests or situations) and purely random sampling error σ^2, two main sources of unreliability (or interaction) entering into predictions made on the basis of total test scores or the "trait" model. We identify unreliability due to the interaction of subjects and individual test items ($\sigma^2_{S \times I}$), that is, the inconsistency of the test, and we may recognize unreliability due to the interaction of subjects with occasions of testing ($\sigma^2_{S \times O}$), that is, lack of repeatability of the test.

Typically, we may distinguish between "fixed" and "random" effects in the analysis of variance (see e.g. Snedecor and Cochran, 1980). We shall always treat subjects as if they are "random" because we wish to generalize from our particular subjects to the population from which our subjects are a sample. The choice of items, however, may be regarded in two ways. In one application we wish to regard the chosen items as a sample from a universe of similar items to which we wish to generalize our findings. In this case we should consider "items" to be "random" effects. In another application, we might be interested only in the particular set of items chosen for study because they represent a questionnaire that is to be used repeatedly without alteration (perhaps in a diagnostic setting). In this case there is no intention of generalizing beyond the particular set of items, so they are regarded as "fixed" effects. The implications of the distinction between fixed and random effects for the design and analysis of psychometric investigations are considered in greater detail by Cronbach *et al.* (1963).

Clearly, whether or not particular components of unreliability contribute to the variation among the test scores of a random sample of subjects will depend on the generalizations that we wish to make from a particular study. If we regard both occasions and items as fixed effects then the only source of unreliability in subjects' test scores will be error in the strict sense, σ^2. In such a case, corrections for unreliability that involve $\sigma^2_{S \times I}$ and $\sigma^2_{S \times O}$ could be misleading. If, however, we regard occasions and/or items as random effects then corrections for σ^2 and/or $\sigma^2_{S \times I}$ will be necessary in addition to that for σ^2. In either case we could be interested in a more detailed analysis of the interaction, since it may reveal additional behavioral traits that could become the objects of further investigation. A preliminary study, such as ours, could identify at least the major areas of concern by distinguishing interactions that are predominantly genetic in origin from those that have an environmental basis. Should we wish to regard occasions and/or items as random effects, our approach gains further force because no correction of genetic parameters of the variation between subjects is possible until it has been established whether the interaction components are themselves genetic or environmental. It is sometimes supposed that unreliability contributes to environmental sources of variation between subjects and thus leads to the underestimation of heritability. This is necessarily the case only when unreliability represents error in the strict sense. In other cases, where unreliability is estimated from interaction components, "genetic" unreliability will be confounded with genetic variation between subjects' scores, and "environmental" unreliability will be confounded with the environmental differences between subjects, unless an experiment is specifically designed to estimate the relevant components of the interaction in addition to the variation between subjects.

9.2 THE DATA

Prior to administering the EPQ to the twins in the London study, an earlier investigation had used an earlier form of the EPQ (the "Personality Inventory", PI) in the same population. Many twin pairs received and returned both questionnaires. Although the questionnaires differ considerably for most of their items, we could identify 11 items related to neuroticism (N) that were formulated identically in the two questionnaires. The interval between the administration of the EPQ and PI was approximately two years. Altogether, 441 pairs of twins completed both questionnaires. The structure of the twin sample is given in Table 9.1.

The 11 N items analysed in this study are given in Table 9.2. Since the scale is such that "Yes" answers are keyed for neuroticism, a simple analysis of the individual responses that identified sources of variation due to subjects

Table 9.1 Structure of the twin sample: numbers of pairs completing both questionnaires.

	Monozygotic	Dizygotic	Total
Male	51	25	76
Female	202	104	306
Opposite-sex	—	59	59
Total	253	188	441

Table 9.2 Neuroticism items common to both questionnaires and analyzed in this study.

Item no.		
PI	PQ	
1	10	3	Does your mood often go up and down?
2	14	7	Do you ever feel "just miserable" for no good reason?
3	22	12	Do you often worry about things you should not have done or said?
4	25	20	Are your feelings rather easily hurt?
5	28	94	Are you sometimes bubbling over with energy and sometimes very sluggish?
6	35	28	Are you often troubled by feelings of guilt?
7	38	44	Would you call yourself tense or "highly strung"?
8	58	16	Are you an irritable person?
9	61	40	Do you worry about awful things that might happen?
10	70	32	Would you call yourself a nervous person?
11	78	52	Do you worry about your health?

and their interaction with items and occasion will partition the total variation into that attributable to neuroticism and the various components of
unreliability.

9.3 ANALYSIS OF VARIANCE

In Table 9.3 we present the mean squares of the analyses of variance of the
five twin groups, in which we recognize the hierarchical classification of subjects into pairs and individuals within pairs. The within-pairs items for the
opposite-sex dizygotic twins (DZ_{OS}) have been corrected for the appropriate
effects due to sex. There is a significant difference between sexes for N and
the sexes differ in their mean responses to the particular items. However,
there is no significant interaction of sexes and occasions of testing.

We now consider in more detail the implications of the analyses of
variance in Table 9.3. We could proceed directly to fitting a genetic model to
the mean squares, but some preliminary considerations will assist in deciding
on an appropriate model for the data.

There are ten independent mean squares involving the triple interaction of
subjects, items and occasions. These are the mean squares I × P × O and I
× W × O for each of the five groups of twins. They all appear remarkably
consistent, and, in fact, are so when tested for heterogeneity (X^2_9 = 11.96, P
= 0.22). This finding supports our interpretation of the triple interaction as

Table 9.3   Mean squares of analyses of variance within twin groups.

Item	MZ_f df	MZ_f ms	MZ_m df	MZ_m ms	DZ_f df	DZ_f ms	DZ_m df	DZ_m ms	DZ_{mf} df	DZ_{mf} ms
Between items (I)	10	29.4103	10	6.4225	10	14.4911	10	2.1527	10	9.5270
Between pairs (P)	201	1.9752	50	1.6501	103	1.1340	24	1.0608	58	1.4417
Within pairs (W)	202	0.5811	51	0.5673	104	1.0490	25	1.2382	58	1.0314
Between occasions (O)	1	2.4979	1	0.0218	1	5.4554	1	0.1782	1	0.7458
I × P	2010	0.2937	500	0.2716	1030	0.2653	240	0.2654	580	0.2451
I × W	2020	0.1631	510	0.1791	1040	0.2158	250	0.2202	580	0.1999
I × O	10	0.5199	10	0.0767	10	0.2593	10	0.1062	10	0.2186
P × O	201	0.1935	50	0.2382	103	0.2083	24	0.2255	58	0.1642
W × O	202	0.1837	51	0.1555	104	0.1888	25	0.1691	58	0.1974
I × P × O	2010	0.1075	500	0.1051	1030	0.1058	240	0.1056	580	0.0983
I × W × O	2020	0.0939	510	0.1006	1040	0.1056	250	0.0991	580	0.0995
Subjects × occasions (pooled over all groups)					876	0.1913				
Items × subjects × occasions (pooled over all groups)					8760	0.1020				

Mean squares within pairs are corrected for sex difference.

error in the strict sense, since heterogeneity would be detected if there were any genetic component of the interaction because the within-pairs items for the DZ twins would exceed those of MZ twins or if there were environmental effects common to members of each pair. In the latter case, we would expect there to be equal and significant components for the triple interaction between pairs, irrespective of zygosity. The only remaining doubt is whether environmental influences specific to individuals could inflate our estimate of error. Of this we have no test with the present design, so we have pooled our ten mean squares to give a joint estimate of σ^2.

It is legitimate to combine the ten interactions of subjects and occasions since these too are homogeneous ($X^2_9 = 4.06$, $P = 0.91$). The pooled S × O interaction mean square, however, is significant when tested against our estimate of σ^2 ($X^2_{876} = 1643.07$, $P < 10^{-6}$), so there is a real interaction of subjects and occasions. Because the magnitude of the interaction components depends neither on zygosity nor on grouping of subjects into pairs, we can interpret such interaction as the result of experiences or endogenous behavioral fluctuations that are specific to individuals irrespective of their genotype or the shared experiences of twin pairs.

For the interaction between subjects and items, and for the variation between subjects, we obtain a different result. A preliminary investigation of the S × O interactions suggests that they are all significant when tested against the pooled error, but that they are not homogeneous ($X^2_9 = 216.39$, $P < 10^{-6}$). We see that the mean squares between MZ pairs are consistently greater than those between DZ pairs and that the reverse is true for the mean squares within pairs. This is consistent with the interaction having at least some genetic basis. A similar pattern emerges for the variation between subjects (i.e. for "neuroticism"), but this can be interpreted without reference to the inconsistency of the test only if we are prepared to regard test items as fixed effects.

A simplified statistical model for the mean squares of a typical analysis is given in Table 9.4. The expected mean squares take into account the kind of generalizations that we might make from the study. In our treatment we have assumed throughout that both subjects and occasions represent random effects, but we have indicated the expectations on both random and fixed models with respect to items. In Table 9.5 we provide the estimated components of variance of the individual responses calculated on the basis of both models for the five groups of twins. Bearing in mind the large errors inevitably associated with estimated components of variance, the estimates that are expected to be similar are quite consistent.

Table 9.4 Expectations of mean squares on statistical model.[a]

Items	σ^2		$+2n\sigma^2_{I\times O}$	$+2\sigma^2_{I\times W}$	$+4\sigma^2_{I\times P}$		$+4n\sigma^2_I$
Pairs	σ^2	$+11\sigma^2_{S\times O}$		$\mathbf{2\sigma^2_{I\times W}}$	$\mathbf{4\sigma^2_{I\times P}}$	$+22\sigma^2_W + 44\sigma^2_P$
Within pairs	σ^2	$+11\sigma^2_{S\times O}$		$+2\sigma^2_{I\times W}$		$+22\sigma^2_W$
Occasions	σ^2	$+11\sigma^2_{S\times O}$	$+2n\sigma^2_{I\times W}$				$+22n\sigma^2_O$
I × P	σ^2			$+2\sigma^2_{I\times W}$	$+4\sigma^2_{I\times P}$	
I × W	σ^2			$+2\sigma^2_{I\times W}$		
I × O	σ^2		$+2n\sigma^2_{I\times O}$			
Pooled subjects × O	σ^2	$11\sigma^2_{S\times O}$				
Pooled error (S × I × O)	σ^2					

[a] Components in bold type appear only if items are not considered fixed.

Table 9.5 Neuroticism: components of variation of individual responses to 11 items on two occasions.

	Item	MZ_f	MZ_m	DZ_f	DZ_m	DZ_{os}
Pooled	σ^2		0.10197			
	$\sigma^2_{S \times O}$		0.00812			
Items fixed or random	$\sigma^2_{I \times O}$	0.00103	—	0.00076	0.00008	0.00099
	$\sigma^2_{I \times W}$	0.03057	0.03857	0.05692	0.05912	0.04897
	$\sigma^2_{I \times P}$	0.03265	0.02313	0.01238	0.01130	0.01130
Items fixed	σ^2_O	0.00520	—	0.00230	—	0.00043
	σ^2_W	0.01772	0.01709	0.03899	0.04759	0.03819
	σ^2_P	0.03168	0.02461	0.00193	—	0.00933
	k^2_I	0.03552	0.03015	0.03382	0.01883	0.03884
Items random	σ^2_O	0.00043	—	0.00223	—	0.00034
	σ^2_W	0.01494	0.01359	0.03381	0.04221	0.03374
	σ^2_P	0.02872	0.02169	0.00081	—	0.00830
	σ^2_I	0.03552	0.03015	0.03382	0.01883	0.03884

9.4 GENETIC ANALYSIS

The estimates in Table 9.5 still do not represent the most parsimonious summary of the data. We may reparametrize our expectations of mean squares for the five analyses of variance in terms of a simple genetic model that makes explicit certain theoretical relationships existing between the components of variance of different analyses if the model is appropriate. In this case we specify genetic parameters for variation in neuroticism (V_{AN}) and for the interaction of subjects and items ($V_{AS \times I}$), and we specify environmental parameters for variation in neuroticism (E_{WN}), for the interaction of subjects and items ($E_{WS \times I}$), for the interaction of subjects and occasions ($E_{WS \times O}$), and for error (σ^2). In Table 9.6 we give expectations for the components of variance in terms of our simple model and in Table 9.7 the expectations for the relevant mean squares of the analyses of variance in terms of our genetic and environmental components. The assumptions that are made in this model are given in detail elsewhere (Chapter 5). The principal assumptions are: random mating; additive genetic effects; environmental influences specific to individuals rather than common to twin pairs; no effects of sex-linkage or sex-limitation.

The model that we are fitting is therefore the simplest possible model for the combined action of genetic and environmental influences. The method of weighted least squares was used to obtain estimates of the parameters and to provide a test of the model (see Chapters 4 and 5). Examination of the raw mean squares (Table 9.3) suggests that, while the data are generally consistent over sexes and for twins of both types, there are anomalies such as the negative intraclass correlation for DZ males. The X^2_{16} for testing the goodness of fit of the model was 19.03 ($P = 0.27$), indicating that our simple model gives quite an economical account of the variation in individual responses to the questionnaire in spite of the minor anomalies. The estimates of the parameters and their standard errors are given in Table 9.8. In Table 9.9 we summarize the contributions of the different sources to the variation in individual responses.

From the appropriate components of the fixed- or random-item models, we may estimate any desired reliability coefficient and obtain values of the heritability of the trait and its inconsistency. The "true" heritability of neuroticism is thus

$$h^2_N = \hat{V}_{AN}/(\hat{V}_{AN} + \hat{E}_{WN}).$$

This ratio represents the proportion of variance in subjects' responses to a randomly chosen item that is due to genetic differences in the neuroticism "trait". We may also estimate the contribution of genetic factors to the interaction of subjects and items from

$$h^2_{S \times I} = \hat{V}_{AS \times I}/(\hat{V}_{AS \times I} + \hat{E}_{WS \times I}).$$

We see that the contribution of genetic factors to responses at the item level reflects both "trait" and "interaction" effects and that the contribution of genetic effects to P × S interaction is comparable to the contribution of genes to differences at the trait level.

9.5 DO GENES AFFECT SHORT-TERM PERSONALITY CHANGE?

In Chapter 7 we showed that the expression of genetic differences apparently changed with age. The analysis of changes over a two-year period described in this chapter suggests that short-term fluctuations are not affected by genetic differences because all the interaction of subjects and occasions is caused by environmental effects specific to individuals (E_W). This analysis, however, takes account of both the magnitude and *direction* of change. If the effects of genes on personality change were mediated by making some individuals more sensitive to their environment than others then we might expect the behavior of some of the genotypes to fluctuate more widely than others, even though there might be *no correlation in the direction of change*. Our data allow us to test for genetic effects on sensitivity to the environment. In Table 9.10 we present the mean squares and intraclass correlations for the

Table 9.6 Genetic and environmental model for variance components.

Component	Parameter					
	V_{AN}	E_{IN}	$V_{AS \times I}$	$E_{WS \times I}$	$E_{WS \times O}$	E_W
σ^2	.	.	.	.	.	1
$\sigma^2_{S \times O}$	.	.	.	.	1	.
$\sigma^2_{I \times MZW}$	.	.	.	1	.	.
$\sigma^2_{I \times MZP}$	.	.	1	.	.	.
$\sigma^2_{I \times MZW}$	.	1	.	.	.	.
σ^2_{MZP}	1	.	.	.	.	.
$\sigma^2_{I \times DZW}$	1	1	$\frac{1}{2}$	1	.	.
$\sigma^2_{I \times DZP}$	.	.	$\frac{1}{2}$	.	.	.
σ^2_{DZW}	$\frac{1}{2}$	1	.	.	.	.
σ^2_{DZP}	$\frac{1}{2}$	.	.	.	.	.

Table 9.7 Expectations of relevant mean squares on simple genetic and environmental model.[a]

Mean square	Observed	df	Information	E_W	$E_{WS \times O}$	$E_{WS \times I}$	$V_{AS \times I}$	E_{WN}	V_{AN}
MZP_f	1.9752	201	25.7599	1	11	**2**	**4**	22	44
MZP_m	1.6501	50	9.1816	1	11	**2**	**4**	22	44
DZP_f	1.1340	103	40.0480	1	11	**2**	**3**	22	33
DZP_m	1.0608	24	10.6639	1	11	**2**	**3**	22	33
DZP_{os}	1.4417	58	13.9524	1	11	**2**	**3**	22	33
MZW_f	0.5811	202	299.1022	1	11	**2**	.	22	.
MZW_m	0.5673	51	79.2346	1	11	**2**	.	22	.
DZW_f	1.0490	104	47.2555	1	11	**2**	**1**	22	11
DZW_m	1.2382	225	8.1532	1	11	**2**	**1**	22	11
DZW_{os}	1.0314	58	27.2611	1	11	**2**	**1**	22	11
$I \times MZP_f$	0.2937	2 010	11 650.8650	1	.	**2**	**4**	.	.
$I \times MZP_m$	0.2716	500	3 389.0695	1	.	**2**	**4**	.	.
$I \times DZP_f$	0.2653	1 030	7 316.9945	1	.	**2**	**4**	.	.
$I \times DZP_m$	0.2654	240	1 703.6462	1	.	**2**	**3**	.	.
$I \times DZP_{os}$	0.2451	580	4 827.3788	1	.	**2**	**3**	.	.
$I \times MZW_f$	0.1631	2 020	37 967.6268	1	.	**2**	.	.	.
$I \times MZW_m$	0.1791	510	7 949.6683	1	.	**2**	.	.	.
$I \times DZW_f$	0.2158	1 040	11 166.0730	1	.	**2**	**1**	.	.
$I \times DZW_m$	0.2202	250	2 577.9553	1	.	**2**	**1**	.	.
$I \times DZW_{os}$	0.1999	580	7 257.2554	1	.	**2**	**1**	.	.
$S \times O$	0.1913	876	11 973.6293	1	11	.	.	.	.
Error	0.1020	8 760	421 239.677	1	.	.	.	.	.

[a] Coefficients in bold type do not apply if items are regarded as fixed.

Table 9.8 Estimates of genetic and environmental components.

Parameters	$\hat{\theta}$	$\sigma_{\hat{\theta}}$
V_{AN}, items fixed	0.0247	0.0030
V_{AN}, items random	0.0439	0.0030
E_{WN}, items fixed	0.0186	0.0022
E_{WN}, items random	0.0155	0.0023
$S \times I$	0.0030	0.0021
$E_{WS \times I}$	0.0338	0.0022
$E_{WS \times O}$	0.0081	0.0008
E_W	0.1020	0.0015

Table 9.9 Percentage of total variation in individual responses attributed to random sources.

	"Neuroticism"	Inconsistency	Unrepeat-ability	Error	Total
	Items fixed				
Genetic (V_A)	11.36	13.97	—	—	25.33
Environmental (E_W)	8.54	15.52	3.73	46.88	74.67
Total	19.90	29.49	3.73	46.88	
h^2	0.57	0.47	—	—	—
	Items random				
Genetic (V_A)	10.37	14.35	—	—	24.72
Environmental (E_W)	7.32	15.94	3.84	48.17	75.27
Total	17.69	30.29	3.84	48.17	
h^2	0.59	0.47	—	—	—

Table 9.10 Analyses of variance of absolute-change scores.

	MZ_f		MZ_m		DZ_f		DZ_m		DZ_{os}	
Item	ms	df	ms	df	ms	df	ms	df	ms	df
Between pairs	1.66	201	2.22	50	2.19	102	1.06	24	1.92	58
Within pairs	2.23	202	1.66	51	2.71	103	3.00	25	1.59	59
Total	1.95	403	1.94	101	2.45	205	2.04	49	1.75	117
$r_{intraclass}$	−0.147		0.144		0.106		−0.478		−0.094	

five groups of twins. The total variances of the groups are homogeneous ($X^2_4 = 5.47$).

The correlations suggest that no simple model is likely to fit the change scores, since, apparently, two of the correlations are significantly negative. We must proceed tentatively because of non-normality, but the correlations are nearly heterogeneous at the 5% level ($X^2_4 = 9.32$), and the best-fitting pooled correlation is -0.09, which again approaches significance at the 5% level. The results clearly provide no suggestion that sensitivity to short-term environmental effects is under genetic control. Indeed, if we take the negative correlations seriously, we have to adopt an explanation in terms of social interaction between twins. The simplest and most conservative interpretation of the small average correlation in the absolute change scores of twins is that they are the result of random environmental effects that are unique to the individual.

9.6 DISCUSSION

The analysis illustrates two important points. First, even the interaction between subjects and situations may have a genetic component. Secondly, the interaction may have a genetic basis for some situations (in our case "items") and environmental for others ("occasions"). Only a genetically informative design can resolve the alternative mechanisms underlying the interaction. The study is deficient in two respects. First, as an illustration of P × S interactions, the selection of items is poor because it is based purely on those neuroticism items that were included in both versions of a personality inventory. A better questionnaire, employing more explicit situations along the lines suggested by Endler and Hunt (1976), would yield more challenging results. Secondly, the analysis ends with the detection of "interaction" and does not address the ways in which the description of interaction and its genetic basis may be parametrized more economically. Our defense of the latter weakness is simply that there are limits to which we should push the analysis of interaction in the analysis of variance of dichotomous items.

The fact that V_{AS} and $V_{AS \times I}$ are significant indicates that there is significant genetic variation both for neuroticism itself and for the interaction of subjects and items. The former conclusion confirms the replicated finding of Chapter 5. The latter conclusion suggests that a strictly unitary model for the genetics of neuroticism measures may not be appropriate. In our case we find that the particular pattern of "symptoms" revealed by responses to a questionnaire is itself partly genetically determined, as well as the overall predisposition to neuroticism. For other aspects of behavior we might expect a different picture. For social attitudes, for example, in which some genetic determination of the factors is indicated (Chapters 14–16), we should not be surprised if the inconsistency of subjects' responses reflected cultural and specific environmental influences.

Our finding that the interaction of subjects and items is partly under genetic control may surprise some social psychologists at first, but such findings are by no means unusual in quantitative genetics. It is regularly seen (see e.g. Mather, 1953; Jinks and Mather, 1955; Paxman, 1956) that variation between measurements of replicated structures in the same organism is a function of genetic differences and may be subject as much to the influence of natural selection as any other genetically determined trait.

From a more formal aspect, our study shows that attempts to correct heritability estimates for unreliability of measurement that ignore the genetic component of unreliability may lead to overestimation of the genetic component of variation among subjects' true scores. Whatever the source of such interaction, the usual correction for unreliability will overestimate the contribution of error to the variation of subjects' scores on a fixed set of items. If, in addition, the inconsistency has a genetic component, it will be wrong to assume that inconsistency in the responses of subjects to a random collection of items contributes only to our estimate of environmental variation. Similar considerations may apply to lack of test repeatability of traits in which developmental factors operate between occasions of testing. Wilson (1972) has illustrated the genetic control of developmental profiles for one behavioral trait. Misleadingly high heritability estimates could result from inappropriate corrections for unreliability. There is therefore no substitute for an appropriate complete experimental design and genetic analysis if the variation in test responses is to be assigned to the appropriate genetic and environmental sources.

Given measurements that were continuous at the item level, rather than dichotomous, we could have examined in greater detail the genetic and environmental contributions to individual components of the interaction. Such an attempt is ill-advised with our data. However, in the next chapter we shall consider how an appropriate analysis of continuous multivariate data can resolve the genetic and environmental components of common and specific factors.

Chapter 10

Genetic and Environmental Covariance Between Traits

Multivariate methods, particularly factor analysis, have played a major role in the study of individual differences in behavior. Such methods have not always been used to the best advantage, for want of a psychological theory to govern the choice of measurements and generate testable predictions about the outcome of a particular study.

Eysenck's personality theory makes a number of types of predictions involving, for example, measurement; physiology; learning; psychiatric disorder; inheritance (see e.g. Eysenck, 1952a; 1967a, b; Chapter 2). The predictions of such a theory are both genetic and inherently multivariate. In this chapter we therefore consider how hypotheses about genetic and environmental causes of variation can be formulated in a multivariate context and show how they may be tested with a variety of personality measures on twins.

As early as 1951, Eysenck and Prell argued that the hypothesized dimension of neuroticism, which accounted for much of the variation and covariation between objective personality tests, was a fundamental property of the nervous system. The differences that created this dimension were encoded in the genotype rather than acquired by experience. Thus, not only did the model for neuroticism predict a common factor underlying a wide range of physiological and behavioral measures, it also predicted that the correlation between measures was genetic and therefore that the common factor would be significantly more heritable than the component measures. A simple version of the mathematical argument behind this prediction is presented in Chapter 8. Eysenck and Prell attempted to test this prediction in a pioneering study, but the samples were too small to yield convincing results. If the underlying common factor were genetic, they argued, then the first common factor underlying a set of tests should be amenable to a rotation in which the factor loadings were proportional to the heritabilities of the component tests. The same idea was used in Chapter 8, when we correlated the loadings on the extraversion factor with the estimated contributions of dominance to the individual items of the EPQ.

The recognition that there is a "genotype–environmental" facet to the multivariate study of individual differences is also found in some of Cattell's work. As long as we restrict multivariate analysis to samples of unrelated individuals then any model that we devise for the correlation between tests is merely describing phenotypic covariation. The factor loadings that we derive are aggregates of genetic and environmental causes. Cattell argued that the aspects of behavior included in a multivariate study may be divisible into "source" and "environmental mold" traits (see e.g. Cattell, 1960). This distinction between genetic and environmental sources of cognitive variation is also implicit in Cattell's theory of "fluid" (inherited) and "crystalized" (acquired) intelligence (Cattell, 1963b).

If the underlying factor structure is so simple that some variables are affected only by genes and others only by environment then it is possible for a factor analysis of the intertest correlations to recover the genetic and environmental components as separate factors. If genes and environmental causes both operated on the same pair of underlying variables then we could get an identical pattern between tests at the phenotypic level. In either case, we shall not be able to identify "genetic" and "environmental" factors without a genetic study, and the only way to distinguish these two hypotheses for the same set of variables is to conduct a study in which the genetic and environmental causes of test covariation can be resolved.

10.1 GENETIC AND ENVIRONMENTAL COVARIANCE: A BIVARIATE STUDY

10.1.1 A bivariate model for genetic and environmental effects

Animal and plant breeders commonly speak of "genetic correlation" between measures. The measures that we make do not correspond directly to specific genetic effects, but rather many genes may have effects on several variables and different loci may have effects on different variables. In so far as genes affect more than one variable, there is the opportunity for genetic correlation between measures. In reality, things may not be so simple. We have described genetic correlations as if they were only caused by the pleiotropic effects of genes on more than one variable. However, correlations between the effects of different genes may not tell us anything fundamental about the mechanism of gene action so much as tell us that there are factors at work in a population creating associations between genes affecting different traits. Assortative mating for a combination of traits that are under the control of quite distinct genes, for example, may nevertheless produce a genetic correlation between the measures (see e.g. Eaves *et al.*, 1984). Linkage disequilibrium resulting from selection can do the same; so can unrecognized racial heterogeneity within a population. Some of these causes of genetic correlation can be distinguished. For example, if genes are unlinked then the effect of assortative mating on the genetic correlation is felt only in genetic differences between sibships rather than within families.

In addition, the pleiotropic effects of genes on a pair of variables may not be consistent over loci. Thus some genes that increase one trait may increase expression of another. Other genes may also increase expression of the first trait and decrease expression of the second. A zero genetic correlation does not therefore necessarily imply that the genes do not have pleiotropic effects, merely that the effects of different loci cancel one another out. In analysing genetic covariation, we are therefore able only to examine the *net* correlation between gene effects over variables.

For any pair of variables i and j, we may define the additive and dominance effects of each locus as follows:

genotype	AA	Aa	aa
effect on trait i	$m_i + d_{ai}$	$m_i + h_{ai}$	$m_i - d_{ai}$
effect on trait j	$m_j + d_{aj}$	$m_j + h_{aj}$	$m_j - d_{aj}$

The model is identical in form with that used in representing the sex-limited effects of genes (Chapter 5) and the developmental consistency of gene effects (Chapter 7). In the case of sex limitation the genetic covariance was defined across sexes. In the case of developmental changes the genetic covariance was defined over time. Now, in the context of multivariate analysis, the genetic covariance is defined across traits.

Eaves and Gale (1974) show that we may define the contribution of all loci to the genetic *covariance* exactly as Mather (1949) and his colleagues (see e.g. Mather and Jinks, 1982) or Falconer (1960) have done in terms of the frequencies and effects of all the loci affecting the traits. Thus we may define additive and dominance components of variance for the two traits:

$$
\begin{array}{lll}
 & \text{additive} & \text{dominance} \\
\text{trait } i: & V_{Ai} = \sum_a 2u_a v_a [d_{ai} + (v_{ai} - u_{ai})h_{ai}]^2 & V_D = 4u_a^2 v_a^2 h_{ai}^2, \\
\text{trait } j: & V_{Aj} = \sum_a 2u_a v_a [d_{aj} + (v_a - u_a)h_{aj}]^2 & V_{Dj} = 4u_a^2 v_a^2 h_{aj}^2.
\end{array}
$$

Similarly, the cross-products of the gene effects on each trait, accumulated over all loci, yield expectations for the additive and dominance components of genetic *covariance* between the traits:

additive genetic covariance:

$$V_{Aij} = \sum_a 2u_a v_a [d_{ai} + (v_a - u_a)h_{ai}][d_{aj} + (v_a - u_a)h_{aj}],$$

dominance genetic covariance:

$$V_{Dij} = \sum_a 4u_a^2 v_a^2 h_{ai} h_{aj}.$$

If the product terms, e.g. $d_{ai} d_{aj}$, are generally positive over all loci then there will be a positive genetic covariance; if they are generally negative then the genetic covariance will be negative; if the products are not consistent over all loci but the positive products are offset by negative products then the genetic covariance will be zero even if the genes have widespread pleiotropic effects.

An analogous argument to the above may be used to express the relationships between variables caused by environmental effects. We may define environmental components of variance and covariance within and between families:

	environmental variances		covariance
	trait i	trait j	traits i and j
within families:	E_{Wi}	E_{Wj}	$E_{Wij} \leq (E_{Wi}E_{Wj})^{1/2}$,
between families:	E_{Bi}	E_{Bj}	$E_{Bij} \leq (E_{Bi}E_{Bj})^{1/2}$.

10.1.2 Summarizing multivariate twin data

In this chapter we shall be considering only the analysis of multivariate twin data. Thus each kinship is of a constant structure (i.e. twin pairs). The summary of twin data for pairs of variables is therefore analogous to the univariate data summary employed in the analysis of the individual variables. In the examples presented in this chapter we began with the multivariate analogue of the one-way analyses of variance described in Chapter 4. For each group of twins, matrices of mean products were computed within and between twin pairs. The mean squares formed the diagonal elements of the mean-products matrices, and the mean products between variables formed the off-diagonal elements. For a typical pair of variables, the structure of the analysis is illustrated in Table 10.1.

Mean squares corresponding to those in the table may be computed for each variable, and mean products for each pair of variables. Just as the expected mean squares could be written for the univariate case (Chapter 4), so the mean products within pairs and between pairs have expectations in terms of the population covariance matrices within and between pairs. However, we have already shown how the ten components of variance represented in the standard components of variance model for the five groups of

Table 10.1   Expected mean squares and mean products from analysis of bivariate twin data.

Item	df	Variable i mean squares (MS)	Variable j mean squares (MS)	Variables i and j mean products (MP)
Between pairs	$n-1$	$\sigma_{Wi}^2 + 2\sigma_{bi}^2$	$\sigma_{Wj}^2 + 2\sigma_{bj}^2$	$\omega_{Wij} + \omega_{bij}$
Within pairs	n	σ_{Wi}^2	σ_{Wj}^2	ω_{Wij}

Table 10.2   The bivariate extension of the V_A, E_W model for twin data.[a]

Item	Expected MS$_i$	Expected MS$_j$	Expected MP$_{ij}$
Between MZ pairs	$E_{Wi} + 2V_{Ai}$	$E_{Wj} + 2V_{Aj}$	$E_{Wij} + 2V_{Aij}$
Within MZ pairs	E_{Wi}	E_{Wj}	E_{Wij}
Between DZ pairs	$E_{Wi} + \frac{3}{2}V_{Ai}$	$E_{Wj} + \frac{3}{2}V_{Aj}$	$E_{Wij} + \frac{3}{2}V_{Aij}$
Within DZ pairs	$E_{Wi} + \frac{1}{2}V_{Ai}$	$E_{Wj} + \frac{1}{2}V_{Aj}$	$E_{Wij} + \frac{1}{2}V_{Aij}$

[a] Parameters are defined in the text.

twins may often be reduced very effectively to only one or two when a genotype–environmental model is specified. In an exactly analogous fashion, we may specify the contribution of genetic and environmental variance components to the mean squares and mean products derived from multivariate data. For the bivariate case typical expectations of mean products are given for MZ and DZ twins in Table 10.2. It is assumed, for simplicity, that the genetic and environmental effects are constant over sexes. Other, more complicated, models may be specified and will be described subsequently.

Given such a data summary, and an appropriate model of the type illustrated in Table 10.2, it is relatively straightforward to employ the methods of maximum likelihood or weighted least squares to obtain estimates of the genetic and environmental covariances and their standard errors. Likelihood-ratio tests may be performed to compare alternative models, just as we did for the univariate case, and to provide tests of goodness of fit.

In applying the method of weighted least squares, the only important factor to bear in mind is that the statistics in the original data summary are no longer independent. Thus, although the different mean-products matrices are independent, the mean squares and mean products within a matrix are not because each observation contributes both to mean squares and a number of mean products. While in the univariate case the weight matrix is diagonal, in the multivariate case it may be represented only in block-diagonal form. The variances and covariances of variances and covariances needed to compute the weight matrix may be obtained from standard statistical theory (see Kendall and Stuart (1977), and Eaves and Gale (1974) for a multivariate genetic application).

10.2 A BIVARIATE EXAMPLE: THE NATURE OF EXTRAVERSION

One of the central problems in personality research has been the question of whether such higher-order factors as extraversion can be regarded in any meaningful sense as *unitary* or whether there are several independent factors, such as "sociability" and "impulsiveness", which should not be thrown together artificially. Carrigan (1960) concluded her survey of the literature by saying that "the uni-dimensionality of extraversion/introversion has not been conclusively demonstrated" (p. 355); she further pointed out that several joint analyses of the Guildford and Cattell questionnaires show that at least *two* independent factors are required to account for the intercorrelations between the extraversion–impulsiveness variables. These two factors, she suggested, may correspond to the European conception of extraversion, with its emphasis on impulsiveness and weak superego controls, and the American conception, with its emphasis on sociability and ease in interpersonal relations. Eysenck and Eysenck (1963) have reported quite sizable correlations between sociability and impulsiveness, a conclusion replicated by Sparrow and Ross (1964); this would suggest that there is a close connection between the two conceptions (H.J. Eysenck & S.B.J. Eysenck, 1969). Furthermore, Eysenck and Eysenck (1967, 1969) have shown that the correlations of extraversion items (whether sociability or impulsiveness) with subjects' reactions on a physiological test devised on theoretical grounds were proportional to their loadings on the extraversion factor. The recognition that extraversion is a unitary factor in behavior is thus vindicated by prediction from a psychological theory as much as by a correlation between primary factors (Eysenck, 1967a).

   We now develop a model for the genetic and environmental determinants of extraversion and of its primary components, sociability and impulsiveness. Our intention is to analyze the phenotypic variation and covariation of sociability and impulsiveness into their genetic and environmental components in order to compare the unitary and dual models of extraversion with regard to their relative contributions to the representation of both genotypic and environmental determinants of variation among the responses of subjects to a personality inventory.

The research reported in Chapter 4 was mainly concerned with the analysis of extraversion as a unitary trait. Claridge *et al.* (1973) reported analyses of extraversion, sociability and impulsiveness, but their samples were too small to justify the kind of analysis that we attempt here.

10.2.1 Data and data summary

The analysis is based on the responses of 837 pairs of adult volunteer twins to the 80-item personality inventory (PI). Of these items, 13 formed a scale of sociability, and 9 were scored to provide a measure of impulsiveness. The relevant items are given in Table 10.3.

Table 10.3 Personality inventory items measuring sociability and impulsiveness.

Item	Key[a]
18. Do you suddenly feel shy when you want to talk to an attractive stranger?	− S
23. Generally, do you prefer reading to meeting people?	− S
27. Do you like going out a lot?	+ S
30. Do you prefer to have few but special friends?	− S
36. Can you usually let yourself go and enjoy yourself at lot at a gay party?	+ S
40. Do other people think of you as being very lively?	+ S
44. Are you mostly quiet when you are with other people?	− S
48. If there is something you want to know about, would you rather look it up in a book than talk to someone about it?	− S
56. Do you hate being with a crowd who play jokes on one another?	− S
66. Do you like talking to people so much that you never miss a chance of talking to a stranger?	+ S
69. Would you be unhappy if you could not see lots of people most of the time?	+ S
75. Do you find it hard to really enjoy yourself at a lively party?	− S
77. Can you easily get some life into a rather dull party?	+ S
1. Do you often long for excitement?	+ I
4. Are you usually carefree?	+ I
8. Do you stop and think things over before doing anything?	− I
12. Do you generally say things quickly without stopping to think?	+ I
16. Would you do almost anything for a dare?	+ I
20. Do you often do things on the spur of the moment?	+ I
33. When people shout at you, do you shout back?	+ I
59. Do you like doing things in which you have to act quickly?	+ I
62. Are you slow and unhurried in the way you move?	− I

[a] S denotes an item scored for sociability, I for impulsiveness; + indicates that "Yes" scored 1, and − indicates that "No" scored 1 for the scale under consideration.

Table 10.4   Structure of twin sample in study of sociability and impulsiveness.

Twin type	Number of pairs
Monozygotic female	331
Monozygotic male	120
Dizygotic female	198
Dizygotic male	59
Unlike-sex dizygotic	129

Table 10.5   Mean sociability and impulsiveness scores of twin groups.

		Mean	
Twin type [a]	N	Sociability	Impulsiveness
MZ_f	662	6.5045	3.7039
MZ_m	240	5.7875	3.8125
DZ_f	396	6.6869	3.7525
DZ_m	118	6.6441	4.0678
DZ_{os}	258	6.4884	3.7054

[a] Abbreviations are as follows: MZ_f, monozygotic female; MZ_m, monozygotic male: DF_f, dizygotic female; DZ_m, dizygotic male; DZ_{os}, unlike-sex dizygotic.

Composition of the sample by sex and zygosity is given in Table 10.4.

The mean sociability and impulsiveness scores of the five groups are given in Table 10.5. An analysis of the variation between and within groups revealed highly significant (but substantively fairly small) differences between groups with respect to the sociability scores. The groups did not differ with respect to their mean impulsiveness scores. We shall regard the groups as representative of the same population as far as their means are concerned. The pooled standard deviations within groups were homogeneous, being 3.00 and 1.76 for sociability and impulsiveness, respectively.

Since we wished to minimize the possibility of spurious interaction between subjects and tests, we standardized the raw scores of the twins on both sociability and impulsiveness by dividing the scores by the corresponding average within-group standard errors. For each group of twins separately, the mean squares within pairs and between pairs were calculated for each of the standardized scales. The analogous within-pairs and between-pairs mean products were also calculated. The mean squares and mean products form the basic statistical summary for the analysis to follow (see Table 10.6)

We studied the inheritance of extraversion by analyzing the mean squares

Table 10.6 Mean squares and mean products within and between twin pairs for standardized sociability and impulsiveness scores.[a]

Item	df	Mean square S	I	Mean product S − I
Between MZ$_f$ pairs	330	1.5339	1.3777	0.6517
Within MZ$_f$ pairs	331	0.5394	0.6403	0.1762
Between MZ$_m$ pairs	119	1.5595	1.2904	0.3126
Within MZ$_m$ pairs	120	0.4817	0.6630	0.1497
Between DZ$_f$ pairs	197	1.0855	1.1804	0.3069
Within DZ$_f$ pairs	198	0.8380	0.8408	0.2918
Between DZ$_m$ pairs	58	1.3919	0.9799	0.6309
Within DZ$_m$ pairs	59	0.6693	0.8441	0.1516
Between DZ$_{os}$ pairs	128	0.9457	1.2839	0.3638
Within DZ$_{os}$ pairs	129	0.9290	0.7697	0.3581

[a] Abbreviations are as follows: MZ$_f$, monozygotic female; MZ$_m$, monozygotic male; DZ$_f$, dizygotic female; DZ$_m$, dizygotic male; DZ$_{os}$, unlike-sex dizygotic; S, sociability; I, impulsiveness. No correction for the main effect of sex was necessary for the DZ$_{os}$.

derived from the subjects' total scores on the two standardized tests. The mean squares for the twins on the measure of extraversion (E) may be derived directly from the mean squares and mean products (MP) of Table 10.6, since $MS_E = MS_{(S+I)} = MS_S + MS_I + 2MP_{(S,I)}$, where S and I refer to sociability and impulsiveness.

Just as we obtained an E score for each subject by summing over tests, so we may obtain a difference (D) score for each subject by taking the difference between his scores on the standardized tests. The MS derived from these differences summarizes the variation arising because subjects do not perform consistently on the two tests. We may obtain the MS for the D scores directly from the raw MS and mean products of Table 10.6, since

$$MS_D = MS_{(S-I)} = MS_S + MS_I - 2MP_{(S,I)}.$$

The mean squares for E and D are given in Table 10.7.

Clearly, since the mean products are all positive, the MS_Es are larger than the corresponding MS_Ds. Since we are only concerned with these particular tests, the mean squares between subjects for E contain none of the interaction variation. Thus the fact that the MS_Es are approximately twice as large as the MS_Ds is an indication that E accounts for more of the total variation of the two tests than D.

We analyze the MS_E to provide a genetic model for variation in extraversion, and we analyze the MS_D to find how much genetic and environmental factors contribute to the resolution of E into sociability and

Table 10.7 Mean squares for extraversion and interaction of subjects and component tests.[a]

Item	df	Mean square	
		E = S + I	D = S − I
Between MZ$_f$ pairs	330	4.2150	1.6082
Within MZ$_f$ pairs	331	1.5321	0.8273
Between MZ$_m$ pairs	119	3.4751	2.2247
Within MZ$_m$ pairs	120	1.4441	0.8453
Between DZ$_f$ pairs	197	2.8797	1.6521
Within DZ$_f$ pairs	198	2.2624	1.0952
Between DZ$_m$ pairs	58	3.6336	1.1100
Within DZ$_m$ pairs	59	1.8166	1.2102
Between DZ$_{os}$ pairs	128	2.9572	1.5020
Within DZ$_{os}$ pairs	129	2.4749	0.9825

[a] Abbreviations are as follows: MZ$_f$, monozygotic female; MZ$_m$, monogyzotic male; DZ$_f$, dizygotic female; DZ$_m$, dizygotic male; DZ$_{os}$, unlike-sex dizygotic; E, extraversion; D, interaction of subjects and components tests; S, sociability; I, impulsiveness.

impulsiveness. Finally, we show that the covariation of sociability and impulsiveness reflects both genetic and environmental factors by an analysis of the raw mean squares and mean products of Table 10.6.

10.2.2 Model fitting to component scores

Genotype–environmental models, with and without sex-limited effects, were fitted by the weighted-least-squares method to the component sociability and impulsiveness scores, to the total extraversion score and to the standardized difference scores (S – I). When the V_A, E_W model was fitted there was no evidence of statistically significant sex differences in genetic and environmental effects, although the correlation between gene effects in males and females was very small (r_{AMF} = 0.04) for sociability scores, and the test of heterogeneity (X_3^2 = 7.9) approaches statistical significance. Table 10.8 presents the parameter estimates and tests of significance for the sex-limited V_A, E_W model.

Tables 10.9–10.12 give the results of fitting models that assume homogeneity of genetic and environmental effects across sexes for both of the components of extraversion, the total extraversion score and the differences between standardized sociability and impulsiveness scores.

For all the scales apart from impulsiveness (Table 10.10) the environmental models both fail. In all four cases the V_A, E_W model fits well.

Table 10.8   Results of allowing for sex limitation in V_A, E_W model for components of extraversion.

Parameter	Sociability	Impulsiveness	Extraversion	S − I
E_{WM}	0.47***	0.67***	1.43***	0.83***
E_{WF}	0.54***	0.65***	1.54***	0.84***
V_{AM}	0.53***	0.30***	1.14***	0.54***
V_{AF}	0.48***	0.37***	1.21***	0.43***
V_{AMF}	0.02	0.48**	0.46	0.59**
r_{AMF}	0.04	1.44	0.39	1.22
χ_5^2	4.4	0.7	4.2	7.3
$P\%$	49	98	52	20
Improvement:				
χ_3^2	7.9	1.2	3.2	1.9

Table 10.9   Model fitting to sociability scores: no sex limitation.

Model	E_W	E_B	V_A	V_D	V_{AS}	χ^2	df
E_W	1.01***	—	—	—	—	137.0***	9
E_W, E_B	0.67***	0.34***	—	—	—	44.3	8
E_W, V_A	0.54***	—	0.48***	—	—	11.3	8
E_W, E_B, V_A	0.51***	−0.23***	0.72***	—	—	7.0	7
E_W, V_A, V_D	0.51***	—	0.04	0.46	—	7.0	7
E_W, V_A, C_{AS}	0.52***	—	0.56***	—	−0.04$^+$	9.4	7

Table 10.10   Model fitting to impulsiveness scores: no sex limitation.

Model	E_W	E_B	V_A	V_D	C_{AS}	χ^2	df
E_W	1.00***	—	—	—	—	71.8***	9
E_W, E_B	0.73***	0.28***	—	—	—	9.4	8
E_W, V_A	0.64***	—	0.36***	—	—	1.9	8
E_W, V_A	0.65***	0.01	0.34***	—	—	1.9	7
E_W, V_A, V_D	0.65***	—	0.38*	−0.03	—	1.9	7
E_W, V_A, C_{AS}	0.65***	—	0.35***	—	0.00	1.9	7

Significance levels: *$P < 0.05$; **$P < 0.01$; ***$P < 0.001$

Although the environmental model fits the impulsiveness data well, the fit of the two-parameter genotype–environmental model is much better. In no case does the addition of a between-families environmental component (E_B) to the genetic model lead to a significant positive estimate of E_B. In the case of the sociability scores the estimate of E_B is negative and significant at the 5% level, suggesting a hint of dominance, but comparison with the sex-limited model in Table 10.8 suggests that the effects of dominance cannot be

Table 10.11   Model fitting to extraversion (S + I) scores: no sex limitation.

Model	E_W	E_B	V_A	V_D	C_{AS}	χ^2	df
E_W	2.71***	—	—	—	—	111.0***	9
E_W, E_B	1.86***	0.85***	—	—	—	31.0***	8
E_W, V_A	1.54***	—	1.17***	—	—	7.4	8
E_W, E_B, V_A	1.49***	−0.42	1.63***	—	—	5.1	7
E_W, V_A, V_D	1.49***	—	0.37	0.84	—	5.1	7
E_W, V_A, V_{AS}	1.50***	—	1.34***	−0.09	—	6.2	7

Table 10.12   Model fitting to difference (S − I) scores: no sex limitation.

Model	E_W	E_B	V_A	V_D	C_{AS}	χ^2	df
E_W	1.30***	—	—	—	—	84.6	9
E_W, E_B	0.94***	0.36***	—	—	—	19.0*	8
E_W, V_A	0.83***	—	0.47***	—	—	9.2	8
E_W, E_B, V_A	0.83***	−0.02	0.49**	—	—	9.2	7
E_W, V_A, V_D	0.83***	—	0.42+	0.04	—	9.2	7
E_W, V_A, V_{AS}	0.83***	—	0.48***	—	0.01	9.2	7

distinguished from the low correlation for unlike-sex twins, which could result from marked sex limitation of gene expression. In view of the discussion of dominance in Chapter 5, it is of some interest to note that the estimate of the dominance component of the extraversion scores (Table 10.11) is quite large, though admittedly not statistically different from zero. Again, it is difficult to disentangle the apparent effects of dominance from those of sex limitation in these data.

Overall, the results for the extraversion scores from the Personality Inventory correspond well with those reported for the EPQ extraversion scores in the London twins (see Chapter 5). The main determinants of individual differences appear to be genetic effects and within-family environmental differences. The possibility that the dominance/competition component of extraversion might be confined to the "sociability" component is consistent with an explanation in terms of competitive social interaction between twins based on availability of resources of social reinforcement, but the evidence is weak. The heritability of the total extraversion score turns out to be 0.42, but it is important to notice that the S − I difference score also has a significant genetic component (h^2 = 0.34) so the distinction between sociability and impulsiveness described by several authors cannot be attributed to environment alone, but the aspects of personality are somewhat distinct genetically. The fact that the heritability of the E scores is slightly higher than that of the S − I scores implies that the genetic covariance between measures of sociability and impulsiveness exceeds the environmental covariance. That is, genes are the main determinant of the phenotypic correlation between these aspects of personality.

10.2.3 The causes of correlation between the scales

The analysis of the sums and differences of the standardized sociability and impulsiveness scores reflects the underlying genetic and environmental covariance structure of the two subscales of extraversion. Rather than combine the scales according to *a priori* criteria, however, we may directly examine the variation and covariance of the component traits. Since the V_A, E_W model gave the best fit to the composite scores, we used the method of weighted least squares to estimate six parameters: V_{Ai}, V_{Aj}, V_{Aij}, E_{Wi}, E_{Wj} and E_{Wij}, from the 30 mean squares and mean products given in Table 10.6. The WLS estimates of the parameters, their estimated standard errors and associated tests of significance are given in Table 10.13.

All the parameter estimates differ very significantly from zero, and the model gives a good fit to the data ($X_{24}^2 = 29.4$, $P = 0.2$). Since the covariance components are both significant, we conclude that the correlation between sociability and impulsiveness is caused both by genetic and environmental factors. The heritability of the sociability scores is estimated to be 0.46, and that of impulsiveness is 0.36.

Our estimates of the genetic and environmental variances and covariances may be combined, if desired, to give estimates of the genetic and environmental variances of composite scores. If we combine sociability and impulsiveness into a single extraversion score by adding the two subscales with equal weight (as we did originally) then the additive genetic component of the extraversion score should be the sum of the additive genetic components of the component scales and twice the genetic covariance between scales. The environmental component should likewise be the analogous sum of the environmental variances and covariance for the two subtests. For the combined extraversion score, therefore, we find the additive genetic variance to

Table 10.13   Estimates (± s.e.) of genetic and environmental variances and covariances for sociability and impulsiveness.

	Genetic (V_A)	Environmental (E_W)
Sociability	0.461 ± 0.041	0.541 ± 0.032
Impulsivity	0.356 ± 0.042	0.644 ± 0.038
Covariance	0.179 ± 0.031	0.176 ± 0.026

be $0.46 + 0.36 + 2 \times 0.17 = 1.16$. Similarly, the environmental variance within families is $0.54 + 0.64 + 2 \times 0.18 = 1.54$. Both of these values are very close to the estimates derived directly from the total extraversion scores under the V_A, E_W model. The genetic component of the difference score, $D = S - I$, is the sum of the additive genetic components of the two separate scales *minus* twice their genetic covariance. Substituting the parameters estimated from the bivariate model, we obtain an estimate of $0.46 + 0.36 - 2 \times 0.17 = 0.48$ for the additive genetic variance in the difference score. Similarly, the environmental variance in the difference score is estimated to be $0.54 + 0.64 - 2 \times 0.18 = 0.82$. Once again, these estimates are very close to those obtained directly from the difference scores.

The genetic correlation between sociability and impulsiveness is

$$r_{Aij} = V_{Aij}/(V_{Ai} V_{Aj})^{1/2}.$$

A similar expression yields the environmental correlation

$$r_{EWij} = V_{EWij}/(E_{Wi}E_{Wj})^{1/2}.$$

Substituting our estimated genetic and environmental covariances in the above expression yields for the genetic correlation between sociability and impulsiveness $r_{Aij} = 0.42$. For the environmental correlation we obtain a value of 0.32. Thus, if we consider the two scales as they are measured, the phenotypic correlation between them is due to genetic *and* environmental causes. The genetic correlation between the scales is slightly greater than that due to the environment, but not markedly so.

Since the test scores are only *estimates* of the subjects' true scores, it may be argued that the heritability of the test scores, and the environmental correlation between them, is less than that of the true scores, since errors of measurement will contribute to estimates of E_W. Thus there may be some justification for correcting heritability estimates for unreliability of measurement. Before making such adjustments, however, it is important to consider the context in which such adjustments are made. First, as we have already shown (Chapter 9), the variance components often used to yield estimates of unreliability (i.e., persons $\times$ items interaction components) are often as much genetic as they are environmental. Under these circumstances, it is wrong to assume that reliability corrections will only reduce estimates of non-genetic variance. Secondly, before making a correction for unreliability, we have to consider the purpose for which the actual test scores are being used. In so far as the test is being used as it stands for predicting the resemblance between relatives, the unadjusted heritability is the correct one for practical purposes. On the other hand, there may be certain processes in the population (of which mate selection is one) that depend not on the test scores but on the latent trait of which the test is but an unreliable index. The genetic consequences of mate selection, i.e. the correlation between the genetic deviations of spouses, A (see Chapter 16), will depend on the heritability of the latent trait and not on the heritability of the raw scores.

Using the estimates of internal consistency given above, and assuming that they indeed have no genetic component, we conclude that 54% of the environmental variation in sociability is "reliable" variation, and the remainder is due to error. For impulsiveness, about 38% of the estimate of E_W is due to real long-term environmental effects and the remaining 62% is due to errors of measurement. Subtracting the fraction of E_W that is assumed to be error leads to estimates of the "real" environmental component of variance. Estimates of the genetic component are unchanged. The heritability estimates of the true scores may now be computed using the original estimates of V_A and the corrected estimates of E_W, yielding estimated heritabilities of 0.61 for sociability and 0.60 for impulsiveness.

The corrected estimates of E_W for the two traits may also be used to compute an estimate of the correlation between the environmental influences on the true scores. The corrected estimate of r_{EWij} is 0.66. This value is substantially higher than the uncorrected value and larger than the genetic correlation. If our unreliability correction is justified then the data imply that such long-term environmental effects as affect the traits of sociability and impulsiveness exercise a significant joint influence on both traits. Overall, when the effects of the short-term errors of measurement are subtracted, the greater part of the reliable variation in both traits, and in the combined extraversion score, is due to additive genetic factors. However, the genetic correlation between sociability and impulsiveness is significant but far from perfect. Thus the genes create much of the common factor underlying the component tests, but there is substantial genetic variation unique to the two traits, which justifies our regarding them as genetically distinct aspects of personality.

10.3 THE COMPONENTS OF IMPULSIVENESS

10.3.1 Models for more than two variables

There are limits to what can be said about the covariance between two measures. A covariance may be estimated, and the methods we have described so far may be used to partition phenotypic covariance into its genetic and environmental components. As long as we only consider pairs of measurements, however, there is no way to partition their variances and covariance into common and specific factors without placing unwarranted constraints on the specific variances. The method that we used in the analysis of sociability and extraversion can easily be extended to the multivariate case (see e.g. Eaves and Eysenck, 1975, 1980), to yield estimates of genetic and environmental covariance between every conceivable pair of variables. While this approach tells us much about the sources of trait covariation in genetic and social terms, it does not address certain kinds of question that naturally arise when we consider multiple variables. For example, having decided that there is substantial genetic covariation between traits, we want to know whether there is a single set of genetic effects common to all measures, or whether a significant part of the genetic variation is specific to individual variables. As far as environmental effects are concerned, we should like to answer similar questions. Given estimates of the genetic and environmental covariances, it does not require much imagination to see that these might be factor-analysed using some of the standard methods to obtain estimates of factor loadings and specific genetic and environmental variances. In practice, it turns out to be not quite as simple, because estimates of genetic and environmental covariance matrices obtained from linear combinations of mean-product matrices are often not positive-definite nor even positive-semidefinite (see e.g. Loehlin and Vandenberg, 1968). Cantor *et al.* (1983) and Martin *et al.* (1984) show how maximum-likelihood estimates of genetic and environmental covariances can be obtained that do not suffer from this mathematical problem. However, the main issue is not mathematical, but hangs on what kinds of hypothesis we want to test given that we have multivariate data on twins that can be used to resolve the structure of genetic and environmental covariance. Some of these hypotheses cannot be tested if our ideal analysis is confined to factor analysis of genetic and environmental covariances. For example, Figures 10.1 and 10.2 show two models for the way in which common and specific genetic and environmental components may contribute to the covariance structure of multiple measures. In Figure 10.2 a common genetic factor and a common environmental factor contribute directly to the correlation between traits, but there is no common intervening pathway. Conceptually, it is as if the measured traits were the outcome of two distinct processes, one that was under genetic control (for example the biochemical pathways underlying "sensitivity" to the environment) and the other that reflected the covariance structure of the information which individuals acquire from their environment. With this model, the ratios of genetic factor loadings to environmental factor loadings will change from one variable to the next. In the model shown in Figure 10.1 there are no separate pathways from the underlying genetic and environmental factors to the variables. Rather, both genes and environment affect a single process P, which then has a joint effect on the multiple outcomes. In this case, the ratio of genetic factor loadings to environmental loadings will be constant for all variables. Conventional unconstrained factor analysis of genetic and environmental covariances does not allow us to make this fundamental distinction between two quite different ideas about how genes and environment create the structure of phenotypic correlations. Rather than simply estimate genetic and environmental covariance matrices as a prelude to conventional factor analysis, we therefore try to specify certain explicit structural hypotheses in advance of the analysis, obtain maximum-likelihood estimates of the parameters, and conduct likelihood-ratio tests to discriminate between alternative hypotheses wherever possible. The approach is illustrated by data on four measures of impulsiveness.

10.3.2 The measurements and sample

Eysenck & Eysenck (1977) have suggested that impulsiveness is capable of resolution into four correlated primary factors, which have been termed "impulsiveness in the narrow sense" (IMPN), "non-planning" (NONP), "risk taking" (RISK) and "liveliness" (LIVE). Questionnaire measurements of these factors are shown to correlate differently with questionnaire measurements of "psychoticism", "extraversion" and "neuroticism".

Copies of an experimental questionnaire were mailed to adult twins on the Institute of Psychiatry Twin Register. Completed questionnaires were received from 587 pairs of twins, for whom the distribution of zygosity and sex is summarized in Table. 10.14. From the 52 impulsiveness items in the questionnaire 40 were selected that best represented the four component factors of impulsiveness.

Three illustrative items from each scale are given in Table 10.15. The numbers of items contributing to scores on each factor were IMPN (12), RISK (10), NONP (12), LIVE (6). An angular transformation was applied to the raw scores for each factor to secure additivity, but the improvement was not marked, probably on account of the relatively small number of items contributing to each of the component scales.

Table 10.14 Composition of twin sample employed in the analysis of impulsiveness.

	Number of pairs	
	Monozygotic (MZ)	Dizygotic (DZ)
Male pairs (M)	144	52
Female pairs (F)	233	83
Unlike-sex pairs (OS)	—	75

Table 10.15 Illustrative items for four impulsiveness factors.

Items	Keyed responses (Yes/No)
Factor: impulsiveness (narrow sense)	
Do you often do things on the spur of the moment?	Y
Do you often get involved in things you later wish you could get out of?	Y
Do you usually think carefully before doing anything?	N
Factor: risk	
Would you prefer a job involving change, travel and variety even though it might be insecure?	Y
Would you enjoy parachute jumping?	Y
Would you enjoy fast driving?	Y
Factor: non-planning	
Do you think an evening out is more successful if it is unplanned or arranged at the last moment?	Y
Would you make quite sure you had another job before giving up your old one?	N
When you go on a trip do you like to plan routes and timetables carefully?	N
Factor: liveliness	
Do you usually make up your mind quickly?	Y
Are you usually carefree?	Y
Are you slow and unhurried in the way you move?	N

10.3.3 Data summary

The mean squares and mean products between and within pairs were computed for the four variables for each of the four like-sex twin groups separately. For the male–female pairs the mean vector corresponding to the overall sex difference was also extracted from the intrapair variation. For each of the five twin groups the linear regression on age was partialled out of the variation and covariation between pairs.

The corrected mean squares and mean products are given in Table 10.16. The df take account of the corrections made for age and sex. Recognizing the constraints imposed by the symmetry of the ten matrices, the basic data summary comprises 100 statistics (ten covariance matrices, each with 10 free statistics).

Table 10.16 Age-corrected mean-product matrices within and between twin pairs for four impulsiveness scales.[a]

Twin type	df		IMPN 1	RISK 2	NONP 3	LIVE 4	df		IMPN 1	RISK 2	NONP 3	LIVE 4
			Between-pairs mean-product matrices						Within-pairs mean-product matrices			
MZ$_f$	231	1	0.12041	0.05049	0.03926	0.04673	233	1	0.05344	0.01945	0.01381	0.02299
		2	0.05049	0.12487	0.04456	0.03785		2	0.01945	0.05960	0.01312	0.01472
		3	0.03926	0.03785	0.03286	0.17454		3	0.01381	0.01312	0.03072	0.01545
		4	0.04673	0.03785	0.03286	0.17454		4	0.02299	0.01472	0.01545	0.09528
MZ$_m$	142	1	0.08964	0.03427	0.01683	0.02782	144	1	0.07445	0.03766	0.02829	0.02863
		2	0.03427	0.10136	0.02636	0.02756		2	0.09791	0.09791	0.02760	0.01528
		3	0.01683	0.02636	0.05672	0.01235		3	0.02829	0.02760	0.05021	0.02387
		4	0.02782	0.02756	0.01235	0.12959		4	0.02863	0.01528	0.02387	0.11747
DZ$_f$	81	1	0.10729	0.04840	0.03176	0.03082	83	1	0.04787	0.01216	0.01178	0.02489
		2	0.04840	0.08159	0.02951	0.01655		2	0.01216	0.03764	0.00877	0.02283
		3	0.03176	0.02951	0.06455	0.02149		3	0.01178	0.00877	0.03307	0.00579
		4	0.03082	0.01655	0.02149	0.19970		4	0.02489	0.02283	0.00579	0.08473
DZ$_m$	50	1	0.09077	0.04154	0.02206	0.01277	52	1	0.07478	0.01300	0.01298	0.00996
		2	0.04154	0.07101	0.02986	0.02515		2	0.01300	0.06791	0.01651	0.00583
		3	0.02206	0.02986	0.04510	0.03040		3	0.01298	0.01651	0.02939	0.01024
		4	0.01277	0.02515	0.03040	0.13465		4	0.00996	0.00583	0.01024	0.08983
DZ$_{os}$	73	1	0.07418	0.02009	0.01796	0.01044	74	1	0.08180	0.03712	0.02634	0.02307
		2	0.02009	0.07473	0.02242	0.02658		2	0.03712	0.07394	0.01977	0.02212
		3	0.01796	0.02242	0.05869	0.01858		3	0.02634	0.01977	0.04465	0.00214
		4	0.01044	0.02658	0.01858	0.10020		4	0.02307	0.02212	0.00214	0.12258

[a] Key to labels for primary factors: IMPN, impulsiveness in the "narrow sense"; NONP, non-planning; RISK, risk-taking; LIVE, liveliness.

10.3.4 The model

The above analysis of impulsiveness and sociability showed that the pattern of individual differences in male and female monzygotic (MZ) twins and in like-sex and unlike-sex male and female dizygotic (DZ) twin pairs was consistent with a model that assumed that the variation and covariation of the two traits were due primarily to the additive effects of many genes and the effects of environmental influences that were largely specific to individuals rather than common to families. Furthermore, the consistency over sexes of particular estimates, and the ability of the model to encompass data on unlike-sex pairs without additional parameters, implied that the causes of variation in extraversion and its components did not depend substantially on sex. We suppose that the phenotypic variation for the four traits may be explained by a model invoking a single factor common to the four variables (impulsiveness in the broad sense) and components specific to each of the variables. However, by including twins in the study we are able to go beyond a simple treatment of the structure of phenotypic variation into an analysis of its causal basis. Thus, combining the simple causal model for impulsiveness described in Section 10.2 with the simple factorial model proposed by Eysenck & Eysenck (1977), we can discover whether the factorial unity of impulsiveness applies with equal force to both the genetic and environmental determinants of the trait.

We assume that individual differences in impulsiveness are due to additive genetic effects with random mating (V_A) and within-family (E_W) environmental effects. For a single variable, we may write our expectation for the total phenotypic variance in terms of our simple model as

$$V_P = V_A + E_W \quad \text{(cf. Chapter 4).}$$

The basic model for individual differences can be extended in many ways for the univariate case, as we have already shown. For multiple variables, the same model can be used for the phenotypic covariance matrix, which is now expressed as the sum of an "additive genetic" covariance matrix, Σ_A and a "within-family environmental" covariance matrix Σ_W. We also assume that the genetic and environmental covariance matrices may themselves be decomposed in terms of the conventional factor model into one or more common factors and a number of specific variances:

$$\Sigma_P = \Sigma_A + \Sigma_W$$
$$= \mathbf{GG}' + \mathbf{D}^2 + \mathbf{HH}' + \mathbf{E}^2.$$

This model may be made more complex if the need arises. As long as we only have unrelated individuals, as is the case in most studies of individual differences, there is no hope of separating the genetic and environmental components of phenotypic covariance. However, once we have kinship data, we can estimate genetic and environmental factors separately. In the twin data we have matrices of mean products within and between pairs for the different types of twins. The contributions of the genetic and environmental factors to the different matrices differ, so we may try to resolve genetic and environmental components of phenotypic covariation. For twins we have the following expectations:

$$\Sigma_{BMZ} = 2(\mathbf{GG'} + \mathbf{D}^2) + \mathbf{HH'} + \mathbf{E}^2,$$
$$\Sigma_{WMZ} = \mathbf{HH'} + \mathbf{E}^2,$$
$$\Sigma_{BDZ} = \tfrac{3}{2}(\mathbf{GG'} + \mathbf{D}^2) + \mathbf{HH'} + \mathbf{E}^2,$$
$$\Sigma_{WDZ} = \tfrac{1}{2}(\mathbf{GG'} + \mathbf{D}^2) + \mathbf{HH'} + \mathbf{E}^2.$$

The subscript B denotes a matrix of mean products *between* pairs, W denotes a matrix of *within-pair* mean products.

There are simple extensions of the expectations for the mean squares for a single variable given in Chapter 5. We have made no distinction between male and female twins, nor between like-sex and unlike-sex dizygotic twins, in writing the initial expectations, since our basic model assumes that the genetic and environmental components do not depend on sex. Allowing for sex limitation in gene expression in the multivariate model is quite straightforward.

For the present case there are four variables. The factor model anticipates that there will be only one common factor. The matrices **G** and **H** will therefore be four-element column vectors of genetic and environmental loadings. $\mathbf{D}^2$ and $\mathbf{E}^2$ will be 4×4 diagonal matrices containing the corresponding specific variances. Our first model will thus attempt to explain the 100 raw statistics by reference to 16 parameters. A further simplification is proposed. We begin by assuming that genetic and environmental effects on the phenotype are mediated through a common underlying process (cf. Figure 10.1). We thus constrain the genetic loadings to be a constant multiple of the environmental loadings thus:

$$\mathbf{G} = b\mathbf{H}.$$

If this model fits then the scale factor b is related to the heritability of the common factor through $h^2 = b^2/(1 + b^2)$. In summary, the simplest form of the model has 13 parameters: four factor loadings (**H**); one scalar factor (b); four genetic specifics ($\mathbf{D}^2$); and four environmental specifics ($\mathbf{E}^2$). In principle, further reductions might be possible by imposing constraints on the relative values of specific components, but this does not seem appropriate in the absence of any theoretical justification.

A second model for the relationship between genes, environment and phenotype in a multivariate system is represented in Figure 10.2. In this case

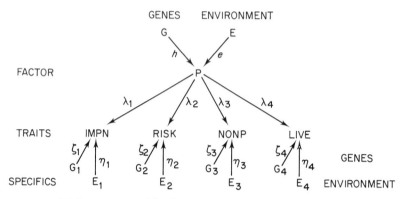

Figure 10.1 Multivariate model when genes and environment operate through a common underlying variable.

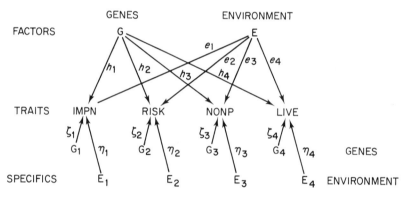

Figure 10.2 Multivariate model when genes and environment affect different pathways.

a common set of genetic effects (**G**) and a common environmental factor (**E**) each affect the multiple variables through different pathways. Here it would be impossible to scale the genetic factor loadings to be constant multiples of the environmental loadings.

Although we have considered only the simplest experimental design and specified only the most basic genotype–environmental factor models, the approach is general in that virtually any model we can write for a single variable can be cast in a form applicable to multiple variables, so that hypotheses can be tested and parameters estimated, given adequate data.

10.3.5 Statistical method for testing the model

Joreskog has developed a conceptual and statistical methodology for problems very similar to ours, (see e.g. Joreskog, 1973, 1978). We have used an approach that is very similar to his, adapting it somewhat to the needs of our particular class of problem. The approach is described fully by Martin and Eaves (1977). Generally, our data will consist of k matrices of mean products. We may write S_i, for the ith matrix, having N_i df. Given some model for the S_i, we may compute the expected values Σ_i, being positive-definite, for particular values of the parameters of the model. When the observations are multivariate normal we may write the log-likelihood of obtaining the k observed independent S_i as

$$L = -\frac{1}{2} \sum_{i=1}^{k} N_i [\log|\Sigma_i| + \mathrm{tr}(S_i\Sigma_i)]^{-1} + \text{constant.}$$

For a given model, we require the parameter estimates that maximize log L. Given maximum-likelihood estimates of our parameters, we may test the hypothesis that a less restricting model (i.e. one involving more parameters) significantly improves the fit by computing $X^2 = 2(L_0 - L_1)$, where L_1 is the log-likelihood obtained under the restricted hypothesis (H_1) and L_0 is the log-likelihood obtained under the less demanding hypothesis (H_0). The H_0 that we shall adopt in practice assumes that as many parameters are required to explain the data as there are independent mean-squares and mean-products in the first place, i.e. $\Sigma_i = S_i$ for every i. In this case we have simply:

$$L = -\frac{1}{2} \sum_{i=1}^{k} N_i [\log |S_i| + p] + \text{constant.}$$

When we have k matrices the X^2 has $\frac{1}{2} kp(p+1) - m$ df, where m is the number of parameters estimated under H_1 and p is the number of variables. The likelihood may be maximized numerically using algorithms for unconstrained optimization as long as the model is parametrized in a form that always yields positive-definite expected covariance matrices. In our application, this restriction is satisfied by the following conditions: (1) we estimate D and E rather than D^2 and E^2; (2) the expected genetic and environmental covariance matrices each parametrized as the products of a (possibly different) matrix and its transpose; (3) the specific environmental components are not zero. We have never encountered a case where the last condition is not satisfied; and we should not expect one to occur, since all variables have some specific environmental variation even if it is only due to errors of measurement.

In order that the estimates of a satisfactory model might be interpreted more rigorously, their covariance matrix is required. This is the inverse of the matrix of the second derivatives of the log-likelihood with respect to the maximum-likelihood estimates of the parameters. Joreskog (1973) gave formulae for the derivatives in problems involving single covariance matrices. We followed his approach, constructing first the matrix of second derivatives with respect to all the parameters, fixed and free, then striking out the rows and columns corresponding to fixed parameters, and finally combining the information on those parameters which are constrained to be equal. Martin and Eaves (1977) described the method for obtaining the covariances of the estimates in further detail.

10.3.6 Results

Table 10.17 gives estimates obtained from fitting the simple model that assumed a single genetic factor with loadings proportional to those of a single common environmental factor. These loadings are scaled to reproduce the covariance matrices. The log-likelihood under this hypothesis was 3867.10 for 13 parameters. If we were to estimate the parameters of a model involving the maximum of 100 parameters (a perfect-fit solution) then we would obtain a value of 3919.37 for the log-likelihood.

The goodness-of-fit test yields $X^2_{87} = 104.54$. Although this value is not significant ($P = 0.097$), the fit is relatively poor. Standard errors of the estimates are not cited at this stage because we are not satisfied with this model. We first consider some modifications of the basic model that might improve the fit. There are several possibilities. We could seek additional common factors. This would seem unwise with only four variables! We could seek explanations that involve effects other than additive genetic and within-

Table 10.17 Maximum-likelihood parameter estimates assuming identical structure and environmental factors.

Trait	Genetic		Environmental	
	Loading[a] ($h\lambda$)	Specific[b] (γ)	Loading ($e\lambda$)	Specific (η)
IMPN	0.115	0.128	0.142	0.186
RISK	0.114	0.143	0.141	0.150
NONP	0.085	0.105	0.105	0.135
LIVE	0.089	0.197	0.111	0.249

[a] Genetic loadings are a constant multiple (0.810) of the environmental loadings. Loadings are scaled to reproduce the phenotypic *covariance* matrix, not the correlation matrix.
[b] Specific standard deviations, not variances, are tabulated.

family environmental effects such as more complex environmental effects or more subtle genetic effects. One possibility would be to relax the constraint that $G = bH$ and allow the genetic and environmental loadings to vary independently. Such a model implies that genes and environment affect the measures of impulsiveness directly and independently (cf. the model in Figure 10.2) and are not mediated through some common underlying latent variable. With this model, a factor analysis of the *phenotypic* correlation matrix could never adequately represent the joint action of genetic and environmental effects because there is not just one underlying factor but two different factors, one of which is substantially genetically determined, and the other of which reflects the structure of environmental influences. When we fitted this model we obtained a value of 3870.40 for the log-likelihood, giving a goodness-of-fit X^2 of 97.94 for 84 df ($P = 0.142$). The three additional parameters led to a reduction in chi-square of 6.60 for 3 df ($P = 0.086$), suggesting a slight but not very significant improvement in fit.

The slight, but not very striking, evidence of some differences between genetic and environmental factor structure led us to revert to the previous model for the factor structure, and we started to examine the specific variation. By a process of tentative model fitting to the data on sexes separately, but leaving out the unlike-sex pairs, we obtained an indication that, although the factor loadings seemed fairly consistent over sexes, the values obtained for the specific variances, especially the specific genetic variances, differed quite markedly between males and females. This suggested that the genetic determinants of trait-specific variation were different in the two sexes. If this were the case then we should expect the common factors to contribute to the covariation of male–female pairs, but we should expect the specific genetic variances to take different values in males and females and to make no contribution to the covariance of unlike-sex twins. Thus a final model was fitted that differed from the initial model in only the following features. Specific genetic variances were fitted that depended on sex, with the further specification that these were genetically quite distinct in the two sexes. This amounts to saying that the genetic component of the trait-specific variation can be best approximated by a model that assumes that quite different genes are expressed in males from those expressed in females (non-scalar sex limitation). The model is thus that described above, except that we have slightly different expectations for the opposite-sex (OS) pairs, as follows:

$$\Sigma_{BOS} = \tfrac{3}{2}\, GG' + \tfrac{1}{2}\, (D_M^2 + D_F^2) + HH' + E,$$
$$\Sigma_{WOS} = \tfrac{1}{2}\, GG' + \tfrac{1}{2}\, (D_M^2 + D_F^2) + HH' + E,$$

where D_M^2 and D_F^2 denote the specific additive genetic variances for males and females respectively. In the expectation for like-sex pairs (above) we merely substitute D_M for D in the males and D_F for D in the expectation for female twins (cf. Table 5.11 for the univariate case).

Table 10.18   Parameter estimates for multivariate model for impulsiveness
(standard errors in parentheses).

			Specific standard deviations		
	Factor loadings		Genetic (c)		
Trait	Genetic (h)	Environmental (e)	Male	Female	Environmental (n)
IMPN	0.114	0.142	0.133	0.127	0.181
	(0.008)	(0.010)	(0.015)	(0.021)	(0.009)
RISK	0.112	0.140	0.134	0^a	0.200
	(0.008)	(0.010)	(0.016)	—	(0.009)
NONP	0.083	0.105	0.101	0.087	0.149
	(0.006)	(0.008)	(0.012)	(0.018)	(0.007)
LIVE	0.090	0.113	0.173	0.201	0.274
	(0.009)	(0.011)	(0.191)	(0.026)	(0.011)

^{a} Parameter is fixed on the boundary.

The model now has four factor loadings, with one constant relating genetic and environmental factors, four specific environmental components and eight specific genetic components (four for each sex), making 17 parameters in all. The log-likelihood is now 3875.12, giving an overall $X^2_{83} = 88.50$ ($P = 0.319$) indicating a good fit and a marked improvement ($X^2_4 = 16.04$, $P = 0.003$) on the original model that assumed that the specific genetic variances were the same in both sexes.

The estimates of the parameters of this model are given in Table 10.18. The standard errors of the estimates are also tabulated where appropriate.

10.3.7 Discussion

The analysis suggests that the covariance structure of impulsiveness is due to a single underlying factor, which is affected jointly by genetic and environmental effects. By showing that the genetic and environmental loadings are proportional to one another, we have in effect showed that the ratio of variation due to common genetic factors to that due to the common environmental factor is consistent over all variables. Thus there is a common factor, which we may call impulsiveness in the broad sense, whose heritability is a simple function of the ratio b of the genetic and environmental loadings. The fact that b differs significantly from zero ($\hat{b} = 0.80$) indicates that the genetic loadings are jointly significant and justifies our attempts to estimate the proportion of the common-factor variance that is due to genetic factors.

If we write g for any one of the four genetic factor loadings and e for the corresponding environmental loading then we have $g = eb$. Since the model fits, we may estimate the "heritability" of the common factor from

$$h^2 = b^2/(1 + b^2).$$

Substitution for b gives the estimate of the proportion of the *common-factor variance* that is genetically determined as 0.39. This value does not require any correction for unreliability, since we presume that sampling error will contribute only to the specific components of variation in the four measurements.

Similarly, we may use our parameter estimates to asses the relative contribution of genetic and environmental differences to the specific variation of the four measurements. We have established that the sexes differ in the genetic mechanism responsible for specific variation, so we are compelled to give separate estimates of the "specific heritabilities" for each sex.

Writing d^2_i for a typical genetic specific variance and s^2_i for the corresponding environmental specific, we have

$$h_i^2 = d_i^2/(d_i^2 + s_i^2).$$

The estimated heritabilities of the trait-specific variances are given in Table 10.19. With the exception of the risk factor (for which males show no specific genetic variation — in contrast with the females, who show significant genetic specific variation), the values are comparable to those obtained for the common factor. However, errors of measurement do not contribute to environmental variation in the common factor (given independent errors), but are expected to contribute to the specific environmental variation. Since the analysis is based on scores transformed to angles, we may approximate the error variance for each scale from the theoretical error of transformed proportions (see e.g. Snedecor and Cochran, 1980). These values will enable us to estimate how much of the trait-specific environmental variance is due to measurement error. For a scale consisting of n items of equal difficulty and given local independence the theoretical error variance takes the value $(4n)^{-1}$. If the items of a scale are not all equally difficult then this estimate of error will be larger than the true value.

Table 10.19 Proportion of variation specific to four impulsiveness scales attributable to genetic factors.

Trait	Females	Males
IMPN	0.350	0.380
RISK	0.329	0.000
NONP	0.332	0.332
LIVE	0.144	0.146

Table 10.20   Analysis of the contribution of measurement error to specific variation.

Trait	Number of items	"Error" variance[a]	"True" specific environmental variance	Proportion due to "error"	Females	Males
					"Heritability" of specific variance corrected for measurement error[b]	
IMPN	12	0.021	0.012	0.63	0.60	0.57
RISK	10	0.025	0.015	0.63	0.54	0.00
NONP	12	0.021	0.001	0.95	0.91	0.88
LIVE	6	0.042	0.033	0.56	0.48	0.55

Estimated contribution to specific environmental variance

[a] The error variance is estimated as $(4n)^{-1}$ (see text).
[b] Cf. uncorrected values in Table 10.19

Table 10.21   Summary of the relative contributions of common and specific genetic and environmental effects to the total variation in each of the components of impulsiveness.

Sex	Trait	Proportion due to genetic effects			Proportion due to environmental effects		
		Common	Specific[a]	Total	Common	Specific	Total
Female	IMPN	0.155	0.211	0.366	0.241	0.392	0.633
	RISK	0.139	0.218	0.357	0.199	0.444	0.643
	NONP	0.138	0.219	0.357	0.203	0.440	0.643
	LIVE	0.064	0.101	0.165	0.238	0.596	0.835
Male	IMPN	0.158	0.245	0.404	0.197	0.399	0.596
	RISK	0.173	0.000	0.173	0.272	0.554	0.827
	NONP	0.146	0.231	0.377	0.158	0.465	0.623
	LIVE	0.059	0.094	0.153	0.296	0.551	0.847

[a] Error variation has not been deducted from the contribution of specific environmental factors.

Table 10.20 shows estimates of the specific environmental variation for each trait, with the appropriate theoretical errors for scales of the corresponding length. The difference between the two sets of estimates is an estimate of the "true" specific environmental variation, which is due to factors other than errors of measurement. In every case rather more than half of the measureable specific environmental variation within families seems to be attributable to errors of measurement. Indeed, in the case of the non-planning factor we conclude that virtually all the detectable specific environmental variation is due to sampling error in the scores. Finally, Table 10.21 gives a summary of the contributions of the different sources of variation to the four measurements in each sex derived from the parameters in Tables 10.19 and 10.20. For each sex, the phenotypic variance V_{Pi} for the ith trait was calculated by substituting the appropriate parameter estimates in

$$V_{Pi} = (g_i^2 + d_i^2) + e_i^2 + s_i^2.$$

The contributions of each of the components g_i^2, d_i^2, e_i^2 and s_i^2 to the phenotypic variance V_{Pi} are then computed as ratios for each sex and variable separately. Many other summary statistics could be derived from the estimates in Table 10.18, including traditional heritability estimates for the individual variables. Adding together the contributions from the genetic and environmental common factor yields the familiar communality estimate for each variable. Adding the genetic contributions due to common and specific variance for each variable in turn, we have the usual heritability estimate applicable to each variable as it would be derived in any equivalent univariate analysis of the individual scales.

Apart from the substantive findings, the analysis of impulsiveness illustrates how the maximum-likelihood method can be applied to answer questions about why multiple variables are correlated. The analysis shows that the phenotypic correlations between measures of impulsiveness are due to genetic and environmental factors. More than that, however, it seems as if the effects of genes and environment are "channeled" through some common pathway before they affect the different facets of impulsiveness. Although the *correlation* between traits appears to be caused by the same genes in both sexes, there is a suggestion that the trait-*specific* components of impulsiveness may be due to different genetic effects in males and females. Even this simple example therefore illustrates some of the power and flexibility of the method.

10.4 THE COVARIANCE STRUCTURE OF NEUROTICISM

In the same questionnaire used to develop subscales of impulsiveness, 76 items were embedded that were drawn from scales that had been used at some time or other in the measurement of neuroticism and related traits. The responses of the 1174 individual twins to these items were correlated and factored using iterative principal-axis factor analysis. Males and females were combined in the analysis. When a single common factor was fitted it was clear that a general factor of "neuroticism" could be identified, but that a multiple-factor solution was needed to account adequately for the pattern of interitem correlations. Seven correlated factors could be identified, covering the spectrum of neurotic responses, some of which might be better described as "psychotic" since they included items relating to depression and perceived threat from others. The seven factors may be tentatively identified in terms of their item content as "depression", "worry", "insomnia", "paranoia", "inattention", "shyness" and "psychosomatic symptoms". Factor scores were derived for each subject and employed as the basis for subsequent genetic model fitting.

10.4.1 Data summary

Table 10.22 gives the matrices of mean squares and mean products between pairs and within pairs for each group of twins in the sample. The linear trend of factor scores on age was partialled out of the mean products between pairs.

10.4.2 Univariate analysis

Before attempting a multivariate analysis of the covariation between factors, we summarize the results of a univariate analysis, which may help us in building an appropriate multivariate model. The results of fitting the V_A, E_W model to the mean squares for each of the variables are given in Table 10.23. The fit is good for two factors ("insomnia" and "paranoia"), barely adequate for "depression" and "inattention", and bad for the remaining three factors.

Allowing for the effects of sex-limited gene expression and sex differences in the within-family environmental component improved the fit significantly for "depression" and "paranoia", but the overall fit of the model was still poor for the "worry", "shyness" and "psychosomatic symptoms" factors (Table 10.24).

We attempted to explain the failure of the additive model for the two worst cases ("shyness" and "psychosomatic symptoms") in terms of competitive interactions between twins. In view of the evidence for sex differences, we excluded the unlike-sex pairs and tested for the improvement in fit attributable to competition for males and females separately. In three out of four cases (Table 10.25) the inclusion of competitive effects in the model led to a significant improvement in fit to the observed mean squares and resulted in non-significant residual effects.

The results in Table 10.25 have to be treated with certain skepticism for two reasons. First, we know (see Chapter 4) that the resolution of

Table 10.22 Mean squares and mean products for seven neuroticism factors.

	Depression	Worry	Insomnia	Paranoid	Inattention	Shyness	Psychosomatic
Between MZ female pairs, df = 231:							
Depression	1.190						
Worry	0.464	1.094					
Insomnia	0.522	0.395	1.152				
Paranoid	0.469	0.227	0.343	0.886			
Inattention	0.631	0.416	0.445	0.380	1.078		
Shyness	0.172	0.293	0.126	0.136	0.093	0.945	
Psychosomatic	0.711	0.546	0.755	0.289	0.651	0.140	1.509
Within MZ female pairs, df = 233:							
Depression	0.498						
Worry	0.112	0.481					
Insomnia	0.162	0.116	0.521				
Paranoid	0.122	0.070	0.902	0.399			
Inattention	0.172	0.108	0.056	0.102	0.403		
Shyness	0.075	0.105	0.043	0.230	−0.004	0.412	
Psychosomatic	0.198	0.131	0.225	0.084	0.088	0.064	0.479
Between MZ male pairs, df = 81:							
Depression	1.116						
Worry	0.716	1.432					
Insomnia	0.404	0.209	0.792				
Paranoid	0.556	0.391	0.388	0.891			
Inattention	0.761	0.723	0.476	0.663	1.424		
Shyness	0.203	0.291	0.044	0.119	0.159	0.630	
Psychosomatic	0.356	0.428	0.323	0.286	0.329	0.171	0.602
Within MZ male pairs, df = 83:							
Depression	0.307						
Worry	0.136	0.297					
Insomnia	0.868	0.772	0.418				
Paranoid	0.096	0.012	0.013	0.282			
Inattention	0.141	0.078	0.074	0.087	0.354		
Shyness	−0.026	0.050	0.063	0.008	0.159	0.281	
Psychosomatic	0.357	0.074	0.130	0.038	0.141	−0.010	0.304
Between DZ female pairs, df = 142:							
Depression	0.869						
Worry	0.285	0.864					
Insomnia	0.292	0.200	0.944				
Paranoid	0.420	0.223	0.304	0.951			
Inattention	0.350	0.171	0.228	0.315	0.745		
Shyness	0.113	0.297	0.059	0.131	0.009	0.857	
Psychosomatic	0.373	0.386	0.452	0.296	0.270	0.167	0.850
Within DZ female pairs, df = 144:							
Depression	0.803						
Worry	0.220	0.573					
Insomnia	0.156	0.773	0.751				
Paranoid	0.202	0.033	0.273	0.570			
Inattention	0.296	0.141	0.155	0.118	0.597		
Shyness	0.114	0.248	0.064	0.115	0.040	0.774	
Psychosomatic	0.315	0.198	0.306	0.228	0.253	0.247	0.749
Between DZ male pairs, df = 50:							
Depression	0.806						
Worry	0.228	0.749					
Insomnia	0.503	0.176	0.891				
Paranoid	0.384	0.245	0.308	0.566			
Inattention	0.447	0.122	0.351	0.267	0.800		
Shyness	−0.114	0.109	−0.017	0.004	−0.099	0.594	
Psychosomatic	0.421	0.188	0.321	0.240	0.200	0.082	0.493
Within DZ male pairs, df = 52:							
Depression	0.628						
Worry	0.315	0.842					
Insomnia	0.227	0.229	0.675				
Paranoid	0.152	0.239	0.108	0.392			
Inattention	0.366	0.451	0.258	0.196	0.879		
Shyness	0.065	0.200	0.123	0.172	0.027	0.642	
Psychosomatic	0.236	0.412	0.264	0.081	0.277	0.043	0.642

Table 10.23 Results of fitting univariate models to individual primary factors extracted from 76 neuroticism items.

Factor	Depression	Worry	Insomnia	Paranoia	Inattention	Shyness	Psychosomatic
V_A	0.324	0.0324	0.272	0.291	0.336	0.237	0.343
E_W	0.466	0.434	0.516	0.362	0.405	0.437	0.446
h^2	0.41	0.43	0.35	0.45	0.45	0.35	0.43
χ_8^2	15.2	17.6	8.2	12.6	14.5	27.2	46.27
$P\%$	5–10	2–5	30–50	10–20	5–10	<0.1	<0.1

Table 10.24 Results of allowing for sex limitation in genetic model for primary factors of neuroticism.

	Depression	Worry	Insomnia	Paranoia	Inattention	Shyness	Psycho-somatic
V_{AM}	0.37	0.47	0.22	0.27	0.45	0.19	0.13
V_{AF}	0.31	0.28	0.30	0.29	0.29	0.27	0.43
V_{AMF}	0.37	0.23	0.12	0.39	0.19	0.04	0.15
E_F	0.32	0.33	0.46	0.28	0.38	0.34	0.36
χ_5^2	6.4	12.1	3.21	4.4	10.3	13.9	14.3
$P\%$	20–30	2–5	50–70	30–50	5–10	1–2	1–2
Improvement over sex-limited model in Table 10.23:[a]							
χ_3^2	8.8	5.5	5.0	8.2	4.2	13.3	32.0
$P\%$	2–5	10–20	50–70	30–50	5–10	1–2	1–2
$\hat{r}_{MF}$	1.09	0.63	0.46	1.39	0.52	−0.18	0.63

[a] $\hat{r}_{MF}$ is the correlation between the additive effects of genes in males and their effects in females, obtained from $r_{MF} = V_{AMF}/(V_{AM} V_{AF})^{1/2}$.

competitive effects from those of genetic non-additivity is very difficult even in quite large samples. Secondly, the final model did not reflect purely *a priori* considerations but was based on decisions made *post hoc* about where to proceed for individual variables. The goodness-of-fit tests assume that the data have not been inspected prior to deciding on a model. Nevertheless, it is remarkable that the "shyness" factor is one of those for which the effects of competition are implicated.

10.4.3 Multivariate analysis

The univariate analyses have already shown significant heterogeneity over sexes for the genetic and environmental components of variation in the individual factors. Multivariate analysis therefore excludes the unlike-sex pairs and focuses separately on male and female twins. The model-fitting analysis used the LISREL V program for maximum-likelihood estimation and hypothesis-testing (Jöreskog and Sörbom, 1981). A recent special edition of *Behavior Genetics* (Volume 19, 1989) gives detailed worked examples of programs for fitting such structural genetic models with LISREL. Other ways can be found of specifying structural models for twin data in LISREL.

The results of the univariate analysis suggest that we are very unlikely to obtain an adequate fit to the multivariate neuroticism data by a simple model because the total dispersion matrices are heterogeneous in a way that yields no consistent explanation across sexes. We therefore present only a few

Table 10.25 Testing for competitive effects on selected primary factors of neuroticism.

			Estimates					
	Shyness				Psychosomatic			
	Males		Females		Males		Females	
Parameter competition:	Absent	Present	Absent	Present	Absent	Present	Absent	Present
$V_A^{(a)}$	0.18	0.45	0.23	0.50	0.13	0.41	0.43	0.49
$2C_{AS}$	—	-0.27	—	-0.22	—	-0.27	—	-0.31
E_1	0.34	0.28	0.55	0.41	0.37	0.30	0.48	0.47
χ^2	6.64	0.12	7.16	0.13	6.70	0.08	7.21	6.78
df	2	1	2	1	2	1	2	1
P%	2-5	70-80	2-5	70-80	2-5	70-80	2-5	2-5

[a] In the presence of competition V_A reflects both direct and indirect genetic effects (see text).

simpler multivariate models, which, although by no means adequate, may nevertheless capture some of the main informative trends in the data.

Three models were fitted to both sexes. The first model assumed that the only sources of variance and covariance were the additive effects of genes and the influences of the environment within families. We assumed that the genetic covariance between traits was due to a single common factor. Similarly, the environmental covariance was assumed to be due to a single common factor. No attempt was made to scale the genetic and environmental loadings relative to one another. By omitting such scaling, we are effectively assuming the model of genetic and environmental covariation shown in Figure 10.2. In addition, this first model allowed for genetic and environmental variances specific to each trait. The two additional models included parameters for the effects of dominance. The second model assumed that dominance effects were trait-specific. The third model allowed for a single common factor of dominance and seven trait-specific dominance components. Although many of the univariate models implicated some form of competition, we know that the effects of competition and dominance are difficult to resolve in practice, so we assumed that the addition of dominance to the model would capture much of what was assigned to competition in the univariate analyses.

Table 10.26 gives the values of chi-square testing the goodness of fit of the three models in each sex. The chi-squares assume normality in the underlying multivariate distribution. With the possible exception of the third model in males, the models all give a bad fit. The chi-squares, however, may still be informative. In neither sex, for example, does allowing for trait-specific dominance effects significantly reduce the residual chi-square. On the other hand, introducing a general non-additive effect in the third model leads to a highly significant improvement in both sexes.

The parameter estimates obtained with the first model (V_A, E_W) are given in Table 10.27. The parameter estimates from the third model, in which general and specific non-additive effects are incorporated, are given in Table 10.28.

The specific effects in the tables are given as specific *standard errors,*

Table 10.26 Goodness-of-fit tests for multivariate models of neuroticism scales.

Model		Males		Females		
Factors	Specific	χ^2	P	χ^2	P	df
V_A, E_W	V_A, E_W	130.7	10^{-3}	186.7	$<10^{-3}$	84
V_A, E_W	V_A, E_W, V_D	122.8	10^{-3}	179.7	$<10^{-3}$	77
V_A, E_W, V_D	V_A, E_W, V_D	99.0	10^{-2}	144.6	$<10^{-3}$	70

Table 10.27 Parameter estimates for factor loadings F and specific standard errors S from multivariate V_A, E_W model for neuroticism factors.[a]

	Males				Females			
	E_W		V_A		E_W		V_A	
Scale	F	S	F	S	F	S	F	S
DFPR	375	607	522	255	363	454	538	301
WORR	353	615	258	426	358	464	415	575
INSOM	427	621	330	401	330	604	328	306
PARA	228	595	349	406	148	504	459	243
INAT	220	549	495	265	337	512	572	437
SHY	238	648	053	501	060	581	101	415
SOMA	534	491	430	442	408	483	233	205

[a] Estimates have been multiplied by 1000 to remove decimals.

Table 10.28 Parameter estimates for factor loadings F and specific standard errors S from multivariate V_A, V_D, E_W model for neuroticism factors.

	Males						Females					
	E_W		V_A		V_D		E_W		V_A		V_D	
Scale	F	S	F	S	F	S	F	S	F	S	F	S
DFPR	365	434	388	345	347	000	432	572	388	269	323	000
WORR	320	456	718	000	−120	000	366	608	236	107	047	000
INSOM	364	597	132	208	347	004	352	630	405	395	032	000
PARA	133	501	315	226	349	004	262	591	223	001	306	355
INAT	334	495	417	462	367	000	244	582	392	243	259	184
SHY	022	558	195	417	−046	000	253	631	037	512	003	003
SOMA	376	470	228	257	127	000	458	516	646	000	−128	000

[a] Estimates have been multiplied by 1000 to remove decimals.

rather than specific variances, so that they may be compared more easily with the factor loadings. Consideration of the simple additive model first yields a few general observations. The trait-specific environmental components are generally greater than the loadings of the traits on the general environmental factor. Thus the environment within families (which includes many types of measurement error) is mainly trait-specific, although there is some environmental component of trait covariation. This finding is not altered when non-additive genetic effects are included in the model (Table 10.28). Genetic effects are both trait-specific and general across all traits. The specific and common components are generally similar, except in the case of "shyness", which has a much larger specific genetic component and a correspondingly smaller loading on the common genetic factor. Since shyness is related more closely to introversion than neuroticism, it is not surprising that the trait appears genetically distinct from the others and also that it shows some of the marks of competition or non-additivity that we found for the extraversion factor. The results for the additive model are fairly consistent over sexes.

The results for the additive-dominance model (Table 10.28) are much more difficult to interpret. The pattern of the environmental components is largely unchanged. The genetic variation, however, is split unevenly and irregularly between common and specific additive and dominance components. Part of the problem is likely to be the high correlation between estimates of additive and dominance effects in twin data (cf. Chapter 5 in the univariate case), which results in correspondingly large standard errors for genetic parameters in the additive-dominance model. However, within both sexes, the loadings on the general dominance factor are all positive and quite large, with the exception of "shyness", for which the greater part of the non-additive variation is trait-specific. There are numerous very small specific components of the additive genetic variance, which suggests that some of the trait-specific additive variation has been shifted to dominance in the extended model. Generally, however, no clear trend emerges from the multivariate analysis of the neuroticism data.

10.5 CONCLUSION

Any univariate genetic analysis may be extended to the multivariate case. The covariance between measures may be divided into genetic and environmental components, and structural models may be devised and tested by maximum-likelihood methods. The issues addressed by factor-analytic models for phenotypic correlations may be applied equally to the analysis of genetic and environmental components of variance and covariance. For example, genetic factor loadings and specific variances may be estimated for particular sets of data.

A number of illustrations have been provided of how the approach can be applied to the analysis of personality data. In general, we have shown that there is a substantial genetic component to the correlation between primary factors of extraversion and neuroticism. Sociability and impulsiveness are correlated genetically, as are the subscales of impulsiveness and component factors of neuroticism. Generally, the contribution of the environment to trait covariation is more limited, but the trait-specific effects of environment are likely to be overestimated because of measurement error. Correction for unreliability in the case of sociability and impulsiveness leads to a far higher estimate of the correlation due to environmental effects.

In the case of impulsiveness there is some support for a model in which genes and environment operate through the same latent variable. There is strong support for a common genetic factor underlying the many different facets of neuroticism.

Chapter 11

Normal Personality and Symptoms of Psychiatric Disorder: A Genetic Relationship?

The original model of personality advanced by Eysenck (1952a) had its roots in the attempt to account economically for the data then available on psychiatric diseases. Since that time, however, the genetic analyses of personality and psychiatric disorders have assumed lives of their own. We have already remarked (Chapter 8) on the similarity between the dimensional conception of psychiatric disease embodied in Eysenck's model and the "liability" or "threshold" model with which psychiatric geneticists have sometimes sought to account for the inheritance of certain psychiatric disorders. The studies described in the previous chapters point very strongly to the role of additive genetic factors in creating personality differences at the level of individual items and the major personality dimensions. The environment is important, but most often the salient features of the environment for personality development are specific to individuals rather than shared with family members.

The gulf between the genetic study of normal personality on the one hand and the study of psychopathology on the other is not going to be bridged by a single effort, but we believe the attempt to be worthwhile for several reasons. We have already alluded to the use of the multifactorial model in attempts to account for the familial aggregation of certain psychiatric disorders. The liability model implies that it might be possible to devise behavioral or physiological measures that are direct indices of liability to psychiatric disorder and that we can understand much about the origins of disease from the study of normal variation. This possibility is especially important because of the practical difficulty of estimating the precise genetic correlation between psychiatric disorders by the analysis of families ascertained through diseased probands.

The dimensional model almost certainly cannot explain all there is to be said about why some individuals manifest certain types of pathology and others do not. We suppose that many people who score high on particular dimensions of personality never manifest clinical psychopathology as it is currently understood. This finding does not harm the search for indices of liability. Many more people may have personalities vulnerable to disease without ever developing symptoms. The task of understanding the disease can therefore be seen as identifying, measuring and analyzing the genetic basis of the underlying personality dimensions related to vulnerability at the same time as identifying those genetic and environmental events that account for the fact that some vulnerable individuals develop a given disease and others do not, or that some vulnerable individuals develop one set of symptoms and others may develop a quite different set. If this were possible then we should go far to reconciling normal differences in personality with the data on psychiatric disease and help in building a biological basis for the differential diagnosis of disease. In this chapter we illustrate the attempt to bridge the gap between normal personality and disease by considering the analysis of data on twins who had completed both the EPQ and a checklist of symptoms relating to anxiety and depression.

The Australian Twin Study, described in Chapter 5, yields data that illustrate the importance of reconciling studies of normal variation with those of clinical populations. In addition to the EPQ data described in Chapter 5, responses of 3798 adult twin pairs were available to 14 items selected from the Delusions–Symptoms–States Inventory (DSSI) (Bedford *et al.*, 1976). The items included on the questionnaire mailed to the twins were those of the anxiety and depression scales of the DSSI and are given in Table 11.1

The scales were developed and validated by Bedford *et al.*. The self-rating scores on the separate scales correlated significantly with clinical ratings of anxiety and depression in a validation sample and differentiated to a highly significant degree between a sample of psychiatric patients and normal volunteers. The scales are supposedly indices of "state" rather that "trait" characteristics. Bedford *et al.* found that the scores of patients had reduced by one-half one month after release from the hospital. In a sample of 96 twins from the Australian Study who returned two questionnaires at an interval of three months the mean correlation between their scores on the individual items was 0.42 ± 0.10. The reliabilities of the seven-item anxiety and depression scales, however, were both about 0.6.

We describe four analyses of these data. The first, reported in full by Kendler *et al.* (1986), addresses the analysis of the individual items within the framework of the liability model. A second analysis, limited to items related

Table 11.1   Items of the "state anxiety–depression" subscale of the Delusions–Symptons–State Inventory.[a]

Item name	Full text	Abbreviated text
Anx 1	Recently I have been worried about every little thing.	Worried about anything
Anx 2	Recently I have been breathless or had a pounding of my heart.	Breathless or heart pounding.
Anx 3	Recently I have been so worked up I couldn't sit still.	Worked up, can't sit still.
Anx 4	Recently for no good reason, I have had feelings of panic.	Feelings of panic.
Anx 5	Recently I have had a pain or tense feeling in my neck or head.	Pain or tension in head.
Anx 6	Recently worrying has kept me awake at night.	Worrying kept me awake.
Anx 7	Recently I have been so anxious that I couldn't make up my mind about the simplest thing.	Anxious can't make up my mind.
Dep 1	Recently I have been so miserable that I have had difficulty with my sleep.	Miserable, difficulty with sleep.
Dep 2	Recently I have been depressed without knowing why.	Depressed without knowing why.
Dep 3	Recently I have gone to bed not caring if I ever woke up.	Gone to bed not caring.
Dep 4	Recently I have been so low in spirits that I have sat for ages doing absolutely nothing.	Low in spirits, just sat.
Dep 5	Recently the future has seemed hopeless.	Future seemed hopeless.
Dep 6	Recently I have lost interest in just about everything.	Lost interest in everything.
Dep 7	Recently I have been so depressed that I had thoughts of doing away with myself.	Depressed thoughts of suicide

[a] All items were scored on a four-point scale: 0, not at all; 1, a little; 2, a lot; 3, unbearably.

to depression and female twins, considers two alternative models for the genetic component of liability: the polygenic model and the single-gene model. A "mixed" model is also considered that allows for one gene with large effects and the cumulative residual effects of polygenes and normally distributed environmental effects (Eaves *et al.*, 1987). The third analysis (Jardine *et al.*, 1984) addresses the extent to which genetic and environmental effects on the anxiety and depression scores derived from the DSSI items can be explained in terms of the EPQ neuroticism dimension. The final analysis (Kendler *et al.*, 1987) considers the contribution of genetic and environmental factors to the correlation between anxiety and depression items.

11.1 THE THRESHOLD MODEL FOR INDIVIDUAL SYMPTOMS

Kendler *et al.* (1986), show that the self-report scores for the anxiety and depression items are significantly higher in females than males when judged by the Mann–Whitney U-test in all items apart from "Worked up, can't sit still" among the anxiety items and the last three depression items "future seemed hopeless", "lost interest in everything" and "depressed, thoughts of suicide". Differences as a function of birth order and zygosity are not more frequent than might be expected by chance. Scores on many of the items decline significantly with age, but the age–item correlations are uniformly small. Kendler *et al.* state: "The mean absolute value of the 21 significant correlations was 0.082 indicating that for these items, age accounted for less than 1% of the variance in symptom scores".

The Australian questionnaire had information about frequency of contact in the twins. MZ twins had more frequent contact than DZ twins, and females contacted one another more than males. If frequency of contact, which is greater for MZs than DZs, were a correlate of the symptom measures, then we should expect a spurious excess in the MZ correlation, which might wrongly be attributed to genetic factors. With the Australian data, this assumption can be tested. For each item and group of twins, correlations were computed between the frequency of contact and intra-pair difference in item score. The correlations were corrected for age. Only four correlations out of the 70 computed (14 items for each of five twin groups) were significant, which is close to the rate of significant associations to be expected by chance alone. It may be concluded that the degree of contact between pair members, which is greater for MZ twins, cannot account for the effect of zygosity on intrapair variation for symptoms of anxiety and depression.

The threshold model, which was applied to the EPQ items in Chapter 8, was applied to the five contingency tables derived from the twins' responses to each of the 14 items. The only difference between the threshold model employed in the analysis of anxiety and depression symptoms and that described in Chapter 8 is the number of categories. The EPQ items were dichotomous, so a single threshold sufficed to account for the response categories in one dimension. The DSSI items have four categories, so three thresholds will be needed to account for the relationship between response categories and the hypothesized dimension of liability. The fact that there are significant sex differences in item scores for most of the items will make it necessary to allow for sex differences in thresholds when the threshold model is fitted to the contingency tables.

Before fitting genetic models to the contingency tables from the twins, the maximum-likelihood method was used to estimate thresholds and polychoric correlations for each item and group of twins separately. This approach has the advantage of generating correlations that can be inspected (see Table 11.2), and provides a test, for each contingency table, of the adequacy of the assumption that the underlying distribution of liability in twin pairs is bivariate-normal. Widespread failure of this aspect of the model at this stage might lead us to prefer a latent-class model (for example a major-gene model) or a multidimensional model for liability. The correlations were computed without the constraint that the thresholds should be equal in first- and second-born twins, so the model for each table involved seven parameters (the polychoric correlation, three thresholds for the first-born twin and three thresholds for the second-born twins). In the case of unlike-sex DZ twins, separate thresholds were estimated for male and female twins. Some of the cell sizes were very small in the most extreme category ("unbearably"), especially for male DZ twins, for which the tables were based on the smallest sample size. When small samples made it impossible to obtain reliable estimates of the polychoric correlation and its standard error, the "unbearable" category was combined with the "a lot" category, and the estimates obtained for the three-category model. This fact is noted in Table 11.2 where appropriate. For a few tables, even collapsing these two categories did not yield sufficiently large cell sizes. These cases are indicated in Table 11.2.

Of the 67 tables for which polychoric correlations could be estimated individually, the assumption of underlying bivariate normality was rejected at the 5% level in 10 cases, a significant excess over the number to be expected by chance alone. Only two tables gave residual chi-squares significant at the 1% level, which is not significantly different from the chance expectation. Taken as a whole, the results do not point to a general severe failure of the threshold model. This finding is in marked contrast with that for social-attitude items. (Chapter 12).

Apart from a few details, the model-fitting procedure was essentially the same as that described in Chapter 7. Initially, a general model was fitted to the five tables for each item, to establish a baseline against which other simplifications of the model could be assessed. The model used five parameters to represent the five correlations in liability for the five twin groups: V_{AM}, V_{AF}, E_{BM}, E_{BF} and E_{BMF}. Separate thresholds were estimated for male and female twins, but thresholds for the same sex were constrained to be equal across twin groups. The discrepancies between observed cell frequencies and their values predicted from the maximum-likelihood parameter estimates

Table 11.2 Polychoric correlations (+ s.e.) for individual anxiety and depression items for each sex and zygosity group.[a]

Item	MZ females	DZ females	MZ males	DZ males	DZ m-f
Anx 1	0.395 ± 0.032	0.163 ± 0.046	0.333 ± 0.053**	0.148 ± 0.074 [b]	0.087 ± 0.044
Anx 2	0.329 ± 0.053	0.109 ± 0.076	0.344 ± 0.096	−0.007 ± 0.132* [b]	0.005 ± 0.076
Anx 3	0.372 ± 0.042	0.245 ± 0.056	0.393 ± 0.061	0.018 ± 0.090 [b]	0.064 ± 0.054
Anx 4	0.435 ± 0.053*	0.040 ± 0.087	0.555 ± 0.086 [b]	0.305 ± 0.133 [b]	0.047 ± 0.083
Anx 5	0.354 ± 0.037	0.154 ± 0.053	0.351 ± 0.066	0.177 ± 0.092 [b]	0.202 ± 0.052
Anx 6	0.331 ± 0.039	0.237 ± 0.052	0.318 ± 0.063	0.192 ± 0.085 [b]	0.125 ± 0.051
Anx 7	0.458 ± 0.047	0.206 ± 0.073	0.390 ± 0.085	0.239 ± 0.122 [b]	0.233 ± 0.069
Dep 1	0.387 ± 0.042*	0.289 ± 0.058	0.292 ± 0.080	0.117 ± 0.107 [b]	0.122 ± 0.061 [b]
Dep 2	0.313 ± 0.039**	0.238 ± 0.051	0.383 ± 0.065	0.222 ± 0.090 [b]	0.118 ± 0.053
Dep 3	0.504 ± 0.067	0.144 ± 0.109	0.345 ± 0.143*[b]	[c]	0.212 ± 0.098
Dep 4	0.480 ± 0.043	0.232 ± 0.065	0.419 ± 0.073	0.277 ± 0.104 [b]	0.215 ± 0.062
Dep 5	0.395 ± 0.048	0.247 ± 0.068*	0.392 ± 0.078	0.095 ± 0.112* [b]	0.152 ± 0.061
Dep 6	0.430 ± 0.055	0.260 ± 0.081	0.326 ± 0.099	0.241 ± 0.137 [b]	0.104 ± 0.081*
Dep 7	0.377 ± 0.104	0.284 ± 0.133	[c]	[c]	0.045 ± 0.147*

[a] Polychoric correlations were calculated by the maximum-likelihood method, with separate thresholds for first and second twin.

[b] For these tables, because of small expected cell frequencies, it was not possible to estimate accurately the polychoric correlations and its standard error from the original 4 × 4 tables. Therefore the values were obtained from 3 × 3 tables combining the scores for responses of "a lot" and "unbearably". In tables where the polychoric correlation and its standard error could be calculated using both 4 × 4 and 3 × 3 tables, the results were very similar.

[c] For these tables, even if converted to 3 × 3, it was not possible to calculate a meaningful polychoric correlation with standard error. This resulted from either rare or a very small number of pairs concordant for scores of greater than zero.

χ^2 goodness of fit to bivariate normal threshold model: $*$ $P < 0.05$; $**$ $P < 0.01$.

were used to generate an overall chi-square test of goodness of fit for the threshold model. The low endorsement rate for the "suicide" item (DEP7), made such a test impossible. For two of the remaining thirteen items ANX1 ("worried about everything") and DEP5 ("future seemed hopeless"), the cell sizes were small for the four-category table, but acceptably large when the two most severe categories were combined. When these adjustments were made the full model fitted twelve out of the thirteen items for which it could be tested, suggesting that some version of the multiple threshold model should provide a good explanation of the symptom data.

Against the background of what can be achieved with a full specification of the correlations in liability, a more parsimonious model may be sought involving fewer genetic and environmental parameters. The models used by Kendler *et al.* allowed thresholds to be the same in both sexes ("sex-independent") and allowed the thresholds to be different between sexes ("sex-dependent"). With thresholds independent of sex, the likelihoods under three models were compared with that of the full model described above. These models were: additive genetic effects (V_A); between-family environmental effects (E_B); both V_A and E_B. As the total variance in liability is assumed to be unity under the threshold model, the effects of the within-family environment (E_W) are not specified separately, but may be obtained by difference (see Chapter 8). The effects of dominance were detected, for the threshold model as with the model for continuous traits, by the detection of significant *negative* estimates of E_B (see Chapter 5).

Allowing the thresholds to differ between the sexes, but letting the genetic and environmental parameters be constant between males and females, is equivalent under the threshold model to allowing for sex differences in mean and variance in the continuous case while constraining the correlations in liability to be the same in DZ twin pairs, regardless of sex. Four additional models were fitted to allow for sex limitation of genetic and environmental effects, with sex differences in threshold. These corresponded to the sex-limited versions of the V_A and E_B models. Two of the models, V_{AM}, V_{AF} and E_{BM}, E_{BF} allowed for sex differences in the correlation between twins, but assumed that the same genetic or environmental effects contributed to twin resemblance in both sexes (i.e. $r_{AMF} = 1$ or $r_{BMF} = 1$). The remaining two models allowed for the expression of different genes or family-environmental effects in the two sexes ($r_{AMF} < 1$ or $r_{BMF} < 1$) and thus for a reduced correlation in unlike-sex twin pairs. The goodness of fit of each model was compared with that of the full model by a likelihood-ratio chi-square test.

The model-fitting results are given in Table 11.3 for the anxiety items and in Table 11.4 for the depression items. The interpretation placed on the results depends on the limits to which we are prepared to sacrifice parsimony and uniformity over items in favor of heterogeneity among the items with

Table 11.3 Likelihood-ratio test of specific models as compared with full model for anxiety items.[a]

Specific parameters	Thresholds[b]	df of χ^2	Anx 1[c]	Anx 2	Anx 3	Anx 4	Anx 5	Anx 6	Anx 7
V_{AM}, V_{AF}, V_{AMF}	Sex-dep	2	0.36	2.50	2.52	4.50	0.18	1.64	0.24
E_{BM}, E_{BF}, E_{BMF}	Sex-dep	2	18.00**	11.12**	12.08**	18.42**	10.78**	3.70	9.12*
V_{AM}, V_{AF}	Sex-dep	3	4.52	5.70	7.28	9.36*	0.32	2.22	0.28
E_{BM}, E_{BF}	Sex-dep	3	33.76**	16.64**	26.18**	31.50**	13.50**	10.70*	12.04**
V_A, E_B/V_D	Sex-dep	3	2.74	1.18	5.74	4.52	0.30	2.32	0.68
E_B	Sex-dep	4	36.06**	18.76**	29.06*	33.20**	13.62**	11.98*	13.06*
V_A	Sex-dep	4	6.14	6.04	7.50	11.18*	0.32	2.54	0.68
V_A, E_B/V_D	Sex-ind	6[d]	67.48**	29.32**	11.64	33.66**	104.50**	42.00**	13.58*
E_B	Sex-ind	7[e]	103.68**	47.38**	34.34**	63.04**	119.54**	52.64**	25.84**
V_A	Sex-ind	7[e]	70.64**	34.14**	13.24	40.54**	104.52**	42.14**	13.60
Minus log-likelihood of full model	—		7004.92	3641.04	5425.10	3166.54	5938.18	5922.81	3740.40

[a] Twice the difference in log-likelihood of the models under comparison has a χ^2 distribution with df equal to the differences in the number of parameters in the two models. For Anx 3 and Anx 7, more complex models were fitted, with sex-independent thresholds that in no case resulted in a significant improvement in fit. Values for best-fitting model or models are underlined.
[b] Sex-dep: sex-dependent thresholds, 6 for Anx 2–Anx 7 and 4 for Anx 1.
[c] Fit to 3 × 3 tables where item scores 2 and 3 were combined.
[d] Equal to 5 for item Anx 1.
[e] Equal to 6 for item Anx 1.
χ^2 goodness of fit: * $P < 0.05$; ** $P < 0.01$.

Table 11.4 Likelihood-ratio test of specific models as compared with full model for depression items.[a]

Specific parameters	Thresholds[b]	df of χ^2	Dep 1	Dep 2	Dep 3	Dep 4	Dep 5	Dep 6	Dep 7
V_{AM}, V_{AF}, V_{AMF}	Sex-dep	2	1.54	2.18	3.16	0.72	0.72	0.40	—
E_{BM}, E_{BF}, E_{BMF}	Sex-dep	2	4.56	3.24	12.84**	12.52**	10.54**	4.84	0.26
V_{AM}, V_{AF}	Sex-dep	3	<u>2.36</u>	3.52	3.18	0.76	1.06	1.90	—
E_{BM}, E_{BF}	Sex-dep	3	9.68*	13.20*	13.44**	18.58**	14.68**	11.24*	<u>0.44</u>
$V_A, E_B/V_D$	Sex-dep	3	6.56	3.74	3.30	1.12	0.70	3.40	6.36
E_B	Sex-dep	4	45.48**	13.26**	15.44**	19.14**	16.06**	13.86**	7.58
V_A	Sex-dep	4	6.56	<u>3.94</u>	<u>4.62</u>	<u>1.12</u>	<u>1.28</u>	<u>3.62</u>	6.36
$V_A, E_B/V_D$	Sex-ind	6[c]	69.02**	127.24**	11.54	14.38*	6.00	5.42	8.36
E_B	Sex-ind	7[d]	80.50**	139.00**	23.38**	32.30**	20.66**	15.80*	9.74
V_A	Sex-ind	7[d]	60.02**	127.28**	12.66	14.38*	6.00	5.62	8.36
Minus log-likelihood of full model	—		4791.91	5777.80	2198.22	4393.23	4190.71	3201.56	1436.74

[a] Twice the difference in log-likelihood of the models under comparison has a χ^2 distribution with df equal to the differences in the number of parameters in the two models. For Anx 3 and Anx 7, more complex models were fitted, with sex-independent thresholds that in no case resulted in a significant improvement in fit. Values for best-fitting model or models are underlined.

[b] Sex-dep: sex-dependent thresholds, 6 for Anx 2–Anx 7 and 4 for Anx 1.

[c] Fit to 3 × 3 tables where item scores 2 and 3 were combined.

[d] Equal to 5 for item Anx 1.

[e] Equal to 6 for item Anx 1.

χ^2 goodness of fit: * $P < 0.05$; ** $P < 0.01$.

respect to the causes of variation in liability. If we restrict ourselves to the model that gives an adequate fit for the smallest number of parameters (a conservative approach) then we find that the assumption of identical thresholds in males and females fails badly in the majority of items, as we expected from the sex differences in mean item scores, but that the V_A model gives an acceptable fit in all but one of the fourteen items, ANX 4 ("Feelings of panic"). Even here, the residuals are only significant at the 5% level. In contrast, the E_B model gives a uniformly bad fit to all but two items, ANX 3 ("Worked up, can't sit still") and DEP 7 ("Depressed thoughts of suicide"). Only for this last item does the E_B model fit better than the genetic model, and then only marginally so.

It is tempting to accept this conservative description of the data and agree that the multifactorial liability model gives a good fit to most of the items and that there is virtually no convincing evidence that anything other than additive genetic effects is responsible for family resemblance. This view is further supported by the analysis of total anxiety and depression scores to be described in Section 11.3. However, although the V_A model is the single best-fitting model for seven of the symptoms (anxiety items 1, 3, 5 and 7; depression items, 3, 4 and 5), there is some ambiguity in the results for the remaining six items. A significant improvement results from the addition of a dominance parameter to the model for ANX 2 ("breathless or heart pounding") and ANX 4 ("feelings of panic"). In the latter case the effects of dominance could not be distinguished from the reduction in correlation of unlike-sex DZ twins due to the effects of sex-limited gene expression. In four of the items (ANX 6, DEP 1, DEP 2, DEP 6) the fit of the E_{BM}, E_{BF}, E_{BMF} model was as good as that of the V_A model with or without sex limitation. Thus for these items it could be argued that the average reduction in DZ correlation is not primarily a function of genetic effects, but rather of the low correlation between shared environmental effects in males and females (r_{BMF}). Certainly, the value of r_{BMF} is low for these items, ranging from 0.28 for DEP 6 ("Lost interest in everything") to 0.44 for ANX 6 ("Worrying kept me awake"). Therefore if our decision is based solely on statistical criteria for the individual items then a purely environmental explanation is as good as, and arguably better than, a simple genetic model in these cases. However, the genetic model is more parsimonious and provides a better explanation of the results for composite measures. If we are to take the heterogeneity seriously then it might imply that family resemblance in general "liability" has a genetic component but that there are family-environmental effects specific to individual items. Attractive though this proposition might be psychologically, it receives little support from a multivariate analysis (see Section 11.4 below).

The parameter estimates under the best-fitting models are given in Table 11.5 for the anxiety items and Table 11.6 for the depression items. There is some thematic association between the items for which alternatives to the V_A model were most appropriate. The items for which dominance may play some role were both "panic-like" symptoms. A common-environmental effect was likely for two "insomnia" symptoms and two core symptoms of "depression". Thus the heterogeneity detected between the items may reflect genuine etiological heterogeneity rather than merely chance variations on the underlying theme of additive genetic causation.

11.2 TESTING GENETIC MODELS FOR LIABILITY WITH MULTIPLE ITEMS

The weakness of the threshold analysis is twofold. First, the underlying liability to anxiety and depression is assumed to be polygenic. The only test of this important assumption is whether the contingency tables for the individual items can be explained by a bivariate-normal model for liability in pairs of twins. Secondly, the items are treated as if they are independent even though they presumably reflect the same underlying dimension of liability. We now consider how the data from multiple items may be integrated into a single analysis of liability in a way that addresses these two important issues. We assume that the same basic underlying biological or psychological variable is being assessed more or less effectively by a number of different items. Failure to take into account the properties of our measurements may lead to erroneous conclusions about the number and action of genes in the determination of a particular trait. Eaves (1983) has shown, for example, that the effects of a major gene might be inferred incorrectly when no allowance is made for the relationship between the latent dimension on which genetic effects are primarily expressed and the test scores used to summarize behavior.

We distinguish two parts of the model: (a) the "psychometric" model, which describes the relationship between the latent dimension and the responses of individuals to the test items or the symptoms on a physician's checklist; and (b) the "genetic" model, which represents the causes of family resemblance in a hypothetical latent dimension.

Although we describe the model as it would be applied to multiple items of a psychological test, the same approach can be employed in the analysis of any set of multiple-symptom data generated by systematic clinical diagnosis according to a prearranged schedule.

Table 11.5 Parameter estimates for best-fitting models for anxiety symptoms.

Symptom	Thresholds	V_A	V_{AM}	V_{AF}	V_{AMF}	E_B	E_{BM}	E_{BF}	E_{BMF}	E_W	E_{WM}	E_{WF}
Anx 1	Sex-dep	0.344								0.656		
Anx 2	Sex-dep	0.574				−0.241				0.667		
Anx 3	Sex-ind	0.358								0.642		
Anx 4[a]	Sex-dep	0.768				−0.304				0.536		
	Sex-dep		0.552	0.400	0.093[b]						0.448	0.600
Anx 5	Sex-dep	0.351								0.649		
Anx 6[a]	Sex-dep	0.333								0.667		
	Sex-dep						0.272	0.293	0.125[c]		0.728	0.707
Anx 7	Sex-dep	0.437								0.563		

[a] For these symptoms, two best-fit models were found, listed in order of likelihood.
[b] Correlation between V_{AM} and V_{AF} for second-best-fitting model for Anx 4 equal to 0.198.
[c] Correlation between E_{BM} and E_{BF} for second-best-fitting model for Anx 6 equal to 0.443.

Table 11.6 Parameter estimates for best-fitting models for depression symptoms.

Symptom	Thresholds	V_A	V_{AM}	V_{AF}	E_{BM}	E_{BF}	E_{BMF}	E_W	E_{WM}	E_{WF}
Dep 1[a]	Sex-dep		0.249	0.414					0.751	0.586
	Sex-dep				0.227	0.349	0.109[b]		0.773	0.651
Dep 2[a]	Sex-dep	0.334						0.666		
	Sex-dep				0.326	0.282	0.116[b]			
Dep 3	Sex-dep	0.445						0.555		
Dep 4	Sex-dep	0.459						0.541		
Dep 5	Sex-ind	0.385						0.615		
Dep 6	Sex-ind	0.398						0.602		
	Sex-ind				0.303	0.362	0.092[b]		0.697	0.638

[a] For these symptoms, two best-fit models were found, listed in order of likelihood.
[b] Correlations between E_{BM} and E_{BF} for the second-best-fitting model for Dep 1, Dep 2 and Dep 6 were respectively 0.387, 0.383 and 0.278.

11.2.1 The psychometric model

There is already an extensive psychometric literature on "latent-trait" models for psychological test data (see e.g. Lord and Novick, 1968; Bock and Lieberman, 1970; Bock and Aitkin, 1981). Mislevy (1984) used a latent-trait model to test for component normal distributions in the distribution of spatial ability caused by a hypothetical sex-linked locus, but did not employ family data in the analysis.

There are several possible models for the relationship between a subject's score on a latent trait and the probability that he/she will endorse a particular item on a test. We assume that the probability that a given dichotomous item will be endorsed is a cumulative normal function of the subject's latent trait value. Thus, writing X_i for the response (zero–one) to the ith item, and for the trait value of the subject, the probability of endorsement is

$$P_i(\theta) = \int_{-\infty}^{a_i(\theta - b_i)} \phi(\theta)\, d\theta.$$

The parameter a_i is the "discriminating power" of the item. The trait value that results in $P_i(\theta) = 0.5$ is the "item-difficulty" parameter b_i. The regression of $P_i(\theta)$ on θ is steeper at b_i for items with larger a_i (see e.g. Lord and Novick, 1968).

For a subject of given ability θ, the likelihood of a particular vector of k zero–one responses X is

$$l(X|\theta) = \prod_{i=1}^{k} P_i(\theta)^{X_i}[1 - P_i(\theta)]^{1-X_i}.$$

11.2.2 The genetic model

The genetic model is summarized in the (multivariate) frequency distribution of the latent trait in the population of families from which a sample has been drawn. In many psychometric applications the distribution of the latent trait is not known *a priori* and the psychometrician is faced with the (formidable) task of estimating the item parameters and latent-trait values for all the subjects. For many genetic applications, however, it is sufficient (though not necessarily easier) to estimate parameters of the distribution of θ in families sampled from the population. For example, if we are prepared to assume a large number of genes of infinitesimal effect (the "polygenic" model) then the distribution of θ is assumed to be normal and the covariances between relatives simply a function of the heritability and the degree of genetic relatedness. Under the so-called "mixed" model (see e.g. Lalouel *et al.*, 1983) the distribution is assumed to be a mixture of normal distributions, each centered on the average trait value of individuals having a given genotype at a locus of large effect. If the major locus accounts for *all* the genetic variation then the residual trait values are uncorrelated in families. If there are also polygenic effects then the trait values will be correlated within families, even when the effects of the major locus are controlled. The elements of the mixed model are given in Figure 11.1.

The parameters of the mixed model are of two kinds: those that describe the effects of the major locus, and those that describe the contribution of polygenic effects and residual environmental effects to variation in the latent trait around the mean value characteristic of each genotype at the major locus. In the case of MZ twins the residual correlation in the latent trait is assumed to be h^2, i.e. the narrow heritability of the normally distributed residual component, and to be $\frac{1}{2}h^2$ in the case of DZ pairs. The effects of the major gene are specified on the assumption that there are two alleles at the major locus and that the heterozygote is intermediate in expression between the two homozygotes. This assumption can also be relaxed to allow for dominance.

The major-gene component has parameters for the frequency p of the allele that increases liability and for the additive deviation d of the increasing homozygote from the midpoint m of the homozygotes. We assume that the heterozygous effect h is zero in our application. The model allows for the effects of a normally distributed random environmental variable that is uncorrelated between twins (i.e. E_W). The family environment could be

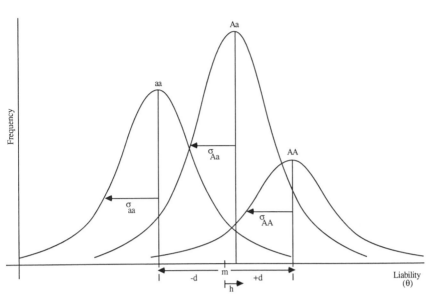

Figure 11.1 Model for effects of single gene with large effect and multifactorial residual effects on liability for depression.

included if desired. In all cases the item parameters and genetic effects are scaled so that the latent trait has zero mean and unit variance. Under models that include the effects of a major-gene model this scaling implies the constraint that $m + (2p - 1)d + 2p(1 - p)h = 0$. The polygenic component of the latent trait is assumed to be normally distributed. The item analysis in Section 11.1 showed that the frequency of "depressive" symptoms declined significantly with age. This effect is included in our models by estimating simultaneously the linear regression z of latent trait on age. Ideally, we should also like to fit the more general "unified mixed model" (Lalouel *et al.*, 1983), which seeks additional likelihood-ratio tests for the Mendelian inheritance of a hypothetical major locus, but this task is not feasible with our current algorithm and computer resources.

We write $\Phi(\theta)$ for the frequency distribution of the latent trait in the population. The unconditional likelihood of a given response, vector **X**, is thus the integral

$$l(X) = \int_{-\infty}^{\infty} \Phi(\theta)l(X \mid \theta) \, d\theta,$$

and the likelihood of a whole sample of unrelated individuals is simply the product of the individual likelihoods. The likelihood may then be maximized with respect to the item parameters and the parameters of the distribution of θ.

In genetic applications, however, unrelated individuals do not yield the information necessary to test alternative hypotheses about the genetic and environmental determinants of θ. For this purpose we require kinship data. For simplicity, we consider only two-member kinships, but the theoretical extension to larger kinships is straightforward. The practical application of the model, even with kinships of two individuals, requires considerable computational resources, so the treatment of larger families may be difficult.

We let θ_1 and θ_2 denote the trait values of pairs of related individuals and the vectors $\mathbf{X}$ and $\mathbf{Y}$ be typical response vectors of the first and second individuals. The likelihood of the responses is then

$$l(\mathbf{X},\mathbf{Y}) = \int\limits_{-\infty}^{\infty} \int\limits_{-\infty}^{\infty} \Phi(\theta_1,\theta_2) l(\mathbf{X}\,|\,\theta_1) l(\mathbf{Y},\theta_2) d\theta_2 d\theta_1,$$

where $\Phi(\theta_1, \theta_2)$ is the bivariate frequency distribution of the trait values in pairs of relatives of the given type. The form of $\Phi(\theta_1, \theta_2)$ will depend on the type of relationship being considered and the genetic model being tested. In applications where the latent trait is dependent on covariates such as sex and age it is conceivable that each pair will have its own unique, expected, trait values, so that Φ will have to be expressed separately for each pair. The likelihood $l(\mathbf{X}, \mathbf{Y})$, may be evaluated for each pair of relatives, and the overall likelihood accumulated and maximized with respect to the item parameters and parameters of the genetic model.

With the mixed model, the bivariate probability density for MZ pairs is the sum of three functions, one for each genotype at the major locus weighted by the genotype frequencies. We thus write

$$l(\mathbf{X},\mathbf{Y}) = \sum_{i=1}^{k} f_i \int\limits_{-\infty}^{\infty} \int\limits_{-\infty}^{\infty} \Phi(\theta_1\theta_2) l(\mathbf{X}\,|\,\theta_1) l(\mathbf{Y}\,|\,\theta_2) d\theta_2 d\theta_1,$$

where f_i is the population frequency of the ith genotype, Φ_i is the joint distribution of θ conditional on genotype at the major locus, which is assumed to be $N[\mathbf{g}_i, \Sigma_i]$. The mean liability of the ith genotype g_i is expressed in terms of the genetic parameters m, d and h. The model assumes that there is no interaction between the effects of the major gene and the effects contributing to the multifactorial component, so that the distribution of the residual effects is identical for the three genotypes. For DZ twins the bivariate density function is the sum of nine components, one for each possible pair of twin genotypes, weighted by the frequency of each genotype pair.

11.2.3 Application to twin data on symptoms of depression

In applying the method to the depression data, we used only six of the seven DSSI items. The item relating to suicide was omitted from this analysis because of the very low endorsement frequency (see Section 11.1 above). Altogether, there were complete data on 1983 pairs of adult Australian female twins (1233 MZ and 750 DZ). Females were chosen because the higher frequency of depression in females makes them more informative. The mean age of the sample of females is 35.45 years.

The threshold analysis of the individual items has shown that an additive genetic model is adequate to account for the polychoric correlations between the responses of twins. There was little evidence of non-genetic familial effects on liability to endorse the items. Table 11.7 summarizes the basic properties of the data. In computing these summary statistics, we included *all* female twins in the sample (including those from male–female pairs). Our subsequent genetic analysis only includes the female like-sex twin *pairs* in the sample. The phenotypic correlations between the items were computed by applying the product-moment formula to the raw responses coded 0, 1, 2, 3 in order of increasing severity. The first two eigenvalues of the correlation matrix are 3.359 and 0.671. There is strong support for a unidimensional model for the latent space. When the latent trait is normally distributed and unidimensional, the item difficulties and loadings on the general factor of the tetrachoric interitem correlations are sufficient to determine the parameters of the latent trait model (Lord and Novick, 1968). This will not be the case if

Table 11.7 Preliminary statistics ($N = 4872$).

Item	Endorsement frequency	Correlations					Loading
		2	3	4	5	6	
1	0.242	0.416	0.406	0.441	0.457	0.451	0.698
2	0.351		0.401	0.471	0.404	0.459	0.692
3	0.073			0.445	0.525	0.563	0.747
4	0.195				0.486	0.538	0.755
5	0.178					0.588	0.778
6	0.120						0.812

a major gene is segregating, since the assumption of normality will be violated, or if genetic and environmental effects do not affect the phenotype through the one common underlying variable (see Section 11.4 below).

For the purpose of genetic analysis, the original responses were recoded as dichotomous by scoring all symptomatic responses as one and all non-symptomatic responses as zero. Although analysis of the raw multicategory data is possible in theory, it is likely to be prohibitive in practice with our current computer resources.

We fitted three genetic models to the data: the "polygenic model", in which we specified no major-gene effects ($p = d = 0$); the "major-gene model", in which $h^2 = 0$; and the "mixed model", in which the major gene effects and the polygenic component were estimated simultaneously. In each case the 12 item-parameters and the age regression were estimated at the same time. The statistical significance of the major-gene component can be judged by comparing the likelihood under the polygenic model against that for the mixed model. A significant increase in the likelihood for the mixed model gives support to the effects of a major gene. The difference between the likelihood for the mixed model and that for the major-gene model is a guide to the importance of polygenic effects.

11.2.4 Numerical method

The maximum-likelihood procedure was implemented in a FORTRAN program. Two main numerical issues have to be faced. The first is the integration of the bivariate expression (11.1) pair by pair to compute the likelihood for a given set of parameter values. The form of $l(X, Y)$ changes markedly with θ from one pair to the next because of the unique response patterns generated by each pair. We have employed Gauss–Hermite quadrature, embodied in the NAG (1982) FORTRAN subroutine DO1FBF. Such methods approximate the integral by a weighted sum of function values for a specified optimal set of abscissae. Preliminary studies suggested that the approximation was very poor unless quite a large number of abscissae were used in each dimension ($N = 12$), necessitating 144 function evaluations to obtain the likelihood for each twin pair. Our finding is consistent with that reported by Bock and Aitkin (1981), who found, for the unidimensional case, that $N = 10$ gave very similar values for the likelihood of 5 LSAT responses of 1000 subjects to those obtained with $N = 40$.

The second numerical problem is that of maximizing the likelihood with respect to variation in the parameters. In our case, the NAG subroutine EO4JBF was used to minimize the negative log-likelihood ($-L$) over all twin pairs in the sample. For the polygenic model and the major-gene model we maximized the likelihood simultaneously with respect to all parameters. Since the item parameters were not greatly different for the two models, we saved some computer time in fitting the mixed model by fixing the item parameters in earlier cycles and maximizing the likelihood with respect to the genetic and environmental parameters only. Other investigators have used an EM algorithm in the unidimensional case to obtain item parameters without assuming any prior distribution for the subjects' trait values (Bock and Aitken, 1981). To enhance the performance of the algorithm, $-L$ was divided by a scale factor s, so that $L^* = L/s$ was in the range 0–1. The procedure was assumed to have converged when parameter estimates were stable to five significant figures. Typically, this criterion resulted in values for the gradients of L^* of the order of 10^{-7}.

11.2.5 Results and discussion

The results of fitting the three models are summarized in Table 11.8 (Eaves *et al.*, 1987). The item parameters do not change much as a function of the genetic model, but have altered rather more than those obtained by Bock and Aitkin in their analysis of LSAT data. The item-difficulty parameters show their expected consistency with the raw endorsement frequencies, and there is broad agreement between the discriminating powers of the items and the loadings of the general factor extracted from the product-moment correlations. The heritability of the latent trait under the polygenic model is estimated to be 0.51. This is remarkably close to the value that Kendler *et al.*

Table 11.8 Parameter estimates from three genetic models for multiple symptoms of depression.

Item	Polygenic		Major-gene		Mixed	
	a_i	b_i	a_i	b_i	a_i	b_i
1	1.038	0.991	0.989	1.009	1.008	1.003
2	0.948	0.568	0.913	0.566	0.928	0.566
3	1.549	1.739	1.430	1.819	1.450	1.770
4	1.337	1.091	1.259	1.116	1.310	1.099
5	1.478	1.133	1.376	1.163	1.429	1.147
6	1.900	1.322	1.751	1.368	1.873	1.330
p	—		0.242		0.045	
d	—		1.091		1.041	
h^2	0.510		[0.437]		0.427	
z	−0.0103		−0.0114		−0.0112	
L	−9073.16		−9074.57		−9072.10	

obtained (see Section 11.4 below) when fitting a single common factor model to the cross-twin interitem polychoric correlations. The value given is the heritability of the latent trait and is expected to be greater than the heritability of scale scores derived from the data because the latter are subject to stochastic error. Jardine *et al.* (1984) report a narrow heritability of 0.37 for the raw depression scores for the same data. Under the major-gene model, the contribution of genetic effects to variation in liability is $g^2 = (1 - 2pq)d^2 - [(2p - 1)d]^2$, from which we find $g^2 = 0.437$ in our data.

The most important result of the analysis is the comparison of the likelihood under the mixed model to those obtained under the other two models for genetic liability. Since the polygenic model and the major-gene model are both special cases of the more general mixed model, we may see if either or both of the former receive significantly less support than the mixed model. It turns out (Table 11.8) that the mixed model is supported slightly better than both the polygenic model ($X^2 = 2.12$) and much better than the major-gene model ($X_1^2 = 4.94$). The particular parameter values for the mixed model are intriguing. Most of the genetic variation, in the mixed model, is due to polygenic factors, which generate a correlation of 0.4 for the residual effects of MZ twins. The parameters for the major-gene component suggest a locus in which the difference in liability between decreasing homozygotes and heterozygotes is 2.32 units on the standardized liability scale. The estimated frequency of the increasing allele is approximately 0.006, which implies that approximately 1% of the population would have this markedly increased liability. This means that our sample of approximately 4000 females contains about 40 individuals who are of such extreme liability that the assumption of underlying normality embodied in the polygenic model cannot account for their response vectors. The general conclusion of this analysis is therefore that most of the liability underlying the depression symptoms in a normal population is apparently polygenic. There is a slight suggestion that a few individuals (<1%) might have very extreme liability characteristic of a syndrome of depression. However, there is no clear statistical evidence to suggest that one gene stands out from the background of normal polygenic variation in liability.

11.3 THE RELATIONSHIP BETWEEN PERSONALITY AND SYMPTOMS

The fact that the data on anxiety and depression are available on twins for whom scores on the EPQ are also known provides a unique opportunity to assess the genetic and environmental basis of the correlation between personality and symptoms of anxiety and depression in an unselected population. We outline the essentials of such an analysis undertaken by Jardine *et al.* (1984).

The items on the anxiety and depression scales of the DSSI were combined with equal weight using a score of 0, 1, 2, 3 for each item scale. The total scores on the anxiety and depression scales (seven items in each) thus range from 0 to 21. The raw anxiety and depression scores are highly skewed, with very few subjects obtaining total scores in excess of 7. A logarithmic transformation $X' = \log(X + 1)$ was used to remove any mean-variance regression for the anxiety scores of the MZ twin pairs. The raw neuroticism scores were subjected to the angular transformation (see Chapter 5). Transformed scores were corrected for regression of age and sex, before being summarized by the matrices of mean products within and between pairs for each of the five groups of twins in the Australian sample (Table 11.9).

A univariate analysis of the transformed anxiety and depression scores (see Jardine *et al.*, 1984) showed that the same genetic and environmental parameters could not be fitted to the depression data from both males and females. There was highly significant heterogeneity between the sexes, even though the models fitted very well within each sex. Models that ignored

Table 11.9 Mean products between and within pairs for transformed anxiety (A), depression (D) and neuroticism (N) scores.

			Mean products					
			Between pairs			Within pairs		
Group	Number of pairs		N	A	D	N	A	D
MZ$_f$	1229	N	330.30	4.4175	4.0954	104.96	1.1982	1.1647
		A		0.1337	0.0951		0.0598	0.0323
		D			0.1280			0.0606
MZ$_m$	566	N	315.18	4.1190	3.4697	115.65	1.1616	1.1665
		A		0.1222	0.0740		0.0571	0.0276
		D			0.1031			0.0540
DZ$_f$	749	N	264.70	3.6399	3.5327	157.57	1.9503	1.8274
		A		0.1168	0.0811		0.0806	0.0500
		D			0.1180			0.0812
DZ$_m$	351	N	263.16	2.9354	2.8184	183.94	1.9195	1.5822
		A		0.0971	0.0535		0.0761	0.0339
		D			0.0912			0.0587
DZ$_{os}$	900	N	228.45	3.1143	2.9562	174.26	2.3166	2.0929
		A		0.1057	0.0711		0.8276	0.0505
		D			0.1079			0.0826

genetic effects altogether failed badly in most cases, whereas the model that includes only additive genetic effects and within-family environmental influences generally gave a good fit. No improvement resulted from adding dominance to the model. In the case of anxiety scores there were no significant sex differences in the genetic and environmental components. Table 11.10 summarizes the univariate findings of Jardine *et al.* with respect to anxiety and depression scores in their sample. A small sample on which repeated measures were obtained yielded an independent measure of error and short-term fluctuations in test scores. The estimated contribution of E_W may thus be partitioned into that due to "stable" environmental effects ("true" E_W) and that due to error and short-term changes. This additional source of variation is recognized in Table 11.10

Anxiety and depression scores on the self-report instrument devised by clinical criteria correlate very highly with the neuroticism scores derived from the EPQ (Table 11.11). The test–retest reliabilities of the neuroticism

Table 11.10   Results of fitting univariate models to anxiety and depression scores.[a]

	Proportion (%) of total variation			
	Anxiety		Depression	
Source	Females	Males	Females	Males
Retest error	38	45	37	29
"True" E_W	23	22	26	38
Genetic (V_A)	39	33	37	33
Goodness-of-fit χ^2		2.15		8.73
df		6		5
$P\%$		91		12

[a] Best-fitting model for depression requires sex differences in V_A and E_W. Model for anxiety requires sex differences in V_A only.

Table 11.11   Age-corrected correlations of transformed anxiety and depression scores with EPQ scores.

	Anxiety		Depression	
	Females	Males	Females	Males
Extraversion	−0.08***	−0.12***	−0.10***	−0.16***
Psychoticism	0.15***	0.12***	0.18***	0.16***
Neuroticism	0.61***	0.60***	0.57***	0.55***
Lie	−0.12***	−0.11***	−0.09***	−0.08***
Anxiety	—	—	0.66***	0.60***

Significance level: *** P < 0.001.

score in the Australian sample is 0.85 in females and 0.83 for males, and the reliabilities of the anxiety and depression scores are 0.63 and 0.66 in females and 0.62 and 0.58 in males. The correlations between traits are comparable to the reliabilities of measurement. Therefore a very high proportion of the *reliable* variation in anxiety and depression symptoms is due to the same common factor measured by the neuroticism scale of the EPQ.

The maximum-likelihood method described in Chapter 10 was used to obtain estimates of the genetic and environmental factor loadings and specific variances for the three tests. Since there are three measures, this model amounts to a full parametrization of the genetic and environmental covariance matrices. Since the univariate analyses had suggested that only additive genetic effects and within-family environmental effects were important, the multivariate analysis made no allowance for dominance or the family environment. However, there were significant sex-limited effects, so the model was fitted separately to male and female twins. Unlike-sex pairs were omitted from the multivariate analysis. Table 11.12 summarizes the multivariate genetic analysis, giving the proportion of variance in each measure, which is explained by the common and specific genetic and environmental factors.

In both sexes most of the genetic variation in the three measures is explained by the common factor. Indeed, the genetic variation in anxiety and depression scores is explained almost entirely by their genetic communality with neuroticism, since less than 10% of the genetic variance in both "symptom" scales is trait-specific. On the other hand, there is a slightly larger specific genetic component to neuroticism. The contributions of the environmental factor and specifics are much more equal. The unique

Table 11.12 Estimates of contributions of common and specific genetic and environmental factors (proportions of variance) from multivariate model for covariation between neuroticism, anxiety and depression.[a].

Sex	Trait	Environmental (E_W)		Genetic (V_A)	
		Common	Specific	Common	Specific
Females	Neuroticism	20	29	35	16
	Anxiety	35	27	35	03
	Depression	33	31	30	06
Males	Neuroticism	22	32	34	12
	Anxiety	31	35	30	04
	Depression	33	35	23	09

[a] All contributions are statistically significant at the 0.001 level. Goodness-of-fit chi-squares (12 df) for the model given are 6.90 (females) and 12.52 (males).

experiences of the individual that contribute to E_W clearly make a substantial contribution to the phenotypic correlation between the measures.

When we interpret the large environmental specifics, however, we have to remember that these are comparable to the error variances assessed by repeated measures. If we subtract measurement error from the estimates of specific variance then we are left with virtually no trait-specific environmental effects. Thus our model for the relationship of neuroticism to symptoms of anxiety and depression in a non-clinical population is comparatively straightforward. Variation is due to additive genetic effects and the environment within families. Variation in self-report symptoms is therefore no different from that of other measures of personality. Furthermore, the phenotypic correlations between neuroticism and the symptom scales are high, indicating that the same factors contribute to variation in all three scales. Anxiety and depression scores are highly correlated with each other and with neuroticism scores. The genetic analysis of trait covariation gives strong support to the view that the same genetic effects that contribute to anxiety symptoms also contribute to mild symptoms of depression. There is virtually no specific genetic variance in either trait. In addition, all that these symptoms have in common with one another genetically is also shared with the neuroticism scores derived from the EPQ. A similar view seems true for the effects of the environment. Short-term changes apart, which contribute to specific environmental variation, virtually all of the environmental variation in neuroticism and scores on the DSSI anxiety and depression scales has a general effect on all scales. Long-term environmental effects contribute to all traits simultaneously. The fact that neuroticism, anxiety and depression are not completely correlated is probably due to short-term fluctuations rather than an underlying difference in the genetic basis of the traits. In so far as neuroticism is a "trait" measure, and anxiety and depression symptoms, as recorded, are "state" measures, we expect that the specific variation in anxiety and depression would largely be due to short-term fluctuations.

The genetic and environmental correlations, derived from the factor loadings and specifics, are given in Table 11.13. The genetic correlations are large because the specifics are small. The environmental correlations are much smaller because of the more substantial short-term fluctuations in the symptom data. In Table 11.14 we give the environmental correlations corrected for unreliability on the assumption that all interaction of subjects and occasions is due to environmental effects (but see Chapter 9 for a critical account of this assumption). Correction for unreliability increases the environmental correlations substantially, making them much more comparable to the genetic correlations. In the case of anxiety and depression scores, the corrected environmental correlation is greater than unity in both sexes. These nonsense values may stem from lack of precision in the esti-

Table 11.13 Estimated genetic and environmental correlations between neuroticism (N), anxiety (A) and depression (D) scores (females in upper triangle, males in lower triangle).

Trait	Genetic			Environmental		
	N	A	D	N	A	D
N	—	0.80	0.76	—	0.47	0.45
A	0.81	—	0.88	0.44	—	0.54
D	0.73	0.79	—	0.45	0.48	—

Table 11.14 Environmental correlations between neuroticism (N), anxiety (A) and depression (D), corrected for short-term environmental fluctuations.

	"Corrected" environmental correlation					
	Females			Males		
	$r^{(a)}$	A	D	r	A	D
N	0.735	0.89	0.76	0.673	0.94	0.73
A	0.377	—	1.27	0.328	—	1.04
D	0.476	—	—	0.567	—	—

[a] r is the estimated proportion of environmental variation due to long-term environmental effects. Note that correlations based on variance component estimates can exceed unity (see text).

mates of unreliability, or they could point to a failure in the model employed in the correction procedure. We have assumed that short-term fluctuations in anxiety and depression scores are uncorrelated and that they may therefore contribute only to *specific* environmental variances. In reality, this may not be the case if anxiety and depression are both subject to the same short-term environmental influences. The fact that we obtain nonsense estimates for the unreliability correction to the environmental correlation may point to correlated short-term changes in self-reported anxiety and depression symptoms, i.e. if a person feels miserable on the day they are as likely to say "yes" to an anxiety item as they are to a depression item.

11.4 CORRELATION BETWEEN INDIVIDUAL SYMPTOMS: A GENOTYPE–ENVIRONMENTAL MODEL

The analyses of the items so far show that: (1) genetic factors can account for much of the variation in response to individual items; (2) there is a common genetic dimension of liability underlying responses to many of the items;

(3) there is a large genetic correlation between scores on the anxiety and depression scales that is shared also by neuroticism; (4) some of the effects of the environment within families are specific to anxiety and depression; (5) short-term environmental effects explain much of the specific environmental variance. These many facets of the model for anxiety and depression are synthesized and extended in a multivariate analysis of the individual item-responses conducted by Kendler *et al.* (1986).

The basic questions addressed are: (1) How many factors were required to explain the genetic and environmental correlation between the individual items? (2) Were the effects of genetic and environmental factors mediated through a common underlying phenotype or was the structure of genetic and environmental covariance different? (3) Are the effects of genes and environment consistent over sexes? Similar questions were addressed in Chapter 10 in relation to the covariance structure of impulsiveness. There we concluded that, specific components apart, the genetic and environmental structures were approximated quite well by a common phenotypic pathway. For neuroticism, however, different genetic and environmental factors were needed.

The data were summarized by computing five 26 × 26 matrices of polychoric correlations between the responses for first and second twins in each zygosity group. The item relating to "suicide" was omitted for lack of informativeness (see Section 11.1 above). Hypotheses about the structure of the polychoric correlations were tested by the method of weighted least squares using the amounts of information about the polychoric correlations as weights. No allowance was made for the correlations between the estimated polychorics. The approximate method was used because an exact treatment by maximum-likelihood is computationally infeasible. The method yields approximate chi-square tests of the goodness of fit of the models, and differences in chi-square may be used to construct tests of significance of alternative hypotheses. More details are given by Kendler *et al.* (1986).

Two basic models were fitted to the data omitting unlike-sex pairs: (1) the "latent phenotypic-factor model", which assumes that genetic and environmental factors all affect a common underlying phenotype that creates the covariation between the items so that the environmental loadings are constant multiples of the genetic loadings (see Chapter 10); (2) the "general model", which allows the genetic and environmental factors to be independent. For each of the two cases we allowed for one, two and three factors with specifics and we constructed tests of heterogeneity of the factor loadings between males and females. When fitting the general model, there are nine possible factor models, allowing for 1, 2 and 3 factors for genetic and environmental correlations independently.

When we use the conservative overall goodness-of-fit criterion, we find

Table 11.15   Goodness of fit of latent phenotypic factor models.

	Females			Males		Combined		
	df	χ^2	P	χ^2	P	df	χ^2	P
3-factor	595	550.04	0.91	638.37	0.11	1245	1355.54	0.01
2-factor	609	729.35	<0.001	742.83	<0.001	1259	1645.77	<0.001
1-factor	623	1333.58	0.0	969.55	0.0	1273	2451.89	0.0

that the three-factor solution for the latent phenotypic-factor model fitted the data for males and females. Neither the single-factor model nor the two-factor model fitted the data in either sex. There was also marked heterogeneity between sexes. Table 11.15 gives the goodness-of-fit chi-squares for the latent phenotypic-factor model.

The loadings for the three-factor version of this model are given in Table 11.16. The rotation is achieved by fixing one loading on the second factor to zero and this loading, plus one other, to zero on the third factor. This solution may be transformed orthogonally, if desired, into other equivalent solutions, including those which would be obtained by assigning the three zero-loadings to other variables (Jöreskog and Sörbom, 1988). Since the environmental loadings are multiples of the loadings of the variables on the corresponding genetic factors, we do not give separate environmental-factor loadings but rather provide the multipliers b needed to generate the environmental loadings from their genetic counterparts. The multipliers are related to the heritabilities of the common factors by $h^2 = 1/(1 + b^2)$.

The first factor is clearly a general neuroticism factor, as is expected from Eysenck's model, and is fairly highly heritable in both sexes. The other two factors are influenced far more by environmental effects (even though the item-specific errors of measurement have been removed into specific environmental variances). The second factor has highest loadings on the "depression" items, and the third is mostly an "insomnia" factor. Although the genetic loadings on these two factors are quite small, the environmental loadings are between 1.6 and 2.6 times as great. For most items the genetic specific loadings are substantial. This surprising finding implies that, even at the level of individual items, gene action is highly specific. In the previous chapter we saw that this was also true at the "primary-factor" level.

Chi-square tests of goodness of fit are given for various forms of the "general" model in Table 11.17. Although a number of models were acceptable by the "goodness-of-fit" criterion, it is clear that the model that allows for three independent genetic and environmental factors is significantly better than almost all the alternatives. The estimated loadings under the

Table 11.16 Factor loadings (×100) under the 3-factor latent phenotypic model.

	Females					Males				
	Genetic loadings			Genetic specifics	Environmental specifics	Genetic loadings			Genetic specifics	Environmental specifics
	I	II	III			I	II	III		
Miserable, difficulty with sleep	49	17	33	0	0	46	16	24	0	31
Depressed without knowing why	53	9	−1	25	63	56	8	2	26	62
Gone to bed not caring	56	21	0	34	38	52	20	1	18	56
Low in spirits, just sat	56	15	−1	39	44	57	15	−2	29	52
Future seemed hopeless	52	26	0[a]	24	44	52	23	0[a]	21	51
Lost interest in everything	60	22	−4	16	37	58	22	−2	0	45
Worried about everything	51	5	13	32	58	44	5	13	32	66
Breathless or heart pounding	39	−8	3	39	73	48	−11	5	24	70
Worked up, can't sit still	52	−2	8	33	60	49	−3	14	31	61
Feelings of panic	58	0[a]	0[a]	26	55	64	0[a]	0[a]	38	42
Pain or tension in head	41	−9	7	40	69	43	−6	8	40	69
Worrying kept me awake	44	11	25	28	51	40	10	19	35	53
Anxious, can't make up mind	61	5	0	28	46	61	8	6	14	56
Environmental multipliers										
b	0.94	1.57	1.78			0.80	2.00	2.57		
$h^2 = 1/(1 + b^2)$	0.53	0.29	0.24			0.60	0.20	0.13		

[a] Parameter is fixed to zero *ex hypothesi*.

Table 11.17 Fit of unconstrained covariance-structure models to twin matrices of polychoric correlations for symptoms of anxiety and depression.

	Goodness of fit					Improvement				Genetic factors[a]			Environmental factors[a]		
	Females		Males			Females		Males		I	II	III	I	II	III
df	χ^2	P	χ^2	P	df	χ^2	P	χ^2	P						
559	470.79	0.99	556.78	0.52						1	1	1	1	1	1
572	506.10	0.98	573.06	0.48	13	35.31	**	16.28	NS	1	1	0	1	1	1
585	542.35	0.90	599.41	0.33	26	71.56	***	42.63	*	1	0	0	1	1	1
572	510.40	0.97	587.80	0.31	13	39.61	***	31.02	***	1	1	1	1	1	1
585	550.20	0.85	611.33	0.22	26	79.41	***	54.55	***	1	1	0	1	1	0
598	673.51	0.02	668.84	0.02	39	116.74	***	112.06	***	1	0	0	1	1	0
585	581.93	0.53	625.77	0.12	26	111.14	***	68.99	***	1	1	1	1	0	0
598	711.59	***	684.76	**	39	240.80	***	127.98	***	1	1	0	1	0	0
611	1070.53	***	848.94	***	52	599.74	***	292.16	***	1	0	0	1	0	0

[a] 1 indicates that parameters are free (unless otherwise stated in text); 0 indicates that all parameters are fixed to zero for a given factor.
*, $P < 0.05$; **, $P < 0.01$; ***, $P < 0.001$; NS, not significant.

three-factor general model are given in Table 11.18. We select one of two rotations discussed by Kendler *et al.* (1986).

The results show strong evidence of general genetic and environmental "neuroticism" factors. These factors are not sufficient to explain the data, however. In both males and females the second genetic factor relates to the physical symptoms of anxiety ("breathless or heart pounding" and "pain or tension in the head"), and the third genetic factor relates to insomnia in females but is less clear in males. In both sexes the second environmental loading is a depression "group" factor, loading primarily on the depression items. The third environmental factor loads primarily on the two "sleeplessness" items in both sexes.

This study has several important implications for our understanding of the causes of anxiety and depression symptoms. It is clear that there is a generalized effect of "vulnerability" to express symptoms of both disorders. The data imply that some genetic and environmental influences will contribute to the development of both types of symptoms. However, the model is more complex than that because there are also environmental effects that precipitate the expression of depression symptoms without affecting anxiety (the second environmental factor in Figure 11.1). Furthermore, some specific symptoms and symptom combinations appear to be affected by their own genetic components (for example the physical symptoms of anxiety).

Taking this finding with the analysis of Jardine *et al.* (see Section 11.3

Table 11.18 Factor loadings ($\times 100$) of symptoms of anxiety and depression on orthogonal genetic and environmental factors.

	Females								Males							
	Genetic factors				Environmental factors				Genetic factors				Environmental factors			
	I	II	III	Specific	I	II	III	Specific	I	II	III	Specific	I	II	III	Specific
Miserable, difficulty with sleep	48	-1	44	0	55	-22	48	0	48	-7	19	10	46	-31	41	50
Depressed without knowing why	52	10	12	21	52	-3	-15	61	56	9	-5	24	50	-17	-3	58
Gone to bed not caring	53	-6	5	43	58	-38	-10	17	58	8	3	3	25	-50	2	59
Low in spirits, just sat	59	12	16	32	50	-17	-16	46	61	8	-27	0	31	-38	4	55
Future seemed hopeless	52	10	0[a]	34	51	-50	0[a]	32	51	9	0[a]	31	32	-60	0[a]	42
Lost interest in everything	62	15	-1	17	55	-34	-10	37	61	8	-9	0	31	-56	-2	45
Worried about everything	51	9	10	33	50	-3	24	56	41	-12	37	0	50	-17	7	63
Breathless or heart pounding	39	-30	-11	25	38	5	3	73	46	-35	-5	0	29	-1	2	76
Worked up, can't sit still	52	0	1	35	50	7	14	59	42	-18	8	37	53	-5	19	57
Feelings of panic	60	0[a]	0[a]	20	49	0[a]	0[a]	60	75	0[a]	0[a]	0	35	0[a]	0[a]	56
Pain or tension in head	40	-26	-2	34	42	12	7	68	34	-44	-16	0	47	-5	6	67
Worrying kept me awake	43	-5	28	29	48	-16	38	50	47	-5	14	29	27	-20	75	0
Anxious, can't make up mind	66	18	-11	0	51	-6	7	51	63	-6	6	8	45	-26	3	57

[a] Parameter value is fixed to zero *ex hypothesi*.

above), we suggest that the specific environmental effects found for the scores on the anxiety and depression scales are caused by short-term environmental effects, which create the environmental "group" factors found in the analysis of covariance at the item level. One major implication of the model is that genetically "vulnerable" individuals may develop symptoms of anxiety or depression at different times in their life as a function of the particular kinds of environmental stress that happen to be operating at the time of follow-up. Taken to its limit, the model that we describe predicts that a person who is anxious on one occasion is equally likely to be anxious or depressed on a second occasion, depending on the particular environment the person has experienced in the interim. Furthermore, the model predicts that the MZ twin of such an individual is hardly less likely to show the symptoms of either anxiety or depression than the individual himself on repeated measurement. Further tests of this model must await the collection of substantial longitudinal data on symptoms and life events in a genetically informative population.

11.5 SUMMARY

As far as the genetic and environmental causes of individual differences are concerned, the results show that there is little difference between the findings for symptoms of anxiety and depression and those for personality items. The same kind of genetic model that we developed for normal variation in personality can account equally for self-report symptoms of anxiety and depression in an unselected twin population. Genetic effects are additive, and there is not much evidence that shared environmental experiences are important. A significant proportion of the variation in liability is caused by unique environmental effects. These could include the specific experiences of a transient nature that lead to temporary changes in liability to endorse symptoms of anxiety and depression, i.e. "precipitating factors".

Kendler *et al.* (1986) have noted that a common perception in psychiatry is that the genetic component of a "severe" disorder is greater than that for "milder" forms. The diverse analyses of the Australian dataset (which deal with a population unselected for psychiatric disorders) suggest that the genetic component of variation in mild symptomatology is no smaller than that for liability to severe psychopathology.

The results we present are consistent with a model for psychopathology advanced by Eysenck a long time ago (1952a) in which patients with diagnosed psychiatric disorders were more extreme than normals on normal dimensions of personality.

We fitted a model that combines the effects of a single gene and polygenic effects with the latent-trait model for item responses in the attempt to resolve the genetic component of liability to depression into its polygenic and major-gene components. We showed that most of the genetic variation was due to polygenic factors, but that the slight non-normality in the latent trait could be explained by a relatively rare gene of large effect imparting very high liability to relatively few individuals. The effects of the "major gene", however, were not statistically significant.

Analysis of anxiety and depression scale scores suggests that there is little genetic variation in the symptom data that is not shared with measures of the neuroticism dimension of normal personality. That is, the same genetic effects create variation in neuroticism scores, depression scores and anxiety scores. The analysis also shows that some environmental effects are specific to anxiety and depression. That is, different environmental effects may contribute to the two kinds of symptoms. By combining the twin data with data on repeated measures, we find that most of the specific environmental variation is attributable to short-term environmental effects.

The findings are strengthened substantially by a multivariate analysis of the symptom data. The common genetic factor extends to all the anxiety and depression items, but a separate group factor is needed to account for the pattern of environmental correlation between symptoms of depression. There are distinct and independent environmental influences that increase the chance of expression of symptoms of depression rather than anxiety or *vice versa*, in individuals of high genetic liability. This means that two individuals with the same high genetic liability will have a higher chance of endorsing symptoms of both anxiety and depression, but that environmental experiences will determine whether they develop the symptoms of anxiety or depression.

Chapter 12

Social Attitudes: A Model of Cultural Inheritance?

The most astounding result of all our analyses of personality is not that genetic factors play a significant part in creating individual differences, but rather that normal variation in the environment between families ("family background") seems to have so little effect. One possible reason for this finding is that personality measures depend on properties of the nervous system that were established long before cultural effects played a predominant role in behavioral evolution. It follows that anyone who is seriously concerned to detect and model the effects of nurture on behavior should concentrate his attention on those areas where there are stronger grounds, *a priori*, for the contribution of language and communication to human variation. For that reason, we devote the remaining chapters of our book to the study of variation in social attitudes. In this chapter we describe the analysis of individual items from two questionnaires dealing with social attitudes.

The first study employed twins from the London sample and employed that 60-item public opinion inventory (POI; Appendix D) devised originally by Eysenck (1951) to yield measures of conservatism and tough-mindedness (see Chapter 13). The second study reports data collected as part of the Australian Twin Study and summarized by Martin *et al.* (1986). The Australian Study used a form of the Wilson–Patterson Conservatism Scale. Both inventories are described in Chapter 13. The London data were collected as part of an earlier investigation using the twin panel that preceded the EPQ study reported in Chapter 5. The Australian data were collected at the same time as the EPQ data described in the same chapter. More detailed accounts, on which our discussion depends heavily, are given by Feingold (1984) for the London data and Jardine (1985) for the Australian Study.

12.1 ANALYSIS OF ATTITUDE ITEMS FROM THE LONDON STUDY

12.1.1 Preliminary statistics

The data were collected from twins on the Institute of Psychiatry Twin Registry in the years 1970–1971. The Personality Inventory (PI) and the Public Opinion Inventory (POI) were administered by mail. The structure of the sample of twin pairs in which complete data were available is given in Table 12.1.

The items of the POI are scored on a five-point Lickert scale. Scoring is as follows: "strong agreement", 1; "agreement on the whole", 2; "can't decide", 3; "disagreement on the whole", 4; "strong disagreement", 5. On the five-point scale, the mean-item scores of the six twin groups are given in Appendix E and the within-group variances are tabulated in Appendix F.

12.1.2 Twin correlations for the items

The raw intraclass correlations for the individual items are given in Table 12.2 for each group of twins. For a few items the study of Loehlin and Nichols (1976) offers parallels. These are included for comparison. It is obvious that there is much variation between items and twin groups with respect to the raw correlations. The correlations for female MZ twins, for whom the sample size is largest, range from 0.137 (item 40) to 0.548 (item 12). The correlations of DZ females range from 0.104 (item 60) to 0.408 (item 37). For 54 out of the 60 items the MZ correlation exceeds the DZ correlation in female twins. The MZ correlation was numerically greater than the DZ correlation in male twins for 48 of the sixty items. The MZ correlation in females exceeded the correlation in MZ males for only 31 items. In 36 items

Table 12.1   Structure of twin sample.

Twin type	Number of pairs
MZ female	325
MZ male	120
DZ female	194
DZ male	59
DZ female–male	61
DZ male–female	66

Table 12.2   Correlations of twin pairs for attitude items.

	MZ		DZ		
	MZ_f $n = 325$	MZ_m $n = 120$	DZ_f $n = 194$	DZ_m $n = 59$	DZ_{os} $n = 127$
Item 1	0.226	0.254	0.203	0.198	0.079
2	0.451	0.129	0.331	0.111	0.082
3	0.318	0.154	0.245	0.188	0.171
4	0.266	0.287	0.230	0.274	0.370
5	0.332	0.277	0.269	0.272	0.160
6	0.277	0.427	0.135	0.219	0.159
7	0.362	0.245	0.172	0.266	0.222
8	0.281	0.448	0.248	0.086	0.330
9	0.175	0.223	0.221	0.075	0.115
10	0.476	0.479	0.350	0.111	0.261
11	0.381	0.389	0.304	0.217	0.112
12	0.548	0.421	0.324	0.198	0.205
13	0.359	0.389	0.293	0.294	0.092
14	0.519	0.403	0.344	0.398	0.093
15	0.430	0.137	0.252	0.040	0.181
16	0.275	0.185	0.222	0.179	0.329
17	0.434	0.221	0.306	0.321	0.254
18	0.496	0.405	0.244	0.204	0.216
19	0.374	0.265	0.282	0.085	0.073
20	0.208	0.273	0.253	0.102	0.001
21	0.348	0.424	0.284	0.389	0.147
22	0.398	0.334	0.371	0.364	−0.052
23	0.360	0.353	0.167	0.318	0.053
24	0.350	0.441	0.315	0.463	0.286
25	0.376	0.399	0.256	0.390	0.091
26	0.335	0.219	0.157	−0.015	0.048
27	0.390	0.508	0.181	0.269	0.224
28	0.332	0.346	0.256	0.252	0.378
29	0.492	0.520	0.392	0.301	0.254
30	0.378	0.039	0.131	0.201	0.146
31	0.413	0.435	0.302	0.273	0.450
32	0.168	0.163	0.164	−0.033	0.012
33	0.518	0.322	0.328	0.345	0.023
34	0.327	0.237	0.355	0.275	0.281
35	0.243	0.191	0.282	0.137	−0.039
36	0.373	0.394	0.239	0.318	−0.019
37	0.392	0.392	0.408	0.359	0.457
38	0.338	0.214	0.332	0.244	0.142
39	0.532	0.422	0.316	0.201	0.153
40	0.137	0.113	0.241	−0.088	0.219
41	0.225	0.242	0.156	0.240	0.340
42	0.382	0.500	0.208	0.311	0.342
43	0.342	0.268	0.310	0.327	0.303
44	0.329	0.118	0.252	−0.076	0.228

Table 12.2 Contd

	MZ		DZ		
	MZ_f $n = 325$	MZ_m $n = 120$	DZ_f $n = 194$	DZ_m $n = 59$	DZ_{os} $n = 127$
45	0.364	0.502	0.323	0.133	0.336
46	0.287	0.264	0.169	−0.053	−0.137
47	0.476	0.434	0.241	0.268	0.167
48	0.396	0.352	0.288	0.120	0.041
49	0.481	0.328	0.352	0.223	0.116
50	0.353	0.427	0.225	0.161	0.148
51	0.448	0.385	0.256	0.117	0.283
52	0.265	0.377	0.185	0.011	0.198
53	0.421	0.414	0.245	0.365	0.328
54	0.442	0.263	0.258	−0.037	0.372
55	0.421	0.225	0.228	0.272	0.066
56	0.414	0.462	0.358	0.311	0.269
57	0.323	0.511	0.235	0.258	−0.123
58	0.336	0.249	0.164	0.094	−0.056
59	0.305	0.375	0.227	0.148	−0.031
60	0.262	0.335	0.104	0.494	0.217
Loehlin and Nichols (1976):					
402 = 31	0.47	0.65	0.68	0.65	
403 = 44	0.30	0.44	0.46	0.31	
404 = 7	0.29	0.24	0.01	0.29	

the correlation for female DZ twins was numerically greater than that for male DZ twins. Preliminary inspection of the data therefore suggests that the resemblance of male twins in attitude items is no different from the similarity of females on average. On the other hand, the similarity of MZ twins of both sexes tends to be greater than the correlation of like-sex DZ twins for a significant majority of the individual items. In the London sample the twin data therefore appear to be consistent with some overall genetic effect on social attitudes at the item level.

This simple comparison of correlations for MZ and DZ twins does not address the more salient issue of whether the shared environment of twins is important. There may be a substantial effect of the family environment over and above any genetic component, and the MZ correlations will still exceed those for DZs on average. However, given additive gene action and random mating, we predict on average that the DZ correlations should be exactly half the MZ correlation (see Chapter 3). Therefore if twice the DZ correlation is greater than the MZ correlation for significantly more than half the items then we should have some evidence that the family environment is moulding family resemblance in individual attitudes. For female twins we find that twice the DZ correlation is greater than the MZ correlation for 51 out of 60 items. For males twice the DZ correlation exceeds the MZ correlation in 40 of the items. Thus, although inspection of the raw correlations suggests that there may be a genetic effect on attitudes, there is also a strong indication that the family environment may contribute significantly to attitude differences. This result would be in striking contrast with the studies of personality described in earlier chapters.

12.1.3 Item analysis using the threshold model

These ideas can be expressed more clearly and tested statistically using the multifactorial threshold model described in previous item analyses (Chapters 8 and 11). Since there are five response categories for each item, there are four thresholds (eight, if we allow for sex differences in threshold). There are marked sex differences in endorsement for 36 of the items, so we restrict our discussion to the models that allow for sex differences in threshold.

In view of the large number of items involved, only four models were fitted to the data for each item, allowing for sex differences in threshold. These were: (1) assuming no twin resemblance at all (the E_W model); (2) assuming that twin resemblance was only due to additive genetic effects (V_A); (3) assuming that twin resemblance was due to the shared environment (E_B); (4) allowing for additive genetic effects and the family environment (V_A, E_B). Since we are dealing with the threshold model, the total variance is constrained to unity and the values of E_W are obtained by subtraction, as before (Chapter 7).

Goodness-of-fit chi-square values were computed for testing the agreement between observed and expected cell frequencies. Since some of the cell sizes are small, it is to be expected that the chi-squares will tend to be biased upwards. In a few cases it was impossible to obtain stable solutions for the parameters of some models (usually model 4). Parameter estimates for the best-fitting model, by the likelihood criterion, are given in Table 12.3.

The fit of the threshold model is often poor, when judged by the chi-square criterion. This result is in sharp contrast with that for personality items (Chapters 8 and 11). The model that assumes no familial association of attitudes never fits, and enormous gains result from allowing for family resemblance in attitudes. The model that excludes genetic factors usually gives a poor fit also. The V_A model gives greater likelihoods than the E_B model in 54 of the 59 items for which solutions could be obtained. The inclusion in the model of a shared environmental component in addition to

Table 12.3 Parameter estimates of best-fitting model for social attitude items allowing for different thresholds in each sex.

Item	Variance Components[a]					Thresholds[b]							
	P	I	V_A	E_B	E_W	t_{1f}	t_{2f}	t_{3f}	t_{4f}	t_{1m}	t_{2m}	t_{3m}	t_{4m}
1	0.12		0.29	—	0.71	−0.91	0.25	0.97	0.17	−0.62	0.49	0.99	0.16
2	0.08	*	0.50	—	0.50	−2.0	−1.2	−0.96	−0.07	−1.8	1.1	−0.91	−0.04
3	0.20		0.35	—	0.65	−1.3	0.31	0.57	1.4	−0.9	0.44	0.63	1.4
4	0.00		—	0.37	0.63	−1.82	−1.26	−1.01	−0.16	−1.64	−1.23	−0.90	0.00
5	0.00		0.21	0.15	0.46	−0.94	−0.14	0.20	0.97	−1.19	−0.28	0.05	0.84
6	0.11	*	0.35		0.65	−1.15	−0.48	−0.01	0.88	−1.13	−0.33	0.13	1.0
7	0.15		0.38		0.62	−1.13	−0.09	−0.21	1.53	−1.27	−0.30	−0.04	1.01
8	0.36		0.40		0.60	−1.11	−0.27	−0.03	0.88	−1.05	−0.04	0.25	1.18
9	0.18	*	0.15	0.09	0.75	−1.92	−1.1	−0.73	0.38	−2.02	−1.14	−0.87	0.28
10	0.01	*	0.55		0.45	−0.89	−0.19	−0.01	0.67	−0.88	−0.29	0.13	0.52
11	0.02		0.19	0.23	0.58	−0.96	−0.07	−0.51	1.43	−0.87	−0.01	0.35	1.24
12	0.07		0.57		0.43	−1.01	0.09	0.27	0.94	−0.64	0.53	0.79	1.34
13	0.00		0.44		0.56	−1.71	−0.55	0.20	1.06	−1.55	−0.70	−0.56	0.89
14	0.11		0.55		0.45	−1.01	0.33	0.51	1.23	−0.62	0.58	0.84	1.51
15	0.00		0.41		0.59	−1.43	−0.71	−0.40	0.71	−1.58	−0.69	−0.38	0.60
16	0.22	*	0.11	0.21	0.68	−0.75	0.42	0.70	1.81	−0.72	0.50	0.84	1.75
17	0.80		0.46		0.54	−1.24	−0.77	−0.09	0.54	−0.90	−0.38	0.31	0.95
18	0.00		0.56		0.44	−0.74	−0.17	0.03	0.82	−0.63	−0.13	0.03	0.61
19	0.00		0.39	—	0.61	−1.62	−0.45	0.16	0.88	−1.58	−0.43	0.11	0.71
20	0.03		0.28	—	0.72	−0.22	0.61	1.09	1.35	−0.68	0.09	0.46	1.35
21	0.68	*		0.45	0.55	−1.73	−1.35	−1.19	−0.44	−1.82	−1.55	−1.37	−0.50
22	0.00	*	0.45		0.55	−0.57	−0.43	−1.77	1.33	−0.50	−0.61	0.87	1.47
23	0.00		0.39		0.61	−1.08	−0.09	0.21	1.13	−0.48	−0.33	0.61	1.50
24	0.00		0.12	0.32	0.56	−1.54	−0.61	0.25	1.20	−1.16	−0.47	−0.05	0.86
25	0.15		0.51		0.49	−2.00	−1.31	−1.08	0.13	−2.17	−1.51	−1.36	−0.23
26	0.02	*	0.34		0.66	−1.26	0.23	0.01	0.83	−1.27	−0.32	−0.08	0.92
27	0.42		0.46		0.54	−1.82	−1.10	−0.70	0.34	−1.39	−0.60	−0.11	0.80
28	0.00		0.43		0.57	−1.27	−0.23	0.02	1.01	−1.35	−0.69	−0.49	0.59
29	0.00		0.62		0.38	0.37	0.06	0.29	0.99	−0.59	−0.18	0.08	0.81

No.		I		P		t_{1m}	t_{2m}	t_{3m}	t_{4m}	t_{1f}	t_{2f}	t_{3f}	t_{4f}
33	0.27		0.53		0.47	−1.20	−0.16	0.16	1.18	−1.31	−0.41	−0.08	0.74
34	0.00		0.38	0.35	0.27	−1.93	−1.22	−0.69	0.38	−1.75	−1.40	−0.97	0.16
35	0.04		0.29		0.71	−1.24	−0.43	−0.17	0.83	−1.06	−0.21	0.12	1.06
36	0.68		0.41	0.39	0.59	−1.65	−0.80	−0.43	0.74	−1.50	−0.54	−0.07	0.91
37	0.01		0.04	0.26	0.57	−1.35	−0.67	−0.51	0.63	−1.68	−1.21	−1.03	0.15
38	0.10		0.11		0.63	−2.15	−1.48	−0.93	0.37	−1.91	−1.19	−0.81	0.52
39	0.13		0.54		0.46	−1.36	−0.53	−0.09	0.96	−1.51	−0.79	−0.35	0.55
40	0.38		0.01	0.25	0.74	−1.44	−0.23	0.39	1.43	−1.23	−0.13	0.41	1.21
41	0.00	*	0.19	0.07	0.74	−1.96	−0.59	0.07	1.09	−1.79	−0.55	0.03	0.96
42	0.88		0.45		0.55	−0.81	−0.34	0.75	1.65	−0.85	0.08	0.43	1.41
43	0.26	*	0.12	0.26	0.62	−0.88	−0.24	0.67	1.80	−0.74	0.33	0.77	1.75
44	0.02	*	0.33		0.67	−0.68	−0.33	0.77	1.68	−0.61	0.27	0.65	1.42
45	0.00		0.47		0.53	−1.25	−0.02	0.19	1.17	−1.43	−0.33	−0.13	0.81
46	0.00	*	0.44	0.01	0.55	−0.25	−1.46	1.94	2.35	−0.23	1.54	2.11	2.59
47	0.11		0.51		0.49	−1.14	−0.28	−0.06	0.91	−0.91	−0.19	0.18	0.87
48	0.23		0.43		0.57	−1.49	−0.37	−0.15	1.37	−1.52	−0.54	−0.17	0.87
49	0.00		0.49		0.51	−1.77	−0.47	−0.30	0.64	−1.65	−0.33	−0.10	0.69
50	0.01		0.41		0.59	−1.67	−0.82	−0.05	0.99	−1.68	−0.80	−0.17	0.71
51	0.70		0.49		0.51	−0.75	0.58	0.83	1.6	−0.51	0.58	0.99	1.63
52	0.03		0.33		0.67	−1.96	−0.82	−0.06	0.76	−1.84	−1.09	−0.38	0.53
53	0.59		0.49		0.51	−0.69	0.33	1.01	1.57	−0.95	−0.06	−0.74	1.3
54	0.32		0.21	0.25	0.54	−1.77	−1.17	−0.87	0.19	−1.89	−1.25	−0.98	0.02
55	0.35	*	0.48		0.52	−2.08	−1.56	−1.21	0.02	−1.68	−1.28	−0.99	0.10
56	0.15		0.28	0.22	0.60	−0.61	0.10	0.84	1.27	−0.95	−0.41	0.46	0.85
57	0.22		0.50		0.50	−0.27	0.48	0.68	1.57	−0.44	0.29	0.51	1.28
58	0.11		0.32		0.68	−1.31	−0.08	0.63	1.44	−1.22	0.06	0.55	1.32
59	0.17		0.36		0.64	−1.91	−1.07	0.48	0.52	−2.10	−1.34	−0.80	0.25
60	0.43	*	0.36	0.29	0.71	−1.52	−0.66	−1.43	0.96	−1.74	−0.78	−1.97	0.79

[a] I, * if thresholds differ significantly between sexes, − otherwise; P, goodness of fit (probability that χ^2 > observed value given model fits).

[b] $t_{1m}, t_{2m}, t_{3m}, t_{4m}$, response thresholds for males; $t_{1f}, t_{2f}, t_{3f}, t_{4f}$ response thresholds for females;

V_A leads to a significantly greater likelihood for 17 of the 60 items. The effects of the family environment, as estimated by E_B, accounted for between 9% and 45% of the total variation in liability. For the other items, the effects of the shared environment were too small to be detectable by our statistical tests. Additive genetic factors could account for between zero and 62% of the total variation in liability.

On the basis of the model-fitting results, we can identify the 10 items for which the contribution of genetic factors was estimated to be greatest, the 10 that showed the biggest E_B component, and the 10 for which the contribution of the within-family environment was largest. To some extent, the decision about the relative contributions of variance components will reflect differences in short-term environmental effects on the items and differences in reliability. These items are listed in Tables 12.4(a–c).

The most "heritable" items relate mainly to religion and the treatment of criminals. Items most affected by the family environment relate to socialism and prejudice, but the distinctions are by no means clear or general. It is well to remember that the "religion" items do not relate to specific religious affiliation (e.g. "Catholic" or "Protestant") but to endorsement of more general religious propositions (e.g. "The idea of God is an invention of the human mind") for which there is expected to be variation even within denominations. A recent study (Eaves *et al.*, in preparation) confirms the overwhelming cultural component of religious affiliation in the strict sense.

Table 12.4(a)   Ten items with the largest V_A component of variation.

Item		V_A
29	Sex crimes, such as rape and attacks on children, deserve more than mere imprisonment; such criminals ought to be flogged or worse.	0.63
12	Men and women have the right to find out whether they are sexually suited before marriage.	0.59
14	The average man can live a good enough life without religion.	0.56
18	The death penalty is barbaric and should continue to be abolished.	0.56
10	Crimes of violence should be punished by flogging.	0.55
21	Birth control, except when recommended by a doctor, should be made illegal.	0.54
33	The Church should attempt to increase its influence on the life of the nation.	0.54
39	Only by going back to religion can civilization hope to survive.	0.54
25	We should believe without question all we are taught by the Church.	0.53
47	Our treatment of criminals is too harsh; we should try to cure them, not punish them.	0.51

Table 12.4(b) Ten items with the largest E_B component of variation.

Item		E_B
37	Sex relations except in marriage are always wrong.	0.46
24	Capitalism is immoral because it exploits the worker by failing to give him full value for his productive labor.	0.42
54	It would be best to keep colored people in their own districts of schools, in order to prevent too much contact with whites.	0.42
34	Conscientious objectors are traitors to their country, and should be treated accordingly.	0.39
11	The nationalization of the great industries leads to inefficiency, bureaucracy and stagnation.	0.38
28	Compulsory military training in peacetime is essential for the survival of this country.	0.38
4	Ultimately private property should be abolished and complete socialism should be introduced.	0.37
43	Nowadays, more and more people are prying into matters which don't concern them.	0.35
38	Asian refugees should be left to fend for themselves.	0.34
60	The practical man is of more use to society than the thinker.	0.30

Table 12.4(c) Ten items with the largest E_W component of variation.

Item		E_W
40	It is wrong to punish a man because he helps another country because he prefers it to his own.	0.83
9	The so-called underdog deserves little sympathy or help from other people.	0.75
41	It is just as well that the struggle of life weeds out those who cannot stand the pace.	0.75
35	Abortion should be freely available on demand.	0.73
1	The nation exists for the benefit of the individuals composing it, not the individuals for the benefit of the nation.	0.72
60	The practical man is of more use to society than the thinker.	0.70
20	The dropping of the first atom bomb on a Japanese city, killing thousands of innocent women and children, was morally wrong and incompatible with our kind of civilization.	0.69
16	People should realize that their greatest obligation is to their family.	0.68
44	All forms of discrimination against the colored races, the Jews, etc., should remain illegal and subject to heavy penalties.	0.68
30	A white lie is often a good thing.	0.68

As long as we are restricted to dichotomous items, the statistics of the threshold model, two thresholds and a correlation, are sufficient to account for any two-way table. This is no longer the case for multicategory data, so that the test of the threshold model is as much a test of the assumed ordering of the items and the assumption of normality as it is a test of whether or not transmission is genetic. If the latent-attitude dimension is not normally distributed or if there are two or more latent dimensions then the multiple-threshold model may not capture adequately the relationship between liability and specific response. For example, there may be two distinct psychological processes underlying responses to multicategory items. Subjects may first decide whether they agree, disagree or are neutral with respect to an item, and then, if they either agree or disagree, evaluate the "strength" of their opinion. Such a two-step decision process would yield an excess of twin pairs in the "strongly agree/strongly disagree" cell and lead to failure of the unidimensional model. If, for example, the causes of agreement or disagreement were purely random but strength of opinion were entirely genetic then a folding of the scale such that all strong opinions were given the same weight and moderate opinions received a lesser weight would yield more marked resemblance between twins.

We fitted such a two-dimensional model to the POI items. The model assumed that there were two uncorrelated dimensions of liability behind observed responses to the individual items. The first was a three-category dimension of agreement, with two thresholds dividing the continuous liability to agreement into "agree," "neutral" and "disagree." The second was the independent dimension of "emphasis", or intensity, having two categories: "weak" and "strong." Separate genetic and cultural effects could contribute to both dimensions. The response process was assumed to occur in two stages. First, the subject decides on a position on the "agreement" scale. If the position is neutral then the "can't decide" category is endorsed. If the position is either "agree" or "disagree" then a further decision is made about the strength of the opinion, and subjects who agree thus divide themselves into those who "strongly" agree and those who agree "on the whole". Similar division is assumed to occur for subjects who are beyond the "disagreement" threshold on the primary dimension. Obviously this is not the only possible model for the response process in giving priority to the more cognitive process of "agreement" over the emotional component of "emphasis". It is also probable that different processes could apply to different items.

Neither the original model nor the agreement/emphasis (a/e) model is a simplified form of the other, so we cannot construct a likelihood-ratio test for detecting the improvement resulting from fitting the a/e model. We have to be content with fairly general statements about relative likelihoods and goodness of fit. In spite of its intuitive appeal, the model only gave an adequate fit for ten items overall. In only two cases was the fit of the new model better than that of the original model. Thus we conclude that for the POI items the problem cannot be identified in terms of a general issue in scaling that can be solved by adopting the same, more intricate, model for every item. In view of the potential heuristic value of such models in future studies of multi-category genetic data, however, we present the parameter estimates for the ten POI items for which the modified model fitted the observed twin data (Table 12.5).

In a final attempt to illuminate the dimensionality and form of the scales underlying the response categories of twins, we used an algorithm for obtaining empirical category weights to maximize the correlation between twins for an item. The method amounts to computing canonical correlations between twins where the weights do not relate to distinct variables but to the response categories of a single item (see Eaves, 1980). If the emergent weights are ordered 1, 2, 3, 4, 5 then we may suppose that the item may legitimately be scored in that direction. If the weights emerge in the pattern 2, 1, 0, 1, 2 then we should assume that the main dimension underlying the item is "emphasis". In practice, more than one set of weights may generate significant twin resemblance, so that a two-dimensional representation may be required for a given item. We know of no statistical model on which to base significance tests. The method was applied to all groups of twins and every item, but the cell numbers are very small for some of the twin groups and there is great inconsistency of results across groups within an item. Such inconsistency could be due to chance, or to genuine differences in the scale of

Table 12.5 Parameter estimates for the 10 items that the two dimensional agreement/emphasis biometrical model fits.[a]

Item	P	V_{AA}	V_{AE}	E_{BA}	E_{BE}	E_{WA}	E_{WE}	a_1	a_2	e
8	0.06	0.40	0.40	—	—	0.60	0.60	−0.20	0.04	−0.42
36	0.14	0.52	0.23	—	—	0.48	0.77	−0.72	−0.32	−0.49
42	0.12	0.57	0.20	—	—	0.43	0.80	0.26	0.65	−0.50
46	0.40	0.43	0.39	—	—	0.57	0.61	1.5	2.0	−1.8
47	0.09	0.58	0.38	—	—	0.42	0.62	−0.25	0.10	−0.31
51	0.07	0.59	0.07	0.00	0.18	0.41	0.75	0.59	0.87	−0.43
53	0.11	0.20	0.66	0.30	0.00	0.50	0.34	0.22	0.92	−0.30
55	0.07	0.50	0.49	—	—	0.50	0.51	1.47	1.14	0.09
58	0.13	0.40	0.31	—	—	0.60	0.69	−0.04	0.60	−0.72
60	0.11	0.23	0.00	0.10	0.58	0.67	0.42	−0.71	−0.17	−0.52

[a] Subscripts A and E denote components for "agreement" and "emphasis" respectively. a_1 and a_2 are thresholds on the three-category "agreement" scale (agree–neutral–disagree). e is the threshold on the two-category "emphasis" scale (weak–strong).

genetic and environmental effects. We do not present all the results, but give examples for the largest twin group (female MZs), which depict two distinct patterns (POI items 2 and 41). Item 2 is an example of an item for which the unidimensional threshold model gave an exceptionally poor fit ("Colored people are innately inferior to white") and item 41 is one for which the threshold model gave a good fit ("It is just as well that the struggle of life weeds out those who cannot stand the pace"). The results (Table 12.6) show that the first canonical correlation for the second item has weights that reflect the hypothesized ordering of the categories from "strong agreement" to "strong disagreement". In contrast, none of the correlations derived for item 2 give clear evidence of strong ordering in the direction assumed in the threshold model. Indeed, the first dimension puts "indecision" at opposite poles from moderate opinion and not quite as far from strong agreement or disagreement. The five categories are plotted in the two dimensions defined by their weights on the canonical variates for each item in Figures 12.1 and 12.2.

Table 12.6   Responses to public-opinion inventory: correlations and category weights for first three dimensions of similarity.[a]

				Category weight				
Item	Twin type	d	r	Strongly agree	Agree	Undecided	Disagree	Strongly disagree
2	MZ	1	0.89	−1.71	−3.72	13.91	−2.21	0.00
	female	2	0.32	1.14	0.42	−4.35	0.41	0.00
		3	0.20	0.94	0.58	−1.20	−0.76	0.00
41	MZ	1	0.23	1.90	1.18	0.45	0.12	0.00
	female	2	0.03	0.05	0.30	0.15	0.50	0.00
		3	0.00	0.00	0.00	0.00	0.00	0.00
41	DZ	1	0.32	1.10	1.04	0.04	−0.40	0.00
	combined	2	0.26	2.63	0.99	1.29	1.27	0.00
		3	0.16	1.69	−0.70	−0.59	−0.51	0.00
2	MZ	1	0.54	1.35	2.35	0.00	1.02	0.00
	female	2	0.24	−0.59	−0.65	0.00	0.83	0.00
		3	0.02	0.61	0.61	0.00	0.04	0.00
4	MZ	1	0.41	0.35	−1.74	−1.19	−1.17	0.00
	female	2	0.22	−0.02	−0.86	1.64	−0.04	0.00
		3	0.14	−2.73	−0.11	−0.12	−0.16	0.00

[a] d, dimension; r, correlation.

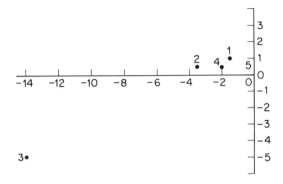

Figure 12.1 Two-dimensional model for response categories to POI item 2 in female MZ twins. Key to categories: 1, strongly agree; 2, agree; 3, undecided; 4, disagree; 5, strongly disagree.

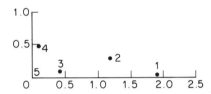

Figure 12.2 Two-dimensional model for response categories to POI item 41 in female MZ twins. For key see Figure 12.1.

12.2 CONSERVATISM ITEMS IN THE AUSTRALIAN SAMPLE

12.2.1 Questionnaire and preliminary statistics

A recurrent problem with the POI administered to the London sample is that the items are complex, often involving more than one proposition. Also, the 1–5 scoring scale may cause problems to some subjects. In the Australian study a less-demanding questionnaire was employed consisting of single-word items to which each subject was asked to respond on a three-point scale by circling "Yes", "?" or "No". The questionnaire was originally developed by Wilson and Patterson (1970) to measure the general dimension of conservatism or "resistance to change". The Australian study (see Martin and Jardine, 1986) employed a slightly abbreviated 50-item form (see Table 12.9) of the Wilson–Patterson Conservatism Scale developed for the Australian population by Feather (1975). In the questionnaire alternate items are designed to be scored in the "conservative" and "liberal" direction.

Jardine (1985) compared the frequencies of responses in the three categories between zygosity groups within sexes by a chi-square test having 2 df. Out of 100 tests conducted, eight were significant at the 5% level, which scarcely exceeds what might be expected by chance alone. Three of these tests were also significant at the 1% level. Overall, there is therefore no marked evidence that the response pattern depends on zygosity in this sample. Table 12.7, however, gives the frequencies of the three endorsement categories by sex for each item. There is evidence of highly significant sex differences in endorsement pattern for all but a handful of the items. These highly significant sex differences in response pattern will translate into marked sex differences in thresholds under the multifactorial model.

12.2.2 Fitting the threshold model

The multifactorial threshold model was fitted to the individual items. Since there are three response categories, there are two thresholds. Different thresholds were fitted to males and females when necessary. The likelihood was computed for each item under a "full" model in which each twin group was assumed to have a separate correlation in liability. Various simplifications of the full model were then tried to arrive at the most parsimonious model that did not lead to a significant reduction in likelihood. Three models were considered for every item: the E_B model, the V_A model and the V_A, E_B model (cf. the analysis of the POI above).

In cases where the fit of the third model was still poor, additional models were fitted that allowed for various forms of sex limitation in genetic and environment effects in the attempt to arrive at a satisfactory fit in comparison with the full model.

Table 12.7 Percentage of individuals giving a liberal, ambiguous or conservative response to Wilson–Patterson items, broken down by sex. Asterisks denote significant differences between male and female response frequencies.[a]

	Liberal		Ambiguous		Conservative		
Item	Female	Male	Female	Male	Female	Male	χ^2
1	33.6	30.4	19.5	10.3	47.0	59.4	149.78***
2	28.3	20.6	34.0	23.7	21.8	19.1	128.28***
3	8.1	24.0	3.3	6.7	88.7	69.3	447.12***
4	22.6	57.8	20.6	15.2	56.8	27.0	979.87***
5	26.9	36.9	23.1	23.5	49.9	39.6	97.85***
6	28.4	34.4	26.3	21.0	45.3	44.6	40.75***
7	9.3	12.0	16.5	12.6	74.2	75.4	30.10***

| | Liberal | | Ambiguous | | Conservative | | |
Item	Female	Male	Female	Male	Female	Male	χ^2
8	48.2	44.1	21.4	21.2	30.5	34.8	16.41***
9	26.1	30.6	26.6	24.8	47.3	44.6	17.44***
10	55.4	52.9	13.9	13.4	30.7	33.7	7.65*
11	52.0	72.0	16.7	12.2	31.3	15.8	306.63***
12	92.3	87.9	3.1	4.7	4.6	7.5	41.09***
13	18.2	26.9	21.4	13.8	60.4	59.2	116.92***
14	91.5	92.3	5.3	5.1	3.1	2.6	1.76
15	18.9	30.4	38.2	30.8	42.9	38.8	134.50***
16	24.0	29.1	30.6	22.6	45.5	48.3	61.07***
17	83.6	81.6	9.3	7.7	7.1	10.7	32.86***
18	14.0	19.1	20.3	24.2	65.8	56.7	65.85***
19	4.0	7.1	9.9	12.9	86.1	80.0	54.22***
20	8.8	11.3	9.8	9.8	81.4	78.9	12.34**
21	56.7	55.1	20.0	20.0	23.2	24.9	2.89
22	61.4	63.6	10.8	11.2	27.8	25.2	5.95
23	43.8	50.5	34.4	22.2	21.7	27.3	126.85***
24	28.2	44.3	19.1	15.9	52.6	39.8	203.31***
25	6.6	13.1	12.9	11.5	80.4	75.4	89.30***
26	28.9	33.4	31.5	24.0	39.6	42.5	49.35***
27	32.9	40.6	26.4	24.3	40.8	35.1	46.66***
28	59.7	58.6	18.6	20.0	21.7	21.4	2.20
29	17.3	30.7	13.3	12.6	69.4	56.6	186.95***
30	94.4	85.3	3.2	6.3	2.4	8.4	191.05***
31	14.2	11.9	15.2	14.5	70.6	73.6	9.70**
32	62.7	70.6	12.3	9.7	25.0	19.7	48.69***
33	65.3	72.2	28.8	19.8	5.9	8.0	79.69***
34	40.9	63.5	18.8	15.2	40.3	21.3	385.46***
35	40.9	50.5	24.4	20.8	34.7	28.7	65.29***
36	40.3	56.2	35.9	17.7	23.8	26.8	297.73***
37	17.3	32.6	13.8	13.3	68.9	54.1	243.08***
38	55.6	51.7	16.2	17.6	28.1	30.8	10.96**
39	54.5	39.7	18.5	15.5	27.1	44.8	252.02***
40	70.7	69.7	12.5	11.5	16.7	18.8	6.37*
41	28.8	33.5	21.4	19.7	49.8	46.8	18.40***
42	69.7	64.2	12.2	14.2	18.1	21.6	24.40***
43	50.9	43.0	29.2	26.2	19.9	30.8	114.55***
44	70.6	79.2	12.1	9.7	17.3	11.1	71.33***
45	48.2	58.9	26.9	22.3	24.9	18.8	81.90***
46	61.9	62.6	12.5	13.4	25.6	24.0	2.98
47	7.7	11.7	29.1	31.0	63.2	57.3	42.39***
48	57.9	54.4	22.0	16.3	20.1	29.3	96.11***
49	20.3	31.6	28.0	26.0	51.7	42.4	126.43***
50	46.5	49.2	20.5	21.0	33.0	29.8	8.59*

[a] Significance levels for χ^2 test of sex differences:
* $P < 0.05$, ** $P < 0.01$, *** $P < 0.001$.

Table 12.8 Parameter estimates ($\times 1000$) for best-fitting models for Wilson–Patterson inventory items.

Item	E_W	E_{WM}	E_{WF}	E_B	E_{BM}	E_{BF}	E_{BMF}	V_A	V_{AM}	V_{AF}	V_{AMF}
1	491	—	—	—	—	—	—	509	—	—	—
3	—	519	485	—	288	515	236	—	193	—	—
4	462	—	—	209	—	—	—	329	—	—	—
5	472	—	—	261	—	—	—	350	—	—	—
6	465	—	—	261	—	—	—	274	—	—	—
7	—	115	588	—	366	248	—	—	519	164	—
8	—	626	514	—	039	257	—	—	335	229	—
9	642	—	—	—	—	—	—	358	—	—	—
10	540	—	—	—	—	—	—	460	—	—	—
11	—	627	504	—	030	195	—	—	343	301	—
12	—	515	338	—	485	465	272	—	—	197	—
13	475	—	—	—	—	—	—	525	—	—	—
14	612	—	—	388	—	—	—	—	—	—	—
15	524	—	—	257	—	—	—	219	—	—	—
16	533	—	—	210	—	—	—	257	—	—	—
17	499	—	—	—	—	—	—	501	—	—	—
18	599	—	—	—	—	—	—	401	—	—	—
19	519	—	—	194	—	—	—	287	—	—	—
21	599	—	—	—	—	—	—	401	—	—	—
22	351	—	—	324	—	—	—	325	—	—	—
23	—	711	603	—	—	—	—	—	289	397	201
24	506	—	—	192	—	—	—	302	—	—	—
25	—	687	855	—	135	145	041	—	178	—	—
26	740	—	—	—	—	—	—	260	—	—	—
27	—	572	521	—	022	398	—	—	406	081	—
28	544	—	—	—	—	—	—	456	—	—	—
29	429	—	—	135	—	—	—	436	—	—	—
30	568	—	—	—	—	—	—	432	—	—	—
31	650	—	—	—	—	—	—	350	—	—	—
32	608	—	—	—	—	—	—	392	—	—	—
33	514	—	—	—	—	—	—	486	—	—	—
34	402	—	—	317	—	—	—	281	—	—	—
35	503	—	—	203	—	—	—	294	—	—	—
36	617	—	—	170	—	—	—	383	—	—	—
37	560	—	—	—	—	—	—	440	—	—	—
38	590	—	—	—	—	—	—	410	—	—	—
39	617	—	—	170	—	—	—	213	—	—	—
40	548	—	—	122	—	—	—	330	—	—	—
41	599	—	—	—	—	—	—	401	—	—	—
42	527	—	—	—	—	—	—	473	—	—	—
43	806	—	—	194	—	—	—	—	—	—	—
44	516	—	—	193	—	—	—	291	—	—	—
45	670	—	—	—	—	—	—	330	—	—	—
46	466	—	—	—	—	—	—	534	—	—	—
47	656	—	—	—	—	—	—	334	—	—	—
48	—	549	570	—	451	235	192	—	—	195	—
49	411	—	—	338	—	—	—	251	—	—	—
50	502	—	—	498	—	—	—	—	—	—	—

The parameter estimates obtained under the best-fitting model are given in Table 12.8 for each item. The contribution of genetic factors to variation in responses to each item (separated by sex where necessary) is summarized in Figure 12.3. The contribution of between-family environmental effects ranges, depending on the item, between zero and approximately 50%. Jardine used the method of weighted least squares to test the heritability estimates for homogeneity, and showed that the items were significantly different ($\chi^2_{43} = 266.7$, $P < 0.001$) in the proportion of variance attributable to genetic factors. Part of this heterogeneity may reflect interitem differences in reliability.

There is a danger in trying to over-interpret the results for individual items in the absence of a clear theoretical reason for such detailed analysis. We note, however, that 27 out of the 50 items show significant evidence of both genetic and social components of twin resemblance since V_A and E_B are both significant by likelihood-ratio tests. In six of these items (3,12,14,15,34,49) the estimate of the shared environmental component exceeds that of the genetic component in both sexes. In all cases random environmental effects within families, including errors of measurement (E_W), account for at least 35% of the variation in liability underlying the attitude items. Nine items show strong evidence of sex differences in genetic and environmental effects. The effects of sex interaction are especially marked for attitudes to birth control, chastity and immigration. In view of the large number of items analysed, however, it is difficult to know whether such patterns would stand replication.

Twenty-one items (1, 9, 10, 13, 17, 18, 21, 23, 26, 28, 30, 31, 32, 33, 37, 38, 41, 42, 45, 46, 47) show support for a genetic component of transmission but no significant support for a cultural component. These include attitudes to issues as important as capital punishment, disarmament and race as well as comparatively trivial items such as computer music and conventional clothes. Three items (14, 43 and 50) showed no significant genetic transmission.

12.2.3 Factor loadings and the shared environment

It helps us to visualize the results of this detailed analysis if we correlate the estimates of V_A and E_B obtained for each sex separately with the loadings of each item on the first general "conservatism" factor extracted by a principal-axis factor of the interitem correlation matrix (Jardine, 1985). The factor loadings on the general factor are given for each sex in Table 12.9.

Factor loadings for conservatism were correlated with the estimated genetic and family environmental components for both sexes (Chapter 7).

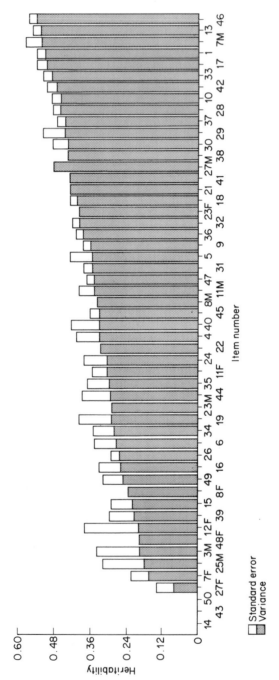

Figure 12.3   Estimated heritability of individual items of Wilson–Patterson conservatism scale.

Table 12.9 Conservatism items used in an Australian study: first principal component.[a]

	Item	Females	Males		Item	Females	Males
1	Death penalty	.08	.19	26	Computer music	− .06	− .12
2	Evolution theory	− .43	− .43	27	Chastity	.48	.48
3	School uniforms	.30	.41	28	Flouridation	.02	.04
4	Striptease shows	− .47	− .46	29	Royalty	.37	.33
5	Sabbath observance	.50	.53	30	Women judges	− .16	− .12
6	Hippies	− .52	− .49	31	Conventional clothes	.31	.30
7	Patriotism	.22	.24	32	Teenage drivers	− .10	− .07
8	Modern art	− .34	− .29	33	Apartheid	.13	.14
9	Self-denial	.29	.27	34	Nudist camps	− .63	− .58
10	Working mothers	− .33	− .34	35	Church authority	.51	.53
11	Horoscopes	− .18	− .15	36	Disarmament	− .22	− .23
12	Birth control	− .24	− .22	37	Censorship	.35	.45
13	Military drill	.38	.43	38	White lies	− .25	− .20
14	Co-education	− .15	− .13	39	Caning	.17	.25
15	Divine law	.53	.58	40	Mixed marriage	− .31	− .28
16	Socialism	− .31	− .32	41	Strict rules	.41	.42
17	White superiority	.18	.14	42	Jazz	− .16	− .12
18	Cousin marriage	− .30	− .28	43	Straitjackets	.04	.04
19	Moral training	.32	.36	44	Casual living	− .44	− .41
20	Suicide	− .38	− .34	45	Learning latin	− .11	− .03
21	Chaperones	.31	.32	46	Divorce	− .45	− .45
22	Legalized abortion	− .45	− .42	47	Inborn conscience	.11	.10
23	Empire-building	.17	.18	48	Colored immigration	− .21	− .19
24	Student pranks	− .38	− .36	49	Bible truth	.60	.61
25	Licensing laws	.10	.20	50	Pyjama parties	− .47	− .45

[a] Modified from Feather (1975) after Wilson and Patterson (1970).

Table 12.10 Correlations of E_B and V_A with factor loadings from the first principal component solution of the C scale.[a]

	Females	Males
E_B	0.58***	0.52***
V_A	−0.08	−0.06

[a] E_B and V_A were estimated from the fit of the E_W, E_B, V_A model.
Significance level: *** $P < 0.001$.

The results (Table 12.10) are very striking indeed. In neither sex is there the slightest evidence that the "heritability" of an item correlates with its loading on the general conservatism factor. However, in both sexes there is a remarkable, consistent and highly significant correlation between the estimates of the between-families environmental component and the factor loadings of the items. That is, items that have the greatest loading on the conservatism factor display, on average, the greatest cultural component. This finding is very important because it indicates that the effects of cultural inheritance, rather than genotype, may be the main cause of the correlation between different conservatism items. In contrast, we might expect a large part of the item-specific variation to have a genetic component. Such a result runs counter to our initial intuition that the best place to look for cultural inheritance would be in the idiosyncratic profile of responses to individual items and that the role of genetic factors in the determination of social attitudes, if any, would be confined to more general dimensions. If anything, the truth *seems* to be the other way round: genetic factors contribute to specific responses, cultural factors contribute to the correlations between items. We shall return to this issue in Chapter 14. All our statements about the contribution of the family environment in this chapter are contingent on the assumption that mating is random. If mating is assortative, the additional genetic variation between families will be confounded with the effects of the family environment in our analyses of twins reared together. The question of assortative mating and its implications for the estimation of cultural effects on social attitudes will be considered in Chapters 15 and 16.

12.3 SUMMARY AND CONCLUSIONS

The analysis of the individual social-attitudes items yields two important conclusions. In the first place, it is surprising that there is evidence for a genetic component to social attitudes at the individual item level in two large studies using different instruments in different populations. On *a priori* grounds we should expect the family environment to play an overwhelming role in the determination of social attitudes. The twin data do not support this conclusion at the individual-item level.

When our surprise at this result has subsided, however, we are left with the second major finding that social attitudes differ consistently from personality measures in both populations in demonstrating that twin resemblance is not due to additive genetic effects alone. For the personality scores, and for the personality items, there was little indication (see Chapters 5–8) that the environment shared by members of a twin pair was a salient feature of personality development. For the social-attitudes items there is consistent evidence for both populations and instruments that the family environment plays a significant role. Estimates of the family environment in this context also reflect any genetic and social consequences of assortative mating. This result is in marked and pleasing contrast to the findings for personality because it suggests that our model-fitting methods, when applied to the right measures in large enough samples, have the capacity to resolve consistently different patterns in the transmission of individual differences.

Generally, the multiple-threshold model does not give a very good fit to the data from the London sample. The unidimensional model of liability does not explain the finer details of the responses to the five-point Lickert scale. Attempts to account for twin resemblance by postulating two independent dimensions met with only marginal success.

Although there are significant sex differences in thresholds between response categories for many items, these do not translate so often into sex differences in the causes of twin resemblance.

The fact that there is a highly significant correlation between estimates of the family environment component and loadings on the "conservatism" factor lends strong support to the idea that the factor is defined significantly by cultural rather than genetic factors. This issue will be pursued in more detail in Chapter 14.

Chapter 13

Conservatism – Radicalism: The Structure of Social Attitudes

The analysis of individual social-attitude items illustrates their use as a basis for exploring models of cultural inheritance in man (see Chapter 12; Cavalli-Sforza and Feldman, 1981; Cavalli-Sforza *et al.*, 1982). Treating fifty or so items as independent indices of cultural or genetic transmission, however, ignores two important psychological issues. The items are not independent but correlated, suggesting that family resemblance may be based less on individual opinions than on underlying common factors that influence how individuals develop the tendency to answer several different items in a consistent uniform direction. In addition, individual differences in such latent common factors may themselves be a function of variation in personality. No less important than the psychological issue is the impact of social attitudes in the formation of political parties and the formulation of social policies.

A widely accepted hypothesis suggests the existence of one major dimension of radicalism–conservatism, ranging from left-wing to right-wing politics through an intermediate stage of liberalism. Communists would be at the extreme left of this continuum, fascists at the extreme right. Such a view, however, cannot explain all the data. There is much evidence to suggest that a second dimension is required in order to do justice to both the expression of social attitudes, and the ordering of political parties (Eysenck, 1954). The need for such a second dimension is seen clearly when it is realized that both extremes, communist and fascist, have many attitudes in common, and oppose the value systems and attitudes of the other parties, particularly liberals (Hayek, 1960; O'Sullivan, 1984).

The following are some of the points on which communists and fascists appear to agree.

(1) The principle of leadership: this often takes the form of a personality cult, but essentially there is a clear distinction between the elite (Fuhrer, Politbureau member) and the great mass of the followers.

(2) Opposition to Democracy: both communists and fascists, whether in power or not, explicitly attack "bourgeois" Democracy.

(3) Subjugation of means to ends: the establishment of a communist/fascist regime is a supreme end, and all means are justified in order to achieve it.

(4) Power as supreme value: internally and externally, the achievement of political power, and the extension of this power to other groups, is the main aim of both communists and fascists.

(5) Expansionism: both communists and fascists, once they are in power, seek expansion through conquest.

(6) Use of the "Big Lie": in the search for power and expansion, both communists and fascists have explicitly advocated the use of the Big Lie (Goebbels, Lenin) and other Machiavellian propaganda techniques.

(7) War and subversion are seen as instruments of the policy of expansionism abroad.

(8) Terror is an important element in the propagation of communism and fascism, both internally and externally.

(9) *Gleichschaltung*: all communal activities, whether in sport, recreation, amusement, education, etc., have to become politicized and appear as organs of the state.

(10) Devaluation of the family. A particularly important aspect of this *Gleichschaltung* is the devaluation of the family, accompanied by the avowal of the superiority of the state.

There are other points of similarity, but these will suffice to suggest that the one-dimensional model of social attitudes and political parties is too simple, and that a second dimension is required. Figure 13.1 shows the simple model under (a), the suggested second dimension under (b), and the combined model under (c). Such a second dimension was first suggested by E. R. Jaensch (1938), in his book on the *"Gegentypus"*, a term denoting the concept of the typical Liberal who opposed the Nazi regime of which Jaensch, a well-known German psychologist, was an ardent supporter. The Jaensch model was adopted by Adorno *et al.* (1950), with certain obvious changes. Naming the opposite pole to the *Gegentypus* the "authoritarian personality", the American group drew a favorable picture of the personality of the liberal, and frowned upon the authoritarian personality, which Jaensch admired. Both, however, made the same error of regarding this dimension as being related to the radical–conservative dimension, with the authoritarian being essentially right-wing, and the *Gegentypus* being essentially left-wing.

Rather different from the models of Jaensch and Adorno *et al.* is that of Kreml (1977), who is concerned with left-wing authoritarianism in particular, which he, too, relates to a variety of personality factors. This is an interesting approach, which would be seen to complement that of the more orthodox studies of the authoritarian personality.

As a contrast, Altemeyer (1981) has published a book on "Right-Wing Authoritarianism", explicitly taking up the implicit message of Adorno's book, but substituting for its psychoanalytic framework a social-psychological one. Social-learning theory could be the explanation of differential social attitudes, and the familial correlation reported by Altemeyer. Genetic factors, however, can equally explain the resemblance between relatives in nuclear families (Chapters 12 and 14–16).

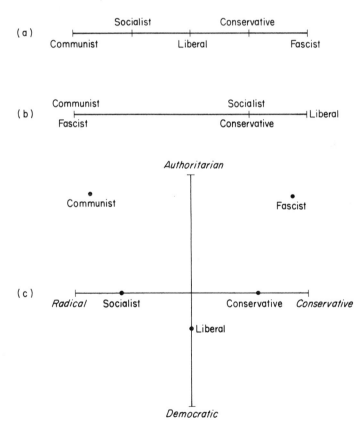

Figure 13.1 Diagram illustrating three hypotheses regarding relative position of five main political groups

Eysenck (1954) and Eysenck and Wilson (1978) have summarized evidence to show that the two dimensions are essentially orthogonal, as shown in Figure 13.1. Evidence for this proposition comes essentially from factor-analytic studies of social attitudes, an example of which is given in Figure 13.2. The questionnaire used in this study is reproduced in Appendix D and was employed in the analysis of the London data reported in the previous chapter. The two dimensions resulting were called radicalism–conservatism (R for short) and tough-mindedness as opposed to tender-mindedness (T for short). The tough-minded attitude is that of the authoritarian, the tender-minded that of the liberal in this system.

As regards political parties, a study by Eysenck (1951) looked at scores on R and T of middle-class and working-class conservatives, liberals, socialists and communists. The results are shown in Figure 13.3, and it can be seen that socialists and communists are high on radicalism, conservatives high on conservatism, and liberals slightly less conservative. Working-class members of these political parties are in each case more tough-minded than middle-class members, who tend to be tender-minded. In another study, Eysenck (1954) showed the intermediate position of the liberals in a more

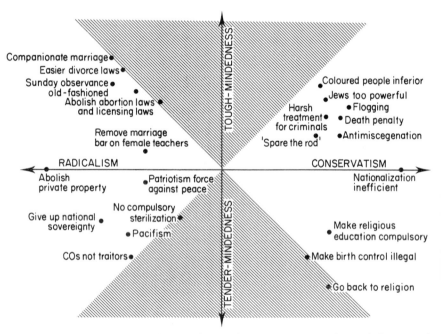

Figure 13.2 Distribution of attitudes with respect to tough-mindedness and radicalism (Eysenck, 1954).

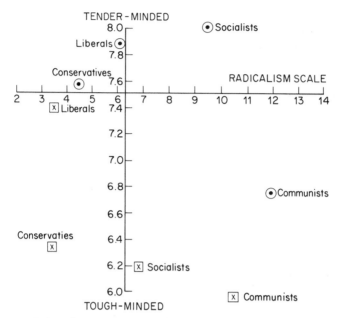

Figure 13.3 Political parties in a two-dimensional framework (Eysenck, 1971). ⊙, middle class; ⊠, working class.

clear-cut fashion; radicalism–conservatism scores of conservatives, liberals and socialists (members of the Labour Party) are shown in Figure 13.4.

Figure 13.5 shows scores on both R and T of communists, fascists and a non-political control group (Eysenck, 1954); it can clearly be seen that communists are radical, fascists are conservative, and that both groups are very significantly more tough-minded than is the control group, which is also intermediate with respect to radicalism–conservatism.

The type of results and analyses reported above is not confined to the United Kingdom, where the original research was carried out, nor is it confined to the time when it was carried out (the early 1950s). A more recent replication by Hewitt *et al.* (1977) showed that the factor-analytic pattern remained practically unchanged after 25 years. Similarly, cross-cultural replications of the original studies in Sweden, Germany, Spain, Italy and Japan, as well as in Canada and the USA, have given very similar results, suggesting that the model is widely applicable in industrialized countries (Eysenck and Wilson, 1978). The concept of authoritarianism, too, has been studied in a different context, by Kool and Ray (1983).

Critism of these studies has largely centered on psychometric objections. Reference to Figure 13.2 will show that there are no items in the

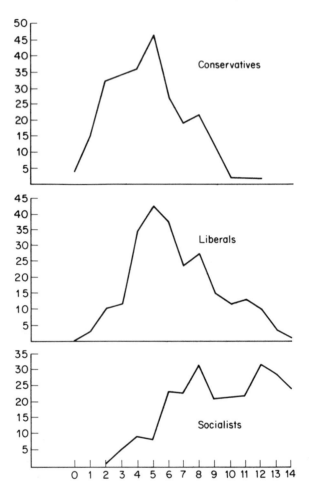

Figure 13.4 Distribution of scores on radicalism factor for three political parties (Eysenck, 1954).

tough-minded and tender-minded quadrants, i.e. those parts of the diagram that have been marked with stippled lines. This suggests rotation of the factors through 45°, resulting in a solution with two factors that might respectively be labelled *religionism* and *humanitarianism* (Ferguson, 1939). Psychometrically there is no doubt that the criticism is justified, and that simple structure would demand a rotation of this kind. However, there are two reasons why the solution shown in Figure 13.2 is preferred.

The first of these is that statistical prescriptions like simple structure are important and may guide otherwise indeterminate solutions and rotations,

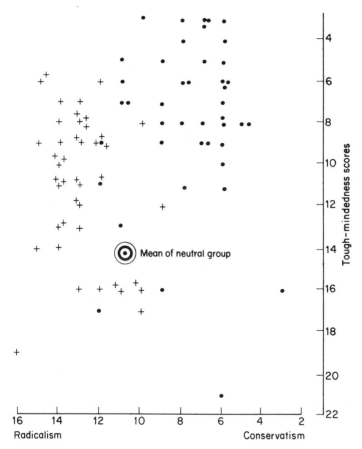

Figure 13.5 Tough-mindedness scores and scores of radicalism–conservatism of communists (+), fascists (•) and neutral group (Eysenck, 1954).

but they do not guarantee psychological meaningfulness of the resulting structure. We are here dealing with a general social and political problem, and it is noteworthy that in all industrialized countries there has been a recognition of a right–left continuum as a fundamental reality ever since the early days of the French Revolution (Meszaros, 1985; Craik, 1984). To consider conservatism a mixture of religionism and anti-humanitarianism, or to consider radicalism a mixture of humanitarianism and anti-religionism, would seem to be a rather absurd interpretation of this fundamental dichotomy. Political parties in all civilized countries have been based on the right–left opposition, and to disregard this fundamental social and political fact in favor of the convenience of a heuristic statistical rule that was never explicitly designed to deal with the situation would seem to ignore political and social reality.

To put the situation in another way, factor analysis may be used to *suggest* theories or to *test* theories. Where no theory exists, rules for rotation, such as the simple structure conventions, are useful. Where a specific and explicit theory is being tested, arbitrary rules of simple structure etc. are irrelevant to the scientific application of structural models of the factor-analytic type (Eysenck, 1952b, 1953; see also Chapters 10 and 11). This brings us to the second point.

The second argument is that the attitude structure actually observed is in fact predicted by Eysenck's theory. According to this theory, there is only one genuine dimension of social attitudes, namely that going from radicalism to conservatism. The orthogonal dimension of tough-mindedness is regarded by him as a *personality* dimension, possibly correlated with extraversion and even more strongly with psychoticism, which finds expression in terms of right- or left-wing attitudes. Hence the middle ground tends to be empty of social attitudes because here we should find personality factors rather than social attitudes. Thus, the shape of the figure is much as theory would predict. The principle of simple structure is irrelevant to a testing of that particular hypothesis.

Tentatively, we may therefore conclude that the suggested two factors of R and T are theoretically and practically superior to the two factors of religionism and humanitarianism, although of course it cannot be disputed that, from the purely statistical point of view, the two solutions are equivalent and can be rotated into perfect agreement.

Altogether, much of the criticism of the Eysenck model (see e.g. Stone, 1974; Altemeyer, 1981) seems to be premised on the priority of statistical and psychometric considerations over psychological and political ones.

Recent work, on the whole, supports the two-dimensional Eysenck model. Recent support for evidence for the R–T theory comes from a study by Ruch and Hehl (1985), conducted in West Germany, using intercorrelations between *scales* rather than *items* to establish the major dimensions involved, and adding measures of *values* and *personality* to direct measurement of *social attitudes*. The scales used are given in Table 13.1.

Figure 13.6 shows the resulting two-factor solution. The two factors are clearly marked. Conservatism–radicalism is highly loaded on Eysenck conservatism, Cloetta conservatism, Q1 conservatism, Wilson–Patterson conservatism, and in addition an intolerance-of-ambiguity, rigidity, and superego strength, as well as by an absence of aesthetic value. Tough-mindedness is marked by Eysenck tough-mindedness, political and theoretical values, disinhibition and Machiavellianism. Tender-mindedness is marked by religious value, social value and Cattell tender-mindedness.

Table 13.1 Scales used by Ruch and Hehl (1985) for factor analysis of social attitudes, social values, personality traits and other variables.

Wilson–Patterson conservatism scale (Schneider & Minkumar, 1972)
Eysenck (1976c) Public Opinion Inventory scale (conservatism, tough-mindedness; capitalism)
Cloetta (1983) conservatism and Machiavellism scales
Q1 Cattell radicalism (inverted) (Schneewind *et al.*, 1983)
Rigidity and intolerance-of-ambiguity scales (Brengelmann and Brengelmann, 1960a, b)
Values: theoretical, economic, aesthetic, social, political, religious
  (Allport–Vernon–Lindzay scales: Roth, 1972)
16 PF scales (Cattell *et al.*, 1970)
  E scale (dominance)
  G scale (superego strength)
  I scale (tender-mindedness)
  M scales (Antia imaginativeness)
Disinhibition scale (Zuckerman, 1979)
Liking for erotic humour (Ruch & Hehl, 1985)

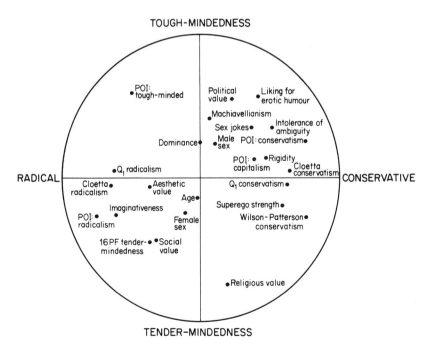

Figure 13.6 Attitude scales in a two-dimensional framework (Ruch and Hehl, 1985).

Males and the young are tough-minded, females and the old are more tender-minded. All of these results replicate previous studies in the USA and the UK, and make a picture perfectly congruent with that presented by Eysenck (1954).

In sum, it is scientifically uninteresting to ask which of the two solutions is "correct" — no meaning attaches to such a question. Scientific concepts are always man-made and constructed for a given purpose. Both models are in accord with the outcome of a statistical analysis of the given data. The question is: "Which is more usful and more congruent with what is known about the social and political structure of attitudes and parties in the countries concerned?" Factor analysis in general, and the principle of simple structure in particular, are excellent tools but bad masters: they cannot prescribe meaningful solutions, but merely suggest certain ways of looking at the data that may or may not be psychologically and sociologically valuable.

One interesting criticism that has often been made of the Eysenck model is that the communist quadrant has many items in it relating to sexual freedom, which it is asserted, are not very relevant to political views. The work of Grossarth-Maticek (1975) gives ample grounds to doubt the value of this criticism: he indicated the central importance of sexual attitudes in the political and social field.

Psychological and factor-analytic studies alone are not the only source of data for crystallizing a model for social attitudes. Discussions by historians and social scientists, for example Thomas (1979), Jupp (1968), Brittan (1968), Berki (1975) and Manning *et al.* (1976a,b), and the writings of Mannheim (1936) and Walford (1979) on ideology are important sources for the significance of terms like "conservatism", "liberalism, and "socialism".

To say that there is some degree of conformity between our psychological analysis and that made by students of politics is not to say that no problems remain. There is, to give but one example, what Eysenck (1972) has called the "paradox of socialism", i.e. the fact that people in the lower SES group tend to take the radical view with respect to economic matters (imposition of taxes, nationalization of industries, etc.), whereas they take a conservative view with respect to cultural and libertarian issues such as the death penalty. Eysenck (1975b) showed, in a factor-analytic study of 88 social-attitude questions, that "Middle-class people tend to be more *radical* with respect to general attitudes, but more *conservative* with respect to economic attitudes, than working-class people" (p. 323). In extreme cases this may lead to the postulation of two independent kinds of conservatism.

The hypothesis that personality factors are important in this connection is borne out by a study reported by Eysenck and Coulter (1972). In this study, working-class fascists, communists and controls were tested with a special version of the Thematic Apperception Test in order to look at two major personality traits that theories suggested would be relevant, namely *aggressiveness* and *dominance*. Aggressiveness is an important ingredient of psychoticism and dominance of extraversion. These two traits seemed to be well adapted to indirect testing of the kind made possible by projective techniques.

Figure 13.7 shows the results for aggression, and it will be seen that communists and fascists have very much higher scores on this trait than does

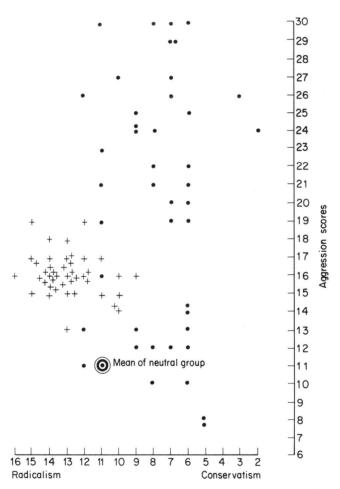

Figure 13.7   Aggression scores of communists (+) fascists (•) and neutral group plotted against radicalism–conservatism (Eysenck and Coulter, 1972).

the control group. (The control group itself was made up of soldiers, who in general show higher aggression and dominance scores than does the general population; hence a different control group might show an even greater differentiation than that observed here.)

Figure 13.8 shows dominance scores for the three groups, and again fascists and communists have much higher scores than do the controls. This study then shows that there is a reliable relationship between membership of tough-minded groups, on the one hand, and aggression and dominance (P and E) on the other.

A number of other studies have also shown that extraversion, and more particularly psychoticism, correlate significantly with tough-mindedness in groups that are not overtly political (Eysenck and Wilson, 1978). The corre-

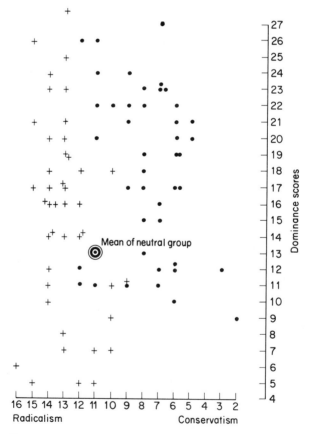

Figure 13.8 Dominance scores of communists (+), fascists (•) and neutral group plotted against radicalism–conservatism (Eysenck and Coulter, 1972).

lations are never very high, but they are always in the predicted direction, thus supporting the view that we may here be dealing with a genuine relationship. Other relevant studies support the view that authoritarianism and aggression are related (Manning *et al.*, 1980), that psychoticism is related to the Ray directiveness scale — a scale that measures behavioral authoritarianism (Ray and Bozek, 1981) — also to dominance and assertiveness (Heavens, 1983), and that authoritarianism is related to intolerance of ambiguity, anti-intraception, dogmatism and other personality traits (Eysenck, 1954; Barker, 1958). Machiavellianism (Stone and Russ, 1976) has also been linked with T, and the authors state that: "Machiavellianism seems to represent Eysenck's tough-minded dimension empirically, as well as conceptually." They found a correlation between Machiavellianism and tough-mindedness of 0.44, but a completely insignificant one of – 0.09 with R.

One further consideration is of importance here. It has been found that tough-mindedness is usually much stronger in males than in females (Eysenck, 1971b), and equally psychoticism has been found to be much more pronounced in males than in females (Eysenck and Eysenck, 1976). This strengthens the case for regarding psychoticism as an important part of the personality structure that underlies tough-mindedness.

Before turning to other models that posit similar dimensions to those originally suggested by Eysenck, such as those of Rokeach, Wilson, Kerling and others, we may note one further score which may be derived from the Eysenck inventory. This is the so-called *emphasis* score, i.e. the tendency to respond with strong support or strong rejection of attitudes in general. This tendency is an expression of the "principle of certainty" first annunciated by Thouless (1935). As he put it: "When in a group of persons there are influences acting both in the direction of acceptance and of rejection of a belief, the result is not to make the majority adopt a lower degree of conviction, but to make some hold the belief with a high degree of conviction, while others reject it also with a high degree of conviction." Thouless worked with religious beliefs, but Eysenck (1954) extended the principle to a variety of social attitudes, using a seven-point attitude scale, and collected 22 208 votes on these issues. The results are shown in Figure 13.9, and it will be clear that the distribution is definitely U-shaped, with a tendency for extreme opinions to be much more widely adopted than intermediate expressions of attitude. This tendency to certainty seems the core of Rokeachs's concept of "Dogmatism" (Rokeach, 1960), and it is interesting to note that Smithers and Lobley (1978) found that on Eysenck's conservatism–radicalism scale both very radical and very conservative subjects scored high on the dogmatism scale, as shown in Figure 13.10.

Of the authors whose models resemble Eysenck's, the most relevant one is perhaps Wilson's (1973), because his scale of measurement of social attitudes

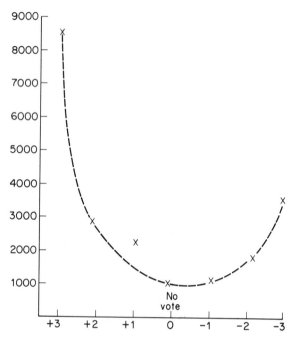

Figure 13.9 Distribution of 22 208 votes on a seven-point attitude scale (Eysenck, 1954).

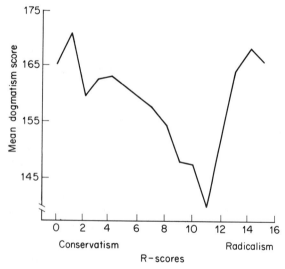

Figure 13.10 Dogmatism and extremity of conservatism – radicalism (Smithers and Lobley, 1978).

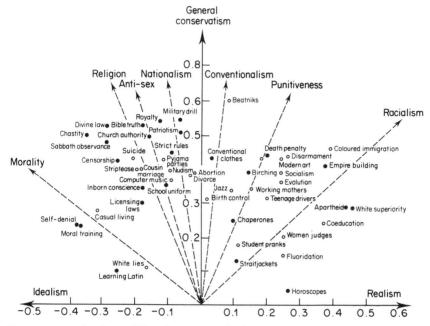

Figure 13.11 Loading of C-scale items on the first two principal component factors averaged over three cultures (England, Netherlands and New Zealand). Coordinates marked (•) represent positive loadings; (○) show negative loadings. The oblique subfactors of conservatism (– – –) are drawn by inspection to assist in the interpretation of the second principal component (Wilson, 1973).

has been used by us in our genetic studies, in conjunction with the Eysenck Public Opinion Inventory. The Wilson–Patterson scale (Wilson and Patterson, 1970) consists of 50 items, such as death penalty, evolution theory, suicide and licensing laws, which the subject is invited to endorse or reject by circling a "Yes" or "No". It is claimed that this form is superior to the usual questionnaire format because it avoids ambiguities due to the sentence structure normally employed. The actual questionnaire used is described in Chapter 12, p. 333.

Factor analysis of the intercorrelations between the items discloses two major factors, which Wilson calls "conservatism" and "realism–idealism", instead of tough-versus tender-mindedness. Figure 13.11 shows the position of the various items on these two factors. They show a clear resemblance to the Eysenck factors, and they too, show evidence of validity. Figure 13.12 shows conservatism scores of various groups known to be radical or conservative respectively.

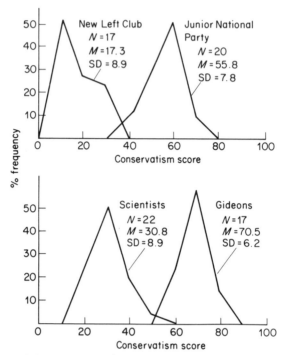

Figure 13.12 Validation comparisons: the distribution of C-scores for four "known groups" (from Wilson and Patterson, 1970).

Eysenck (1976c) has also published a factor analysis of an extended Wilson–Patterson-type questionnaire for a quota sample of 1441 subjects. He found 13 meaningful primary factors, the correlations between which gave rise to two major higher-order factors. "These were closely similar in content to the Radicalism and Tough-mindedness factors previously isolated in research with more orthodox types of items" (p. 463).

Rokeach (1973, 1979), although like Christie (1956) an early critic of Eysenck's *Psychology of Politics* (Rokeach and Hanley, 1956; but see replies in Eysenck 1956b,c), arrived at a model in his own research that is very similar to Eysenck's. He too questions the adequacy of the traditional reliance on a single radicalism–conservatism continuum and proposed that a more sophisticated understanding would ensue if political ideologies were conceptualized in terms of two independent value dimensions — one defined by *equality*, the other by *freedom*. He believed that the extremes of these dimensions could be used to typify four major political orientations. Socialism is characterized by concern for both equality and freedom, while fascism is characterized by lack of concern for both. Capitalism values freedom at the expense of equality, while communism emphasizes equality at the expense of freedom.

As Rokeach (1973) himself acknowledges: "The two-value model presented here most resembles Eysenck's hypothesis (1954) that the four major ideological orientations can best be differentiated in terms of two orthogonal traits — a liberalism–conservatism dimension and a tough-tender-mindedness dimension" (p. 186). While the attempt is interesting, it does not seem to capture the essence of the various political groupings. The ideological position that communism is concerned with equality has not, in general, translated into practical equality of life style for all members of communist countries. Similarly, the notion that freedom necessarily characterizes socialism is one that is often contradicted by history even in the United Kingdom. These various values are frequently espoused in theory by socialists, communists, fascists and capitalists, but they are often irrelevant to the practical policies of the political parties so designated (Boltomore, 1984).

Tests of the Rokeach system by Cochrane *et al.* (1979) in the United Kingdom and Braithwaite (1982) in Australia, as well as others cited by Braithwaite, have not on the whole been kind to the Rokeach model, particularly with reference to the discriminating power of freedom. Factor-analytic studies of values, such as Braithwaite's (1982), also failed to give support to Rokeach.

Kerlinger (1970, 1972) also agreed that liberalism–conservatism cannot be viewed as a single bipolar dimension, but that we have to deal rather with two dimensions, liberalism and conservatism, that are independent of one another. Kerlinger proposes that, for the conservative, private property, religion, educational subject matter and certain other referents are criterial, whereas such referents as social change and civil rights, not normally criterial for the conservative, are criterial for the liberal. His findings indicated that separate measures of liberalism and conservatism were indeed independent of each other. The apparently paradoxical nature of Kerlinger's findings are produced by his use of the term "liberalism" in two different senses. On the one hand, as in the characterization of the radical–conservative dimension, he uses it, instead of the term "radical", as the opposite pole of conservatism, a practice that is not really to be recommended by virtue of the fact that Liberals (i.e. members of the Liberal party) are usually intermediate between radicals and conservatives. He also uses the term "liberal" as we have done, and as Hayek (1960) had done, as the exemplar of the tender-minded, and as opposed to the tough-minded. His general scheme is thus very similar to the one proposed by Eysenck.

This is a brief and compressed account of a very complex topic on which

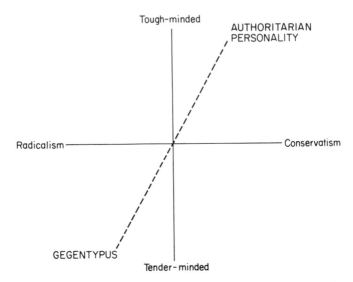

Figure 13.13   Positions of Gegentypus and authoritarianism in a two-dimensional framework.

many hundred articles and a large number of books have been written, with only a modicum of agreement between them. There does seem to emerge, however, a good deal of agreement on certain propositions. The first of these is that the important dimension of radicalism–conservatism seems to account for a good deal of the total variance when intercorrelations are calculated between different social attitudes. The second proposition would be that this dimension is not sufficient to account for the major relationships between social attitudes, and that another dimension is necessary, which has been variously called authoritarianism, liberalism, directiveness, *Gegentypus*, or tough-mindedness versus tender-mindedness. Additional concepts, such as dogmatism or Machiavellianism, have been used as being largely synonymous with authoritarianism or tough-mindedness, and referring particularly to certain personality correlates of authoritarianism. Figure 13.13 illustrates these agreements.

When one looks at the interrelations between political parties, which is easier to do in European countries because they tend to be more diverse politically than in the United States, it is clear that two dimensions are required to account for the observed relationships here also, and the same two dimensions would seem to fulfil this function adequately. The origin and development of these two dimensions, and the position on them of any particular individual, has usually been accounted for in terms of some form of social-learning theory, but the implied rejection of genetic factors has not been based on any empirical evidence, but purely on *a priori* judgements. In this part of the book our concern is to determine the degree to which this almost universal assumption is justified.

In undertaking this study, we have used the Eysenck model and have used the social-attitudes questionnaire employed by him, as well as the Wilson inventory. It seemed desirable to use two rather distinct measuring instruments, different from each other in many ways, in order to determine to what extent any of our findings might be due to instrument variance, rather than true variance.

Hence our study can be looked at from two different points of view. On the one hand, we are assuming the validity of a social-attitudes model and attempting to assess the quantitative contribution made to the variables in this model by genetic and environmental factors. On the other hand, we are testing some of the assumptions of the model by subjecting the scales involved to an analysis that could disclose their lack of homogeneity, reliability, or validity.

Chapter 14

Twin Studies of Social-Attitude Dimensions

Our item analyses in Chapter 12 support the idea that variation in many important attitudes, including certain general kinds of religious attitudes, may not be explicable in purely social terms. Between the experiences of an individual and his behavior there is a nervous system that depends on genes for its development and function. The brain filters and transforms information derived from the environment. If there is genetic variation in the process of transformation then it is hardly surprising that there is genetic variation in behavior. Lewontin *et al.* (1984) attribute to "vulgar hereditarianism" the view that socially important variables such as religious beliefs could be encoded in a particular gene. We can think of no behavior geneticist who seriously maintains such a simple model for the relationship between genotype and phenotype for complex behavioral characteristics. In denying the "one-gene–one-belief" model for social attitudes, however, we should not be forced to affirm out of pure faith that individual differences in social attitudes are entirely immune from genetic influences.

Our data do not show that genetic factors explain all the diversity of opinion in society, and they certainly do not mean that there is a single gene (or even two genes) in which "belief in God", "Protestantism" or "Republicanism" are directly coded. Indeed, other data on twins concerning religious affiliation (Eaves *et al.*, in preparation) confirm that MZ and DZ twins are almost equally concordant for the particular religious tradition with which they see themselves as identified (Catholic, Protestant, etc.). Once we move away from a person's historical allegiance, however, there is great diversity within a cultural tradition (beliefs about the role and influence of the church, endorsement of particular theological or philosophical propostions). Individuals who identify themselves as Jewish, for example, nevertheless display great diversity in the extent to which they subscribe to belief in God or traditional practices. Our study, which focuses mainly on variation *within* certain societies, suggests that genetic factors play a surprising role in maintaining human diversity. Even within families, it is clear that dizygotic twins hold religious and political views that are significantly more diverse than monozygotic twins. Given a "religious" home environment, it may still be partly a matter of genotypic predisposition whether one child or another remains faithful to the beliefs of his parents or chooses a more secular philosophy. Many social and accidental effects may intervene between the individual genotype and the final expression in a particular attitude or belief. In our present Western society, however, the probabilities along the way may be biased as much by the genotype of the individual as they are by the social and educational environment to which he is exposed. If this is a good model of human development then we expect a correlation between genotype and social attitudes. If social attitudes play any part in the behavior of one person towards another then we may find that genetic differences between people are partly responsible for the distinction between godly and ungodly and between liberal and conservative in contemporary societies. By acknowledging the possibility that genes may affect attitudes and beliefs, we are making no judgments about the moral virtues of religion or unbelief nor about the merits of one political system over another. A genetic model for behavior is no more nor less revolutionary and offers no more nor less support for the "status quo" than a purely environmental model.

Our surprise at finding signs of a genetic component of variation in attitudes at the individual-item level should not override the other striking result of the item analysis that sets the social-attitude items apart from the vast majority of the personality items. The item analyses also provided many significant estimates of the family-environmental effects E_B. Furthermore, estimates of E_B were correlated with the loadings of the individual items of the common factor underlying the items, suggesting that the family environment plays a significant part in the causes of correlation between the items. We have already seen (e.g. in Chapter 5) that with twin data alone the effects of the family environment are confounded with the genetic consequences of assortative mating. We shall postpone a detailed consideration of this issue for the next two chapters. The important point now is that the social-attitudes items display a pattern of variation that is quite different from the personality items and that the distinguishing characteristics of the items are related to what the items have in common rather than to their specific characteristics.

In this chapter we consider models for variation in composite scores derived to reflect the factor structure of the social-attitudes questionnaires. First, we outline four separate twin studies of "conservatism", which yield remarkably consistent results for different instruments and samples. Where appropriate, we also consider the determination of "tough-mindedness". Next, we examine the longitudinal consistency of conservatism with particular attention to the genetic and environmental components of stability and change. We then consider analyses of primary factors derived from a public-opinion inventory. In Chapter 15 we consider the contribution of cultural inheritance and assortative mating to family resemblance and describe some results of a preliminary attempt to analyze the mating system in more detail in Chapter 16.

14.1 TWIN STUDIES OF CONSERVATISM

In Table 14.1 we give mean squares from the analyses of variance of three twin studies of conservatism undertaken with the London twin sample.

The first study employed the Public Opinion Inventory, for which item analyses were reported in Chapter 12 (see Appendix D). These data were collected at the same time as the questionnaire data reported as part of the study of sociability and impulsiveness in Chapter 10. The POI (see Chapter 13) was designed in the early 1950s, but Hewitt *et al.* (1977) showed that exactly the same factor structure was obtained for the POI in the twin sample 25 years later. Thus, although the mean responses may have changed on individual items, the factor structure was stable over time. Scores were obtained on the two factors of conservatism and tough-mindedness. The items were combined according to the original scales in Eysenck (1954), but a five-point scale was used for each item to reflect the response categories on the questionnaire. The mean squares for the twin groups were computed by Hewitt (1974). Hewitt's mean squares were corrected for sex but not for age.

The second study comprises a further sampling of the London Twin

Table 14.1   Analyses of variance of conservatism scores of twins from three studies of social attitudes.

Twin type	Item	Hewitt		Martin		Last	
		df	ms	df	ms	df	ms
MZ$_f$	Between pairs	323	8.36	93	338	231	1.13
	Within pairs	324	1.91	95	62	233	0.25
MZ$_m$	Between pairs	119	10.36	37	357	81	0.30
	Within pairs	120	1.53	39	49	83	0.30
DZ$_f$	Between pairs	193	9.51	52	365	145	1.22
	Within pairs	194	2.89	54	101	147	0.39
DZ$_m$	Between pairs	58	9.82	15	272	50	1.25
	Within pairs	59	3.18	17	82	52	0.45
DZ$_{mf}$	Between pairs	127	10.07	39	351	70	1.28
	Within pairs	127	3.23	41	129	72	0.49

Registry in connection with an anonymous study of sexual attitudes reported by Martin and Eysenck (1976). A different instrument was used to measure conservatism (see Eysenck, 1976b) and, since the questionnaire was returned anonymously, we have no way of pairing responses to the first study with those of the second. The mean squares were corrected for age and sex.

The third study was a later study of the London sample using the Wilson–Patterson conservatism scale. The data summary was computed by Last (1978). The mean squares were corrected for age and sex. All three authors conducted their own model-fitting analyses of the data.

In all three studies the authors arrived at the same conclusions, even though the questionnaires differed greatly in format. The causes of variation in conservatism scores were consistent over sexes. In every case, the best-fitting model required within-family environmental effects (E_W), additive genetic effects (V_A) *and* between-family effects (E_B). The results for conservatism are therefore in marked contrast with those we obtained in repeated studies of the main personality dimensions (see e.g. Chapter 5). In the case of personality there was virtually no evidence that the family environment was a component of individual differences. The results for conservatism, on the other hand, as distinguished by a threefold replication, in the London sample, show a major contribution of the family environment.

Table 14.2 gives the estimated proportions of variance explained by the three sources on variation in the twin data. It is clear that, once differences of scale are removed, the proportions are remarkably consistent across the three studies. Approximately 40% of the variation in conservatism scores is apparently due to additive genetic factors, 30% due to environmental differences within families and 30% due to other differences between families. These could be strictly environmental, but could also reflect the genetic effects of positive assortative mating. We are especially struck by the consistency of these findings between studies, and by the consistency of the contrast between the apparent importance of the family environment for

Table 14.2 Summary of sources of variation in conservatism in three twin studies.

Study	Contribution (%)			Test of model	
	E_W	V_A	E_B	χ^2_7	$P\%$
Hewitt	31.4	40.1	28.8	10.08	18
Martin	27.4	44.3	28.4	3.26	78
Last	35.1	37.6	27.3	6.20	52

social attitudes and its comparatively small role in the etiology of personality differences.

14.2 CONSERVATISM IN THE AUSTRALIAN TWIN SAMPLE

It might be argued that the internal consistency of the British results is less impressive because each replication was obtained by resampling from the same registry. Although there are some changes in the composition of the registry over time, there is also a large degree of overlap between pairs completing the different questionnaires. Indeed, this partial overlap provides the basis for an analysis of longitudinal stability to be reported below (Section 14.3).

Replication in a completely different population would be compelling. In Chapter 12 we reported the study undertaken by Jardine *et al.* of the responses of the Australian twins to the shortened Wilson–Patterson conservatism scale. Martin and Jardine (1986) also analyzed the total conservatism scores from this sample. The mean squares from the Australian Study are given in Table 14.3 and the results of model-fitting are summarized in Table 14.4.

When the model-fitting analysis is confined to members of each sex separately, it is clear that both two-parameter models (E_W, E_B and V_A, E_W) cannot fit the data. Models that include all three main sources of variation, however, give an excellent fit within sexes. In both sexes, estimates of V_A, E_W and E_B are all highly significant. In this important respect, therefore, the Australian data replicate the results of the three studies of the London twin population.

Table 14.3 Observed mean squares for conservatism in Australian sample.

Statistic		Degrees of freedom	Mean square
MZ_f	Between[a]	1231	200.39
	Within	1233	43.66
MZ_m	Between[a]	564	250.49
	Within	566	62.44
DZ_f	Between[a]	749	175.62
	Within	751	64.25
DZ_m	Between[a]	350	238.67
	Within	352	85.06
DZ_{os}	Between[a]	904	179.92
	Within[b]	905	76.05

[a] Corrected for regression on age.
[b] Corrected for sex differences in means.

Table 14.4 Summary of model fitting to age-corrected conservatism scores in Australian sample.

	E_W	E_B	V_A	V_D	df	χ^2	h^2
Female:							
E_W, E_B	51.45***	69.79***	—	—	2	41.31***	
E_W, V_A	41.82***	—	77.82***	—	2	21.31***	
E_W, E_B, V_A	43.58***	35.67***	41.92***	—	1	0.11	0.35 + 0.06
E_W, V_A, V_D	43.58***	—	148.92***	−71.33	1	0.11	
Male:							
E_W, E_B	71.11***	87.43***	—	—	2	11.24**	
E_W, V_A	59.59***	—	97.28***	—	2	13.32**	
E_W, E_B, V_A	62.69***	52.71***	43.28**	—	1	0.20	0.27 + 0.09
E_W, V_A, V_D	62.69***	—	201.41***	−105.42	1	0.20	
Female and Male:							
E_W, E_B	57.67***	75.35***	—	—	6	108.31***	
E_W, V_A	47.42***	—	83.99***	—	6	97.73***	
E_W, E_B, V_A	49.58***	40.95***	42.50***	—	5	60.97***	
E_W, V_A, V_D	49.58***	—	165.35***	−81.90	5	60.97***	
Female and male and opposite sex:							
E_W, E_B	62.04***	69.78***	—	—	8	135.90***	
E_W, E_B	62.04***	69.78***	—	—	8	110.94***	
E_W, E_B, V_A	49.45***	34.33***	47.97***	—	7	64.41***	
E_W, V_A, V_D	49.45***	—	150.95***	−68.66	7	64.41***	

Significance levels: ** P < 0.01, *** P < 0.001.

Table 14.5 Parameter estimates from Australian conservatism data.

Parameter	Estimate ± s.e.
V_A	41.5 ± 6.3
E_{WM}	62.1 ± 3.3
E_{WF}	43.4 ± 1.7
E_{BM}	49.4 ± 7.5
E_{BF}	34.6 ± 6.1
E_{BMF}	37.1 ± 5.0
χ^2_4	4.40
h^2_{males}	0.27 ± 0.04
$h^2_{females}$	0.35 ± 0.05

In another major respect, the results for the larger Australian sample are different. Whether unlike-sex pairs are included in the analysis or omitted from it, there is evidence of heterogeneity between the parameters in the two sexes. A single estimate of V_A can explain the genetic component of variation in males and females, but the estimates of the two environmental components differ significantly between sexes. The final parameter estimates are given in Table 14.5, and the estimated proportional contributions of the three sources to variation in conservatism scores are summarized, for both sexes, in Table 14.6. The environmental variation within families is divided into long-term effects and the effects of short-term (three-month) unreliability. Most of the environmental variation within pairs is due to long-term effects.

The proportions of variance attributable to the main sources only differ slightly from those obtained in London. The contribution of E_B is about the same, the average contribution of genetic effects is about 30% rather than 40%, and the overall contribution of the environment within families is increased from about 30% in the London sample to approximately 40% in the Australian data.

In many important respects the results from the much larger Australian study replicate the essential features of the smaller British study. As with personality, so with conservatism, the larger study confirms most aspects of the smaller one, but adds certain glosses. In the case of the conservatism dimension of social attitudes both studies concur in showing a major component of the family environment (or assortative mating) in twin resemblance. Such a finding is gratifying because it confirms that our models and methods are perfectly capable of detecting important environmental effects. Furthermore, given large enough samples, the results of our analyses replicate very well across different Western populations. Consistent similarities appear for the same trait. Consistent differences appear for different traits.

Table 14.6 Sources of variance (%) for age-corrected conservatism scores.

	Females		Males	
E_1 — error	36 ⟨ 18		41 ⟨ 9	
E_1 — individual environment	18		32	
V_A — total genetic	35] 49		27] 38	
E_2 — assortative mating	29 ⟨ 14		32 ⟨ 11	
E_2 — family environment	15		21	

The detection of a genetic component of variation in social attitudes surprised us at first, and it will probably surprise others too. The fact that the social-attitudes data reveal a significant E_B component, while the extraversion and neuroticism scales do not, vindicates the marriage between large genetically informative samples and a few well-defined variables in replicated studies. The variables for which the *a priori* expectation of cultural determination is greatest are those for which the twin study detects apparent cultural inheritance. Traits for which it is less easy to visualize mechanisms of social interaction between relatives (neuroticism, for example) are those for which the apparent effects of social interaction are smallest. The genetic models that result represent an important heuristic for the interpretation of human differences in personality and social attitudes.

The results lead to further predictions. Because of the large samples involved, we have a fairly good idea of how results should look in other types of study that would provide further tests of our models. For example, if we are correct about the lack of a social-environmental effect on personality measures then we should predict that the correlations between unrelated individuals reared together should be approximately zero. The correlation for separated monozygotic twins should be not significantly less than the correlations for twins reared together. In contrast, for conservatism scores we might expect correlations of 0.3–0.4 between unrelated individuals reared together under the environmental model, and the correlations of separated monozygotic twins to be about half those of twins reared together.

14.3 THE CAUSES OF LONGITUDINAL CONSISTENCY IN CONSERVATISM

Two of the London studies involving the POI and the Wilson–Patterson conservatism scale yield a subset of twin pairs for whom both measures of conservatism can be derived. The interval between tests was approximately three years, so we may analyze the genetic and environmental basis of consistency over instruments and occasions of measurement. The study will provide a direct test of whether or not the two instruments are really identifying the same genetic and cultural effects and will determine whether individual changes in the environment over a three-year period, and their interaction with genotype, are a major component of individual profiles of attitude change.

Our analysis focuses only on those twin pairs for whom complete data are available on both occasions. The measurements made on the two occasions are regarded as two distinct variables, which may be correlated genetically or environmentally. The data are summarized, for each group of twins, by a

Table 14.7 Twin covariances and correlations for repeated measures of conservatism.[a]

Twin type	df	Twin 1 Occasion		Twin 2 Occasion	
		1	2	1	2
MZ$_f$	181	69.90	3.91	45.27	3.03
		0.55	0.72	3.80	0.46
		0.59	0.49	84.42	4.64
		0.43	0.64	0.60	0.71
MZ$_m$	53	61.92	4.17	50.63	3.40
		0.63	0.71	3.50	0.40
		0.74	0.48	76.12	3.79
		0.50	0.55	0.50	0.74
DZ$_f$	97	119.27	6.33	60.87	5.01
		0.64	0.81	3.89	0.42
		0.58	0.45	91.96	6.96
		0.47	0.47	0.74	0.97
DZ$_m$	21	123.21	6.52	32.69	2.33
		0.61	0.93	1.95	0.08
		0.31	0.21	89.54	4.07
		0.26	0.10	0.53	0.65
DZ$_{mf}$	39	86.13	4.52	15.23	2.18
		0.53	0.83	2.32	0.25
		0.18	0.28	84.64	4.27
		0.29	0.34	0.57	0.66

[a] Correlations are given in the lower triangle.

4×4 covariance matrix in which the variables are occasions classified by first and second twin. The raw covariance matrices, corrected for age, are given in Table 14.7. That the numbers are smaller than in previous analyses of conservatism is due to the fact that many twins on the registry did not return both questionnaires. The differences in variance between first and second occasions are merely a reflection of the fact that different tests and transformations of the data were used on the two occasions.

The basic model for the covariance structure of the repeated measures assumes that differences in conservatism are ultimately caused by additive genetic differences, within-family environmental effects and between-family environmental effects. We denote the additive genetic variance on the first and second occasions by V_{A1} and V_{A2} respectively. The genetic covariance between occasions is V_{A12}. We may define similar parameters for the environmental variances and covariances within families, E_{W1}, E_{W2}, E_{W12}, and between families, E_{B1}, E_{B2}, E_{B12}.

Table 14.8   Expected covariances for MZ (upper triangle) and DZ (lower triangle) twins measured on two occasions.

		Occasion 1		Occasion 2	
		Twin 1	Twin 2	Twin 1	Twin 2
Twin 1 Occasion 1		$V_{A1}+E_{W1}+E_{B1}$	$V_{A1}+E_{B1}$	$V_{A12}+E_{W12}+E_{B12}$	$V_{A12}+E_{B12}$
Twin 2		$\frac{1}{2}V_{A1}+E_{B1}$	$V_{A1}+E_{W1}+E_{B1}$	$V_{A12}+E_{B12}$	$V_{A12}+E_{W12}+E_{B12}$
Twin 1 Occasion 2		$V_{A12}+E_{W12}+E_{B12}$	$\frac{1}{2}V_{A12}+E_{B12}$	$V_{A2}+E_{W2}+E_{B2}$	$V_{A2}+E_{B2}$
Twin 2		$\frac{1}{2}V_{A12}+E_{B12}$	$V_{A12}+E_{W12}+E_{B12}$	$\frac{1}{2}V_{A2}+E_{B2}$	$V_{A2}+E_{B2}+E_{W2}$

The expected variances and covariances for MZ and DZ twins for the two occasions are given in Table 14.8. The maximum-likelihood method (see Chapter 10) was used to obtain estimates of the components of variance and covariance from the observed covariance matrices.

The correlation in the between-family environment over occasions approaches the upper bound of unity for these data, so we substituted the term $(E_{B1}, E_{B2})^{1/2}$ for E_{B12} in the expectations for the covariances. The chi-square for testing the goodness of fit of the model was 44.8 for 42 df $(0.3 < P < 0.4)$, confirming that the model that fitted each of the separate twin studies can also account adequately for the joint pattern of covariation in the analysis of repeated measures. The maximum-likelihood parameter estimates are given in Table 14.9.

The cross-temporal stability of the causes of variation in conservatism is greatest for the between-families environmental component of variance $(r_{B12} = 1)$, and lowest for the within-families environmental component $(r_{W12} = 0.27)$. The correlation in genetic effects over time, r_{G12}, is 0.72. Thus, in spite of the fact that attitudes are commonly assumed to be very labile, certain crucial determinants seem to exercise a consistent effect over

Table 14.9   Parameter estimates for repeated measures of conservatism different scales.

	Variances		
	Occasion 1	Occasion 2	Cross-temporal correlation
V_A	41 ± 11	0.39 ± 0.09	0.72
E_W	30 ± 3	0.28 ± 0.03	0.27
E_B	16 ± 10	0.10 ± 0.08	1^a

[a] Parameter fixed on upper bound.

time, even when the measurements taken on the two occasions differ greatly in their format and content. The items of the POI are complex statements, requiring a response on a five-point scale. The items of the Wilson–Patterson scale are single words, to which response is on a simpler three-point scale (see Chapter 12). Nevertheless, the only effects for which the cross-temporal correlation is small are the environmental influences within families, with which errors of measurement are confounded. In so far as individuals' overall conservatism scores change with time relative to their peers, therefore, the main causes of change are the unique experiences of the individual, which are not shared, even with a cotwin. The fact that the genetic correlation across occasions is large (not significantly less than unity but significantly greater than zero) implies that the expression of genetic factors that affect conservatism is remarkably consistent across time and questionnaires. The two conservatism scales are assessing the same genetic effects. Similarly, the correlation in the between-family environmental effects was unity. The shared environmental experiences of twins therefore exert a long-term effect on conservatism that is consistent over the two types of measurement. There is no evidence that shared experiences over the three-year period in question caused some pairs of twins to increase their conservatism scores and others to respond in a more radical direction on the second occasion. Furthermore, since the genetic effects are virtually the same on the two occasions, there is no evidence that any cultural effects intervening between the two occasions had changed the ranking of genotypes with respect to the trait. That is, there is no evidence of genetic control of the direction of attitude change over the three-year period in question. If different genotypes had responded differently to the two instruments, or had responded differently to the events of the intervening years, then the genetic correlation would have been significantly less than unity.

The longitudinal study shows that genetic effects and environmental differences between families have a consistent and lasting (i.e. over three years) effect on the overall tendency of individuals to endorse conservative or radical attitudes. These effects are not altered or eradicated by such normal cultural changes as occurred in the three-year period of follow-up. This finding does not mean that attitudes do not change, but rather that there are effects of genotype and the family environment that persist in society in spite of short-term changes. An alternative interpretation, equally consistent with the twin data, is that the genetic effects of assortative mating affect those aspects of social attitudes that display the greatest long-term consistency. As far as individual environmental experiences are concerned, we find that they could explain about 35% of the total variation in conservatism on each occasion. The long-term consistency of these effects is remarkably low. Thus the differences that we see, even within identical twins reared together, are substantial and extremely labile. The low correlation of the within-family environmental effects over time implies that it is a matter of chance and individual experience which identical twin is more conservative than his cotwin on each occasion. In so far as conservatism is a reflection of unique experiences of the individual, these effects are short-lived and cannot be shown to extend over a three-year period. If the unique experiences of the individual had a lasting effect then we should expect the correlation of within-family environmental effects to be much greater than zero.

14.4 ANALYSIS OF PRIMARY SOCIAL-ATTITUDE FACTORS

Although much of the variation in attitudes is explained by the first two principal components, factor analysis reveals a number of primary factors that are stable over sexes and populations. Feingold (1984) identified five correlated primary factors in the POI in British and US samples for which there was reasonable consistency over sexes. The factors were identified by item content to be "authoritarianism", "religion", "socialism", "prejudice" and "permissiveness". Items having salient loadings on each of the primary factors are listed in Table 14.10. The summary statistics for the British twin sample for the five primary factors are given in Table 14.11.

Models were fitted by the maximum-likelihood method, using the General Linear Interactive Modelling ("GLIM") program (Nelder, 1975). Results for models with no sex-limited genetic or environmental effects are summarized in Table 14.12.

Models that excluded genetic effects did not fit the data on any of the variables. Models that included both sources of environmental variation *and* genetic effects gave a good fit to four out of the five factors. None of the three models could explain the data on "prejudice". The "authoritarianism" and "religion" factors showed statistically significant estimates of the between-families environmental component. Parameter estimates under the best-fitting model, assuming no sex limitation, are given in Table 14.13.

The results of allowing for sex differences in genetic and environmental effects (cf. Chapter 4) are seen in Table 14.14. Once again, the purely environmental explanation failed to account for variation in four out of five factors. "Religion" formed the only exception.

As before, the data on "authoritarianism" and "religion" were explained far better by models that allowed for both genetic and cultural components of twin resemblance. Similarly, when allowance is made for sex differences in the expression of genetic and environmental differences, the data on "prejudice" require a joint genetic and cultural explanation. The parameter estimates for the two primary factors for which sex-limited effects improved

Table 14.10 Items loading on the five primary attitude factors.

Item	Primary factor	Item loading
	Authoritarianism	
6	Peace, not national sovereignty	−0.41
10	Flog violent criminals	0.69
13	My country right or wrong	0.44
16	Obligation to family	0.37
18	Abolish barbaric death penalty	−0.59
28	Compulsory military training	0.53
29	Flog sex criminals	0.74
34	Conscientious objectors traitors	0.34
42	Retain independence in world organization	0.57
47	Treatment of criminals too harsh	−0.61
51	Life short and to be enjoyed	0.34
	Religion	
14	Good enough life without religion	−0.56
17	No survival after death	−0.61
23	Sunday observance old-fashioned	−0.43
25	Acceptance of Church's teachings	0.33
31	God an invention of human mind	−0.65
33	Church should increase influence	0.64
36	Religious people hypocrites	−0.36
39	Religion civilization's only hope	0.68
45	Compulsory religious education	0.58
48	Church main bulwark against evil	0.47
53	Christ divine	0.70
56	Universe created by God	0.74
59	Faith in supernatural power	0.48
	Socialism	
4	Introduce socialism	0.43
24	Capitalism immoral	0.45
52	Occupation better than war	0.37
	Prejudice	
2	Coloureds innately inferior	0.50
15	Coloureds shouldn't be foremen over whites	0.46
38	Don't help Asian refugees	0.52
44	Make discrimination illegal	−0.49
54	Racial segregation	0.60
55	Punish homosexuals	0.38
	Permissiveness	
8	Make divorce easier	0.38
12	Trial marriage	0.58
27	Encourage free love	0.48
37	Extramarital sex wrong	−0.47
49	Travelling without a ticket	0.32
51	Life short and to be enjoyed	0.33

Table 14.11 Mean squares between and within twin pairs and correlations for primary-factor scores. [a]

| | | Factor | | | | | | | | | |
Relationship	df	Authoritarianism ms	r	Religion ms	r	Socialism ms	r	Prejudice ms	r	Permissiveness ms	r
Between MZ$_f$	324	1.234	0.69	1.329	0.66	0.969	0.48	1.122	0.61	1.213	0.66
Within MZ$_f$	325	0.228		0.277		0.344		0.301		0.245	
Between MZ$_m$	119	1.569	0.74	1.389	0.58	0.894	0.40	0.931	0.45	1.146	0.59
Within MZ$_m$	120	0.235		0.374		0.382		0.353		0.296	
Between DZ$_f$	193	1.473	0.54	1.095	0.47	0.819	0.26	1.139	0.48	0.834	0.35
Within DZ$_f$	194	0.442		0.391		0.482		0.398		0.406	
Between DZ$_m$	58	1.298	0.44	1.491	0.51	1.050	0.38	0.818	0.22	0.722	0.18
Within DZ$_m$	59	0.507		0.483		0.476		0.526		0.503	
Between DZ$_{fm}$	59	1.444	0.48	1.085	0.35	0.854	0.22	1.010	0.22	0.958	0.21
Within DZ$_{fm}$	59	0.507		0.528		0.417		0.647		0.632	
Between DZ$_{mf}$	65	1.201	0.44	1.120	0.27	1.274	0.37	1.173	0.19	1.129	0.39
Within DZ$_{mf}$	65	0.465		0.650		0.581		0.802		0.497	

[a] All mean squares are corrected for the effect of age, and the DZ opposite-sex pairs are corrected for the main effect of sex.

Table 14.12 Chi-square values for models fitted to five primary factors.

		Factor				
Model	df	Authori-tarianism	Religion	Socialism	Prejudice	Permis-siveness
E_W, E_B	10	55.68***	36.12***	18.78*	49.25***	56.54***
V_A, E_W	10	17.29*	20.09*	11.83	24.07**	11.33
V_A, E_W, E_B	9	9.32	14.07	9.71	23.31**	11.32

* $P<0.05$; ** $P<0.01$; *** $P<0.001$.

Table 14.13 Sources of variance for best fitting model to five primary factors (%).

		Factor			
Parameter	Authori-tarianism	Religion	Socialism	Prejudice	Permis-siveness
V_A	51	41	49	59	63
E_W	27	37	51	41	37
E_B	22	22	—	—	—
χ^2	9.32	14.07	11.83	24.07	11.33
df	9	9	10	10	10
	$0.3<P<0.5$	$0.1<P<0.25$	$0.3<P<0.5$	$P<0.01$	$0.25<P<0.5$

Table 14.14 Chi-square values for models fitted to five primary factors allowing for sex-dependent genetic and environmental effects.

		Factor				
Model[a]	df	Authori-tarianism	Religion	Socialism	Prejudice	Permis-siveness
1	7	45.51**	11.64	16.98*	16.13*	37.66**
2	7	14.88**	12.72	9.36	16.75*	8.37
3	6	7.65	1.75	8.62	9.96	8.15
4	6	8.26	1.92	8.46	8.02	7.85

[a] Parameters are as follows:
(1) E_{WM}, E_{WF}, E_{BM}, E_{BF}, V_{AM}, V_{AF}, V_{AMF};
(2) E_{WM}, E_{WF}, V_{AM}, V_{AF}, V_{AMF};
(3) E_{WF};
(4) V_A, E_{WM}, E_{WF}, E_{BM}, E_{BF}, E_{BMF}.
* $P<0.05$, ** $P<0.01$.

Table 14.15 Sources of variance for models fitted to religion and prejudice factors when genetic and environmental influences depend on sex (%).

	Factor	
Parameter	Religion	Prejudice
V_{AF}	30	28
V_{AM}	26	32
E_{WF}	36	38
E_{WM}	40	54
E_{BF}	34	34
E_{BM}	34	14NS
χ^2	1.75	8.02
df	6	6
	$0.9 < P < 0.95$	$0.1 < P < 0.25$

NS, not significant.

the fit significantly are summarized in Table 14.15. For "religion", the relative contributions of the three sources of variation do not differ greatly between the sexes, but the low correlation between unlike-sex pairs leads to a very small and non-significant estimate of the consistency of gene expression across sexes. We thus conclude that quite different factors are responsible for differences in the "religion" factor in the two sexes.

The sex differences in causation are even more marked for the "prejudice" factor. The family environment plays no detectable part in creating individual differences in males, but accounts for about a third of the variation in females. In contrast, the effects of the environment *within* families is greater for males than females. We should be cautious about reading too much into these results, since they admit a variety of inter-pretations, which could only be resolved by great ingenuity. The greater cultural resemblance of females could be a function of reduced mobility, limiting the range of unique environments to which an individual woman is exposed, or to a more lasting impact of the home environment. The pattern of results obtained for the "prejudice" factor is comparable to those reported by Martin and Eysenck (1976) in their analysis of a "libido" factor extracted from a questionnaire concerning attitudes to sex.

Our analysis suggests that the family environment is equally important for all the primary social-attitude factors. The twin data show that family environment plays a very marked part in creating individual differences in social attitudes. The results also show that the sexes differ in the causes of variation. In the case of "religion" males and females differ in the relative importance of genetic and cultural effects. In the case of the "prejudice" factor the primary difference between the sexes relates to the relative importance of the environment within and between families.

14.5 SUMMARY

The twin analysis of social-attitude dimensions at the level of primary factors and higher-order factors shows that there is a marked and consistent difference between the results for social attitudes and personality. In contrast with the results for personality, the attitude data show that dizygotic twins resemble one another far more than the additive genetic model would predict under the hypothesis of random mating. Only about 70% of the variance in conservatism scores, for example, can be explained by the parameters of the V_A, E_W model. This result is obtained in both the London population and in the large Australian twin sample, and involves effects that have high temporal stability. It is tempting to conclude that social attitudes are much more sensitive to shared environmental effects than personality, but the analysis of twin data alone cannot resolve such effects from the genetic consequences of assortative mating. In the next chapter we consider the effects of assortative mating on social attitudes and how far these might contribute to the excess similarity of DZ twins.

Chapter 15

Attitudes: Cultural Inheritance or Assortative Mating?

The main unanswered question of Chapter 14 was whether the large contribution that we had called E_B in our model for twin resemblance in attitudes was really due to the family environment or whether it was due, wholly or in part, to the genetic consequences of assortative mating. In this chapter we examine spousal resemblance in social attitudes and estimate its contribution to twin similarity.

15.1 A JOINT ANALYSIS OF SPOUSAL RESEMBLANCE FOR PERSONALITY AND ATTITUDES

In Chapter 7 we described personality data from a quota sample of spouses in the London area. The same sample yielded data on an abbreviated POI omitting the 18 items of the original inventory that were scored neither for radicalism nor tough-mindedness in the original study (Eysenck, 1954). The 42 items were scored for radicalism and tough-mindedness using the original scales. Any pair in which an individual missed any of the items was omitted from the analysis. The EPQ was scored similarly for P, E, N and L.

Table 15.1 gives the correlations between husbands and wives for the six questionnaire scores. Also tabulated are the phenotypic correlations between the measures for males and females *and* the cross-trait correlations between spouses (e.g. the correlation between extraversion in husbands and radicalism in wives). The linear regression of test scores on the ages of husbands and wives are partialled out of the correlation matrix.

There are small correlations between the scales in both males and females, including correlations of 0.207 (males) and 0.173 (females) between radicalism and tough-mindedness in this sample. The most important features of the data, however, are the very large positive correlations between spouses for radicalism ($r = 0.540$) and tough-mindedness ($r = 0.571$).

Table 15.1 Age-corrected correlation between attitude and personality scores of husbands and wives ($N = 447$).

		Wives					
		P	E	N	L	R	T
Husbands	P	0.16	−0.07	0.17	−0.01	0.03	−0.14
	E	−0.25	0.06	−0.02	0.03	−0.10	−0.10
	N	0.16	−0.07	0.13	−0.12	0.03	−0.01
	L	−0.03	0.03	−0.05	0.28	−0.03	0.01
	R	0.07	−0.04	0.04	−0.15	0.54	0.09
	T	−0.12	−0.03	−0.13	−0.02	0.17	0.57

Since the correlations are corrected for age, this observed correlation cannot be a simple function of age alone. Some of the resemblance between mates could be due to social interaction within the pair. The ideal way to detect such interactions would be a longitudinal study of spouses before and during marriage. In the absence of such data, however, we have to be content with more oblique approaches. In the British data we find correlations ranging from − 0.06 to 0.15 between absolute differences within spouse pairs and ages of husband and wife, suggesting that *older* spouse pairs are not much more alike than younger pairs. In a sample of 301 spouse pairs in Virginia for whom duration of marriage had a range of 42 years (Feingold, 1984) we found that absolute intrapair differences between spouses correlated − 0.11 and 0.08 for radicalism and tough-mindedness with duration of marriage. Such correlations are difficult to interpret because the data might be censored for less-similar pairs if they are more readily divorced. Taken at their face value, however, they do not indicate that spouses get more alike as they live together longer.

15.2 THE STRUCTURE OF SPOUSAL RESEMBLANCE IN ATTITUDES

In our discussion of assortative mating for social attitudes we have taken it as given that spouses select one another on the basis of the same combinations of characteristics that generate the factors derived from the correlations within individuals. Put another way, the pattern of correlations between mates is assumed to be like that within individuals. There is no particular reason why this should be the case. Indeed, for cognitive and socio-economic variables there is a persistent indication that mates may take a fairly "cavalier" attitude toward the descriptive refinements of sociologists and cognitive psychologists in choosing their partners (Eaves *et al.*, 1984). Thus, even though psychologists recognize specific cognitive abilities in designing tests, it is possible that mate selection and cultural transmission are based on linear combinations of these primary factors. A similar process may operate with social attitudes. Mates may select one another on the basis of higher-order combinations of primary factors, for example, so that there could be cross-correlations between primary factors of spouses that are not reflected strongly in the phenotypic correlations within individuals. Even though it may be possible to identify a large number of factors from the analysis of the phenotypic correlations between the items, it may be that the individual dimensions of attitudes are not recognized in the process of mate selection.

Table 15.2 gives the correlations between spouses for the raw responses to the 42 individual items of the British quota sample. These are all significantly greater than zero, and range from 0.13 (item 22) to 0.42 (item 21). Factor analysis of the items revealed five primary factors that could be matched to those extracted from the London twin sample: "religion", "authoritarianism", "socialism", "prejudice" and "permissiveness".

Do these factors play a part in mate selection, or are different combinations of items involved? The canonical correlations between

Table 15.2 Marital correlations of items and primary factors.

Item	Correlation	Item	Correlation
1	0.34	26	0.27
2	0.29	27	0.21
3	0.37	28	0.24
4	0.18	29	0.15
5	0.36	30	0.17
6	0.25	31	0.25
7	0.36	32	0.21
8	0.18	33	0.15
9	0.37	34	0.27
10	0.27	35	0.29
11	0.34	36	0.33
12	0.34	37	0.26
13	0.18	38	0.24
14	0.22	39	0.37
15	0.26	40	0.39
16	0.35	41	0.33
17	0.19	42	0.15
18	0.19		
19	0.31	Factor:	
20	0.33		
21	0.42	Religion	0.52
22	0.13	Authoritarianism	0.56
23	0.30	Socialism	0.54
24	0.39	Prejudice	0.35
25	0.23	Permissiveness	0.52

spouses define those linear combinations of the items that maximize the correlations between mates. If mate selection were truly based on a single dimension then we should expect only one significant canonical correlation. If the five primary factors operate independently in assortative mating then we expect at least five independent dimensions of resemblance between the attitudes of husbands and wives reflected in at least five significant canonical correlations. Table 15.3 summarizes the results of this analysis. Indeed, if we can accept the statistical tests as appropriate to these data then there are nine or ten independent dimensions on which spouses select one another with

Table 15.3 Canonical correlation analysis of husband and wife responses to social-attitude items.

Number of canonical variates	Eigenvalue	Canonical correlation	Chi-square	df	Significance
1	0.49	0.70	2971.7	1764	0.0
2	0.45	0.67	2624.3	1681	0.0
3	0.41	0.63	2316.9	1600	0.0
4	0.31	0.56	2040.8	1521	0.0
5	0.29	0.54	1848.1	1444	0.0
6	0.26	0.51	1672.9	1369	0.0
7	0.24	0.49	1516.7	1296	0.0
8	0.21	0.46	1375.6	1225	0.002
9	0.20	0.45	1251.3	1156	0.025
10	0.19	0.44	1133.3	1089	0.171

Table 15.4 Marital correlations and cross correlations of primary factors.[a]

	Husband					Wife				
Factor	RL	A	S	PJ	PR	RL	A	S	PJ	PR
RL	1.00	0.05	−0.01	0.04	−0.40	0.52	−0.03	−0.05	0.03	−0.33
A		1.00	−0.23	0.43	0.29	−0.11	0.56	−0.11	0.29	0.19
S			1.00	0.09	0.19	0.07	−0.11	0.54	0.02	0.09
PJ				1.00	0.25	−0.04	0.30	0.10	0.35	0.16
PR					1.00	0.30	0.20	0.18	0.16	0.52
RL						1.00	0.09	−0.03	0.04	−0.44
A							1.00	−0.11	0.41	0.30
S								1.00	0.13	0.25
PJ									1.00	0.20
PR										1.00

[a] Key to symbols: RL, religion; A, authoritarianism; S, socialism; PJ, prejudice; PR, permissiveness.

respect to their attitudes. It turns out that the canonical variates are difficult to interpret, so the analysis was repeated using only the factor scores on the five primary factors that showed cross-cultural stability in Feingold's study. Table 15.4 gives the correlations within and between spouses for these factors.

The pattern of spousal correlations for the five primary factors mirrors the structure of phenotypic correlations very closely. The correlations are very high indeed and comparable to those we reported for the R and T scales in the British sample. The diagonal of the matrix of correlations between mates contains the highest values. These are the correlations for the same variables rather than between different variables. The off-diagonal correlations between spouses, however, also reflect the structure inherent in the attitudes of the individuals in the sample. The correlation between "religion" and "prejudice", for example, is -0.40 for husbands and -0.44 for wives. The correlation between husband's religion and wife's prejudice is -0.33, and the reciprocal correlation is -0.30.

Therefore once again there is much to suggest that mates select one another on the basis of traits that resemble those that factor analysis derives from the item correlations in the population. There is no obvious structure to the spousal correlations beyond that found in the correlations between the primary factors. The canonical correlations derived from this matrix (Table 15.5) confirm the results of our inspection.

All five canonical correlations are highly significant, and the first three are quite large. The pattern of loadings of the factors on the canonical variates is very similar for husbands and wives (Table 15.6). The important feature of the data, however, is that mate selection or the interactions between mates occurs on a "trait-by-trait" basis. In contrast to what has been claimed for cognitive and educational variables (Eaves *et al.*, 1984), mate selection for attitudes does not occur on some one-dimensional combination of the attitudes of potential spouses. Furthermore, since the main correlations arise

Table 15.5 Canonical-correlation analysis of husband and wife social-attitude factor scores.

Number of canonical variates	Eigenvalue	Canonical correlation	χ^2	df	Significance
1	0.37	0.61	735.3	25	0.0
2	0.35	0.59	481.6	16	0.0
3	0.26	0.51	244.8	9	0.0
4	0.09	0.31	77.3	4	0.0
5	0.04	0.20	22.5	1	0.0

Table 15.6   Husband and wife factor loadings on canonical variates.

Factor	Husband					Wife				
	Canonical variate									
	1	2	3	4	5	1	2	3	4	5
Religion	−0.57	−0.05	0.38	−0.85	0.26	−0.58	−0.29	0.36	−0.83	−0.39
Authoritarianism	0.05	−0.76	0.41	−0.65	0.59	0.13	−0.67	0.35	−0.81	0.43
Socialism	0.39	0.34	0.87	−0.36	0.24	0.44	0.40	0.86	−0.24	0.13
Prejudice	0.06	−0.17	0.16	0.19	−1.00	−0.02	−0.22	0.12	0.31	−1.00
Permissiveness	0.46	−0.12	−0.26	1.00	0.32	0.39	−0.27	−0.34	1.00	0.39

between identical factors rather than across different factors, there would seem to be little justification of any principle of "complementary needs" as far as mate selection for social attitudes is concerned.

15.3 RESOLVING ASSORTATIVE MATING AND CULTURAL INHERITANCE

In Chapter 7 we developed a basic model for the joint contribution of additive genetic effects, vertical cultural inheritance and phenotypic assortative mating. A similar model can be applied to social attitudes by combining the data on twins with those on spouses to provide estimates of the genetic and cultural contribution in the presence of assortative mating. There are, however, several other possible models of mate selection and cultural inheritance. For example, mate selection may not be based directly on the phenotype we measure but on a correlated variable or on aspects of the parental phenotype (for a more complete discussion of this issue see Heath and Eaves, 1985). Cultural inheritance may not operate directly from parental attitudes but rather through a latent environmental variable. However, the different types of mate selection and cultural inheritance have fairly similar consequences for data on spouses and twins, so our analysis should yield a first approximation to their separate effects. Alternative hypotheses will be considered in Chapter 16. The model we fitted to social attitudes was formally the same as the one we used in Chapter 6, but was parameterized slightly differently. In the original model the effects of cultural inheritance were represented in the path *b* from the phenotype of parents to the environment of offspring ("P-to-E" transmission). Residual effects were thus allowed to affect E, but the total variance was defined completely by differences in G and E. In the present case (Figure 15.1) we

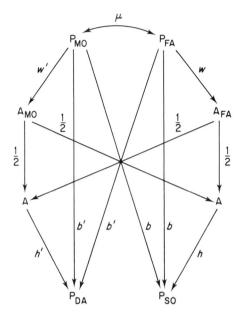

Figure 15.1   Cultural inheritance in the presence of genetic effects: "P-to-P" transmission (after Martin *et al.* 1986). Key: P, phenotypic deviation; A, additive genetic deviation; MO, mother; FA, father; DA, daughter; SO, son; see text for definition of parameters.

allowed the cultural parameter to go directly from phenotype of parent to phenotype of children ("P-to-P" transmission) and assigned all residual effects to random environmental effects E. This simple device means that variation in the phenotype can be represented as regressing directly on three main causes: the individual's genotype G; the phenotypes P of his/her parents and the uncorrelated effects of the random environment E. The regression coefficients are represented by h, b and e respectively. The influences of the genotypes and phenotypes of their parents are correlated. In a random-mating population in which there is no cultural inheritance the correlation is h. At equilibrium under phenotypic assortative mating and vertical "P-to-P" cultural inheritance from both parents the correlation becomes

$$\hat{w} = \frac{h}{1 - b(1 + \mu)},$$

where μ is the marital correlation. This function defines a constraint that is analogous to that we imposed on the genotype–environmental correlation in

Chapter 6. The rules of path analysis may be used to derive expected correlations for MZ and DZ twins and spouses in terms of the free parameters h, b and μ.

The data summary used in the analysis comprises the correlations given in Table 15.7. Twin correlations in conservatism for the Australian sample were augmented by an estimate of the spousal correlation ($r = 0.675$) published for an Australian sample ($N = 103$) by Feather (1978). Because the estimate was taken from published results, we were unable to correct for age. The same model was employed in the analysis of the radicalism and tough-mindedness scores ("R and T") from the London twin sample, augmented by the data on 562 twin pairs. The twin correlations were all corrected for the linear regression of test score on age.

The method of weighted least squares was applied to fit several forms of the basic model for biological and cultural inheritance to the three sets of correlations. The statistical method is described more fully in Chapter 7. Three simple models were fitted, all including a parameter for assortative mating. The "cultural" model included b and e but set the genetic parameter h equal to zero. The "genetic" model included h and e but set b equal to zero. The full model estimated h and b simultaneously as free parameters.

In addition, for each of the three models, we tested whether there were significant sex differences in the transmission parameters by allowing the values of h and b to depend on sex. This step was achieved by letting h and b be the genetic and cultural parameters in male twins and using separate parameters h' and b' for the expectations in female twins. In pairs of unlike-sex twins the terms in h^2 and b^2 were replaced by the products hh' and bb' respectively.

Table 15.8 gives the model-fitting results for the Australian sample. Models that omit the genetic parameters fail badly, whether the cultural effects are assumed to depend on sex or not. In contrast, a model that assumes only biological inheritance gives an excellent fit ($\chi_4^2 = 4.61$), which is scarcely improved by allowing for sex differences in the genetic component ($\chi_1^2 = 1.77$). In addition, the joint model specifying both cultural and biological inheritance gives no better fit than a model specifying biological inheritance alone ($\chi_1^2 = 0.15$), while it represents a very great improvement over the purely "cultural" model ($\chi_1^2 = 68.92$).

The estimate of the cultural parameter b in the joint model is also extremely small ($b = -0.02$) compared with the values of the other parameters, and does not differ significantly from zero. Thus the Australian data suggest very strongly that the additional resemblance of twins that we assigned to the family environment in our earlier analysis *could* be explained genetically in terms of the consequences of assortative mating. This result is counter-intuitive. We wondered if it could be explained by the large value we

Table 15.7 Distribution and correlation for attitude-factor scores in twin and spouse samples.

Sample		MZ_m	MZ_f	DZ_m	DZ_f	DZ_{mf}	Spouses
London	Number of pairs	120	325	59	194	127	562
	Radicalism Mean	0.08	2.97	−2.23	1.61	−1.32	4.05
	s.d.	9.67	9.06	10.23	9.98	10.47	7.88
	r (age corr.)	0.75	0.60	0.52	0.51	0.48	0.51
	Tough-mindedness Mean	−21.5	−28.0	−22.8	−28.7	−26.3	−17.8
	s.d.	13.3	14.0	12.8	12.9	14.5	14.1
	r (age corr.)	0.49	0.69	0.18	0.41	0.28	0.55
Australia	Number of pairs	565	1232	351	750	905	103[a]
	Radicalism Mean	45.3	49.5	45.1	49.2	46.2	—
	s.d.	13.2	12.2	13.9	12.3	12.6	—
	r (age corr.)	0.60	0.64	0.47	0.46	0.41	0.68

[a] Feather (1978).

Table 15.8   Results of fitting phenotype-to-phenotype transmission models to Australian twin and spouse correlations for conservatism scores.

Model	h	h'	b	b'	μ	df	χ^2	P
(1) b,μ			0.40		0.67	4	73.38	<0.001
(2) b,b',μ			0.38	0.41	0.67	3	68.68	<0.001
(3) h,μ	0.79				0.67	4	4.61	0.33
(4) h,h',μ	0.78	0.80			0.67	3	2.84	0.42
(5) h,b,μ	0.81		−0.02		0.67	3	4.46	0.22
(6) h,h',b,b',μ	0.75	0.83	0.02	−0.02	0.67	1	2.70	0.10

Table 15.9   Results of fitting phenotype-to-phenotype transmission models to English twin and spouse data for radicalism and tough-mindedness factor scores.

Radicalism	h	h'	b	b'	μ	df	χ^2	P
(1) b,μ			0.44		0.51	4	15.05	0.005
(2) b,b',μ			0.47	0.43	0.51	3	11.46	0.009
(3) h,μ	0.81				0.52	4	9.55	0.049
(4) h,h',μ	0.87	0.79			0.52	3	3.39	0.34
(5) h,b,μ	0.63		0.13		0.51	3	6.73	0.08
(6) h,h',b,b',μ	0.85	0.49	0.01	0.20	0.51	1	0.05	0.83

Tough-mindedness		h'	b	b'	μ	df	χ^2	P
(1) b,μ			0.41		0.55	4	47.22	<0.001
(2) b,b',μ			0.32	0.44	0.55	3	30.32	<0.001
(3) h,μ	0.79				0.54	4	16.21	0.003
(4) h,h',μ	0.65	0.83			0.54	3	2.25	0.52
(5) h,b,μ	0.96		−0.15		0.55	3	12.43	0.006
(6) h,h',b,b',μ	0.87	0.89	−0.21	−0.05	0.55	1	0.07	0.79

had employed for the marital correlation. Since the published estimate was based on only 103 pairs, there was a reasonable chance that the true value could have been as low as 0.4. We repeated the analysis using this smaller value to set an upper bound on the cultural parameter for this dataset. Using a marital correlation of 0.4 did not change the conclusions substantially and gave a value of 0.07 for the cultural-inheritance parameter.

Table 15.9 gives the analogous results for the R and T scores from the London sample. The samples are smaller than those from Australia, and a different instrument was used, but the results for radicalism compare quite well with the results obtained from the Australian sample. The model that leaves out genetic effects cannot explain the data on conservatism. The genetic model without sex differences in h almost fits ($P = 0.049$), but not as well as the model that allows for sex-differences in the genetic component. Allowing for sex differences in h improves the fit significantly ($\chi^2_1 = 6.16$). The model incorporating both biological and cultural inheritance is a marked improvement on the cultural model ($\chi^2_2 = 11.39$), but is no better than the simple genetic model when we allow for sex-differences in h ($\chi^2_2 = 3.34$). The conservatism data from the London sample still give small values for b and b' compared with h and h'. The cultural parameter in males is only 0.01, compared with a value of 0.85 for h. The value of the female cultural parameter b' is 0.20, compared with $h' = 0.49$.

The results for tough-mindedness are qualitatively similar to the results for conservatism, but the trends are more marked as judged by the values of chi-square. Models without genetic effects do not fit, neither do models that assume that genetic effects are the same in both sexes. The best model out of those we tested was the h, h', μ model ($\chi^2_3 = 2.25$). The addition of parameters for cultural inheritance gave little improvement ($\chi^2_2 = 1.18$), but allowing genetic effects to depend on sex gave a very marked improvement ($\chi^2_1 = 13.96$) over a model that assumed $h = h'$. To what extent further interpretation of the parameters is justified is hard to say. While genetic effects play a significantly greater role in the development of radicalism in males, the contribution of genetic factors to tough-mindedness appears greater in females. However, since the sex-dependent effects do not generalize to a different instrument in a different population, we should not read to much into them. Martin and Eysenck (1976) also report significant heterogeneity of genetic parameters across sexes in their study of tough-mindedness.

It is difficult to interpret the substantive significance of path coefficients because we are used to thinking in terms of contributions to *variation* that are proportional to the square of the path coefficients. Table 15.10 presents the proportions of variance in males and females attributable to the effects of genetic effects, random environmental effects (E_W), and cultural inheritance in the presence of assortative mating. Because of the correlation w between genotype and phenotype, a fraction of the variance remains unassigned

Table 15.10   Proportions of variation (%) in conservatism factors attributed to various sources.

Sample		Genetic	Environment		
			Cultural	Random	Unassigned
London	Male	72	0	38	0
	Female	24	12	43	21
Australia	Male	56	0	44	0
	Female	69	0	31	0

explicitly to either genetic or environmental effects with this model, when $b \neq 0$.

The contributions of the four sources to the total variation in the phenotype are obtained as follows: genetic, h^2; cultural, $2b^2(1 + \mu)$; genotype–environmental ("unassigned"), $2bh\hat{w}(1 + \mu)$; within-families environment $= 1 - h^2 - 2b(1 + \mu)(b + h\hat{w})$. The parameter $\hat{w}$ may be obtained as a function of the other parameters by solving for $\hat{w}$ in $\hat{w} = h + b\hat{w}(1 + \mu)$.

15.4 OTHER DATA

The model assumes that all transmission, genetic or cultural, is from parent to child. The model fits well, but it might be possible to devise an *ad hoc* model for horizontal transmission that could account for the resemblance of twins and spouses without reference to vertical inheritance. Our model makes a number of strong predictions that would not follow from models of horizontal transmission between family members: (1) the correlation in conservatism scores of separated MZ twins should be the same as that for MZ twins reared together (about 0.62); (2) the correlations between biological parents and their adult children should be high (about 0.52); (3) the correlation between siblings should be about the same as that for DZ twins (0.45); (4) the correlation between foster-parents and their adult adopted children should be close to zero. Predictions also follow for more remote relatives.

There is still a large gap in the data on social attitudes. There is only one set of published correlations for conservatism in adoptees (Eaves *et al.*, 1978), and the sample sizes are so small that the data would be consistent with almost any estimate of the cultural parameter. The adoption data do not alter our conclusions for the twins and spouses (see Eaves *et al.*, 1978), but would be incapable of doing so with the sample sizes available.

Two studies show high correlations between parents and offspring for conservatism, consistent with a model of vertical transmission. A study by Feather (1978) shows large parent–offspring correlations for the Wilson–Patterson conservatism scores, but the correlations depend on sex. This finding does not fit our model for conservatism because we find no evidence of heterogeneity in the genetic and environmental parameters over sexes in the twins. The study of Insel (1974) shows that the parent–offspring correlations are large enough to be compatible with our model and that the spousal correlations are close to ours. The exact sample sizes are not given, so we are unable to conduct a more rigorous test of the model. Insel's study also yields the correlation between sibs of unlike sex (0.42), which is close to our own for DZ twins. Correlations for like-sex siblings are not given.

15.5 SUMMARY

There are very high correlations between spouses for social attitudes. These are highest for identical traits, and comparatively small across traits, suggesting that assortative mating operates on a trait-by-trait basis. The correlations for attitudes are in stark contrast to the very small values found for psychoticism, extraversion and neuroticism.

The degree of assortative mating for attitudes is so high that its genetic consequences could account for all the additional resemblance between twins that our earlier analyses had ascribed to the "family environment". When we allow for the joint effects of genes, cultural inheritance and assortative mating in the model for family resemblance in conservatism, estimates of the cultural parameter do not differ significantly from zero. This result does not agree with our initial intuition that cultural factors derived from parents are major determinants of family resemblance in attitudes.

Our model is consistent with the broad features of published family and adoption data for conservatism, but most of the published studies are either too small or inadequately documented to allow a more rigorous test of the model. The provision of new data on adoptees and nuclear families for the same measures is a major goal for future investigation.

Chapter 16

Testing Assumptions About Mate Selection

16.1 ALTERNATIVE MODELS FOR MATE SELECTION

The model that we have used in resolving biological and cultural inheritance assumes that the selection of spouses is based on the phenotype for the principal social-attitude dimensions of radicalism and tough-mindedness. This assumption might be faulted in three main ways: first, mate selection may be based directly on the primary factors rather than higher-order components; secondly, the phenotype measured may only be an error-prone index of an underlying dimension of mate selection; and, thirdly, potential spouses may integrate other information on family background in their assessment of one another prior to mate selection.

Our model of mate selection, which may be called "primary phenotypic assortment", is often used in behavior-genetic analysis. It may be attributed to Fisher (1918) and, in its simplest form, the model assumes that mate selection is based primarily and without error on the trait being measured. However, at the very least, we know that no measurements are completely error-free. The primary-phenotypic-assortment model may allow for errors of measurement (or other random effects) that are not correlated between mates. This model implies that the observed correlation between mates is a pale reflection of a somewhat higher correlation for a latent variable, but that the only effects intervening between the latent trait and the measured phenotype are uncorrelated between spouses. Eaves (1973b) and Loehlin (1979) employ this version of the "phenotypic-assortment" model.

Rao and Morton (1978) introduce what, at first sight, might seem to be an entirely different type of model. They remark that

> . . . we suppose instead that individuals are characterized by social homogamy H, which includes status, tastes, contacts, and academic performance: assortative mating for these factors leads secondarily to the marital correlation for IQ.

It would indeed appear to be unlikely that the "social background" should have no effect on mate selection. The probability that people will marry may depend on such factors as geographical propinquity (see e.g., Vandenberg, 1972) and on common place of education (Jensen, 1978). However, the mere fact of mate selection for a correlated aspect of the measured phenotype does not necessarily imply that mate selection is based on the "social background" of the potential partners. In so far as an individual's tastes, for example, are a function of his genotype or his own experiences, mate selection is better regarded as selection for a correlated variable. Insofar as his parents' phenotypes determine where he lives and goes to school, we may legitimately speak of "social homogamy". This distinction was first made clear by Heath (1983, see also Heath and Eaves, 1985). While it may, at first sight, seem like "hair-splitting", there is a fundamental conceptual difference, which may be important biologically, between mate selection based on aspects of the individuals themselves, whether the salient variables are measured directly or not, and mate selection based on the phenotypes of relatives. We use the term "phenotypic homogamy" to refer to the former and "social homogamy" to refer to the latter. From a biological perspective, there could be aspects of mate selection where the best "information" about a potential spouse is derived from relatives rather than the individual himself. For sex-limited traits the best index of the genotype of a potential partner would either be the parent of unlike sex or the sibling of unlike sex. The long-term earning potential of a possible partner might be gauged better by looking at the social status of his/her parents. The process of "social homogamy" is especially likely to occur in cultures where parents still play an active part in the selection of partners for their offspring without regard to their individual preferences.

In practice, it is likely that both phenotypic and social homogamy operate at the same time to give an integrated process of mate selection that incorporates all the relevant information about a potential mate. The resolution of these different components of mate selection, for variables determined both by biological and cultural inheritance, requires a study design that is even more complex than that employed here. Many complex effects of mate selection and cultural inheritance may be resolved given data on identical and non-identical twins, their spouses, and the parents of both twins and spouses (Heath, 1983; Heath and Eaves, 1985).

In this chapter we first consider some alternative models for assortative mating that could explain the data on spouses without increasing the *genetic* component of family resemblance. One of these models fits the data of the previous chapter almost as well as the model of phenotypic assortment. Finally we shall examine preliminary data on twins and their spouses that may help distinguish between models of mate selection that cannot be resolved by independent samples of spouses and twins alone.

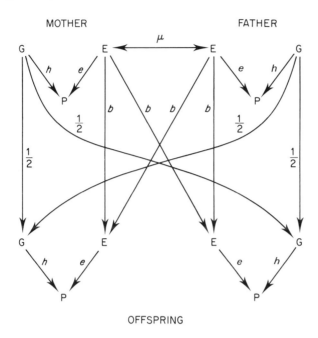

MOTHER   FATHER

OFFSPRING

Figure 16.1 "E-to-E" cultural inheritance with assortment for latent environmental variable.

16.2 ASSORTATIVE MATING FOR ENVIRONMENTAL DETERMINANTS

16.2.1 Unreliable environmental transmission

In Figure 16.1 we give a simple alternative to our previous model that might account for both assortative mating and the apparent effects of the family environment without recourse to a purely genetic explanation. The model assumes that mating is based on the latent environmental component of the phenotype only, so that the correlation we see between the phenotypes of spouses is a secondary consequence of mate selection for environmental factors. We describe the transmission of the environment from parent to child as "unreliable" because the environmental variable on which mate selection is based is not transmitted with complete reliability to the next generation. The model assumes that assortative mating involves all the environmental determinants, but that these are only partly inherited.

Under this model, and assuming equal genetic and environmental effects in males and females, the correlations in our study are expected to be

$$r_{MZ} = h^2 + 2e^2b^2(1 + \mu),$$
$$r_{DZ} = \tfrac{1}{2}h^2 + 2e^2b^2(1 + \mu),$$
$$r_{spouse} = e^2\mu.$$

The effects of this form of mate selection are different from the model assumed in the previous chapter. In particular, assortative mating for the environmental component alone does not inflate the genetic resemblance between relatives. The genetic component of the DZ correlation is still $\tfrac{1}{2}h^2$. The fact that the DZ correlation is greater than half the MZ correlation can thus still be explained in purely environmental terms. Furthermore, the correlation between the *phenotypes* of spouses is only a pale reflection of the correlation between their environments. That is, a modest phenotypic correlation between spouses may point to a substantially higher marital correlation for the latent environmental component.

This model was fitted to the conservatism data from both the London and the Australian samples by the method of weighted least squares. The results are summarized in Table 16.1. The model fits both samples, although the fit is better for the Australian sample than for the London data. In the case of the Australian data the model fits as well as that which assumes phenotypic assortative mating. As we expected, the estimated correlations between the latent environmental determinants of spouses' phenotypes are very high for both samples. In the case of the Australian sample the estimated marital correlation approaches its upper bound of unity.

This model relegates genetic effects on social attitudes to the level of "noise" that has no significance for mate selection. With this model, the genetic component of attitudes is similar to that which contributes to variation in personality. The most striking aspect of the model is the very

Table 16.1 Results of fitting model with assortative mating for environmental effects

	Sample	
Parameter	London	Australia
h	0.54	0.60
b	0.38	0.32
μ	0.72	1.00[b]
e [a]	0.84	0.80
χ_3^2	7.03	4.33
$P\%$	7	23

[a] Obtained as $\sqrt{1 - h^2}$.
[b] Fixed on upper bound.

high marital correlation that has to be presumed for the environmental determinants in order to explain the relatively high phenotypic correlation between mates. With our model, for the Australian sample at least, we have to assume that spouses are perfectly matched for all the environmental factors that influence social attitudes. Such an assumption may not be so implausible for variables such as religious affiliation.

16.2.2 Reliable environmental transmission

The previous model assumes that all the environmental factors are involved in assortative mating but that they are not transmitted with complete reliability between generations. This model predicts a high correlation between the environmental determinants of spouses. It is difficult to conceive of a variable that is so completely correlated between spouses and yet that is not transmitted easily between parents and children. Religious affiliation, for example, is very highly correlated between spouses, but is also transmitted readily to children. We thus consider an alternative model for the cultural component of conservatism that assumes, in effect, that not all the environmental effects on conservatism are transmissible, but those that are transmissible are perfectly correlated between spouses and transmitted with complete reliability to their children. This model approximates the kinds of prediction that would follow if the cultural component of conservatism were a secondary consequence of the non-genetic inheritance of a variable such as religious affiliation.

In Figure 16.2 we give such a modified model for the effects of assortative mating for the environment. We postulate a latent environmental variable R that is perfectly correlated between spouses and between parents and children. The "cultural" path coefficient b is set equal to $\frac{1}{2}$. When the marital correlation for R is unity, this value for b ensures that all members of the nuclear family are completely correlated for R. There are residual environmental effects on the measured phenotype, which are assumed to be uncorrelated between family members. The model predicts the following correlations:

$$
\begin{aligned}
r_{\text{MZ}} &= h^2 + r^2, \\
r_{\text{DZ}} &= \tfrac{1}{2}h^2 + r^2, \\
r_{\text{spouse}} &= r^2.
\end{aligned}
$$

Note that the predicted correlation for DZ twins must be at least as high as that between spouses. This constraint is clearly violated by the small sample of Australian spouses and nearly so by the London sample.

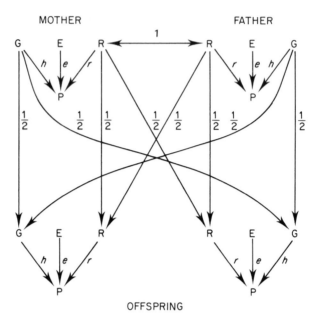

Figure 16.2   Cultural inheritance for "perfectly transmitted latent variable with complete assortment".

Fitting this model by weighted least squares yields the results in Table 16.2. The model fails in both samples, suggesting that the non-genetic inheritance of attitudes cannot be explained simply by postulating a latent variable that is perfectly correlated between the members of nuclear families.

16.3 A STUDY OF TWINS AND THEIR SPOUSES

16.3.1 The data

Feingold (1984) collected data on a US sample of MZ and DZ twins and their spouses using a slightly modified form of the 42-item Public Opinion Inventory employed in the London study. Although the sample sizes are too small to discriminate effectively between all the different components of the mating system, the study is worthy of particular attention for its unique design and for the theoretical models employed in the data analysis.

The study involved a total of 321 spouse pairs in the State of Virginia. The questionnaire necessitated only very slight modification to enable its administration in the USA. The unique feature of the data lies in the fact that

Table 16.2   Results of fitting model assuming complete family resemblance for transmissible environmental effect.

	Sample	
Parameter	London	Australia
h	0.38	0.53
r	0.70	0.57
e [(a)]	0.61	0.63
χ^2_4	0.63	0.63
$P\%$	3.6	$<10^{-3}$

[a] $e = (1 - h^2 - r^2)^{1/2}$.

one member of each spouse pair is a twin, i.e. the responses were obtained from married twins and their spouses. This design should allow more powerful resolution of alternative theories of mate selection than could be achieved from the study of nuclear families alone. Eaves (1980), Eaves and Heath (1981a,b), Heath and Eaves (1985) and Heath *et al.* (1985b) provide a more detailed treatment of the theoretical models for mate selection.

Since we have already shown that mate selection for attitudes is a multi-dimensional process, the analysis focuses on the five primary factors separately. For each group of twins in Feingold's sample (MZ male, MZ female, etc.) there are four distinct relationships: between twins, between spouses, between the spouses of twins, and between a twin and the spouse of his/her cotwin. The correlations are given for each of the five primary factors in Table 16.3.

16.3.2 A model

Given large enough samples, these correlations for the five groups of twins provide enough information to discriminate between a number of distinct models of mate selection which are described most fully by Heath (1983), Heath and Eaves (1985), Heath *et al.* (1985b) and Eaves and Heath (1981a,b). Even though samples are small, the study allows us to test some important ideas about mate selection. Figure 16.3 gives one form of the model for mixed homogamy that allows for four potential complications of the mate-selection process. Errors of measurement are represented by allowing the path r from "true" phenotype to measured phenotype to be less than unity. We assumed that the primary variable on which mate selection is based is the latent variable P_L. The correlation between mates for P_L is μ. The contribution of phenotypic assortment is represented by the path f from the

Table 16.3 Correlations of twin and spouse relationships on five primary attitude factors.

Relationship	Number of pairs	Primary factors				
		Religion	Authori-tarianism	Socialism	Prejudice	Permis-siveness
Twins:						
MZ_m	23	0.763	0.446	0.626	0.412	0.737
MZ_f	19	0.535	0.570	0.271^a	0.683	0.518
DZ_m	11	0.593	0.389^a	0.307^a	0.251^a	0.151^a
DZ_f	8	0.513^a	0.445^a	0.051^a	-0.178^a	0.083^a
DZ_{os}	26	0.488	0.559	0.281^a	0.396	0.623
Spouses:						
MZ	78	0.423	0.559	0.406	0.309	0.625
DZ	80	0.536	0.598	0.451	0.347	0.726
Twin – cotwin's spouse:						
MZ_m	42	0.520	0.540	0.411	0.239^a	0.617
MZ_f	36	0.420	0.421	0.302^a	0.039^a	-0.044^a
DZ_m	18	0.309^a	0.533	0.367^a	0.671	0.001^a
DZ_f	12	-0.299^a	0.039^a	0.006^a	0.029^a	-0.229^a
DZ_{os}	25	0.503	0.506	0.379	0.134	0.709
Female spouse:						
DZ_f	25	0.704	0.451	0.352	0.600	0.566
Male spouse:						
Spouses of twins:						
MZ_m	21	0.551	0.322^a	0.074^a	0.634	
MZ_f	18	0.057^a	0.462	0.608	0.041^a	0.149^a
DZ_m	9	0.007^a	0.060^a	0.452^a	-0.034^a	0.149^a
DZ_f	6	-0.205^a	0.395^a	0.330^a	0.008^a	-0.206^a
DZ_{os}	25	0.415	0.378^a	0.281^a	0.448	0.566

^{a} Not significant correlation.

true pheonotype for the measured trait P_T to P_L. The effects of social homogamy are represented by the regression s of P_L on the parental phenotype. The effect of "asymmetric assortative mating" (i.e. in which the "preferences" of males and females differ) is incorporated by allowing sex differences in the paths f and s. The concept of "asymmetric" assortment has been the cause of some confusion, but Eaves and Heath (1981a,b) have presented a consistent model for the process. The full model allows for the contributions of additive genetic effects h, dominance effects d and cultural inheritance c to the phenotype of interest. Expected correlations under the full model were derived by Heath (1983), but are very cumbersome and can only be solved given data on the parents and parents-in-law of the twins in addition to twins and their spouses.

Table 16.4 Expectations for correlations for twins and their spouses under simplified model (based on Heath, 1983), assuming no measurement error.

Relationship	Expected correlation[a]
MZ twins	h^2
DZ twins	$\frac{1}{2}h^2(1+x)$
Twin–spouse	$\mu\phi^2$
MZ-cotwin's spouse	$\mu\phi\{f[h^2+hs(h+v)]\}$
DZ-twin–cotwin's spouse	$\mu\phi\beta$
Spouses of MZ twins	$\mu^2\phi^2\pi$
Spouses of DZ twins	$\mu^2\phi\delta$

[a] Notation is as follows:

$\phi = f+hs(h+v)$, $\beta = \frac{1}{2}h^2f(1+x)+hs(h+v)$,
$\pi = h^2f^2+2s^2(1+w)+2sfh(h+v)$,
$\delta = \frac{1}{2}h^2f^2(1+x)+2s^2(1+w)+2sfh(h+v)$,
$\omega = \mu\phi^2$, $\nu = \mu\phi\psi$, $\chi = \mu\psi^2$,
where $\psi = hf+s(h+v)$, and $f^2+2s^2(1+w)+hfs(h+v) \leqslant 1$.

In Table 16.4 we give expected correlations under a simplified model that assumes biological inheritance of the phenotype (i.e. $c = 0$), additive gene action ($d = 0$), no sex differences in s and f, and no measurement error ($r = 1$).

The assumption that cultural inheritance plays no role is consistent with our analyses of the larger twin datasets. In practice, with these small samples, purely cultural models give an equally good fit to those data (Feingold, 1984). Whether we assume biological or cultural inheritance, however, has little impact on our main concern with these data — namely the analysis of mate selection. The assumption that mate selection is symmetric is justified by the data on four of the five variables studied. The data on "permissiveness" have an extremely anomalous large and significant sex difference in the correlation between the spouses of twins ($r = -0.525$ for the spouses of female MZ twins). This large inconsistency results in highly significant sex differences in the parameter estimates, but precludes any of our models from fitting the data. We can offer no explanation of these data consistent with the theory.

In the absence of retest data, we cannot estimate reliability r under the full form of the mixed model, so, in fitting the mixed model, we have assumed $r = 1$. However, we may still fit the "phenotypic-plus-error" model, i.e. we can obtain an estimate of r if we put $s = 0$. The fit of this model can at least be compared (though there is no valid significance test) with that of the mixed model to see whether the alternative explanation yields smaller residuals. Without data on the parents, we are unable to estimate the latent marital

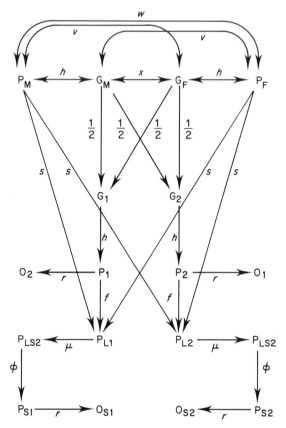

Figure 16.3 Univariate mixed homogamy without cultural inheritance — assuming no sex limitation and additive genetic effects (simplified from Heath, 1983). Key: P, "true" trait value; O, "observed" trait value; P_L, primary trait for which assortment occurs; s, effect of parental phenotype on P_L; f, effect of offspring phenotype on P_L; μ, marital correlation for P_L; w, correlation between true phenotype of spouses ($= \phi^2$); v, correlation between phenotype of one spouse and genotype of the other; x, genotypic correlation between mates.

correlation μ independently of f and s. Instead, we may only estimate $s^* = s\mu$ and $f^* = f\mu$.

16.3.3 Results of model fitting

The non-linear weighted least-squares method was used to estimate the parameters of the various models that we have outlined, and others that are

Table 16.5 Comparison of different models of assortative mating for social attitudes of twins and their spouses (after Feingold, 1984).

Factor	Model	Parameter estimates[a]						
		h	f^*	s^*	r	χ^2	df	$P\%$
Religion	Primary phenotypic	0.87	0.74	—	1^b	22.8	16	12
	Primary + error	0.99	0.74	—	0.81	17.0	15	32
	Social homogamy	0.84	—	0.58	1^b	18.6	16	29
	Mixed homogamy	0.83	0.25	0.40	1^b	17.3	15	30
Authoritarianism	Primary phenotypic	0.80	0.80	—	1^b	15.0	16	52
	Primary + error	0.89	0.91	—	0.81	9.0	15	88
	Social homogamy	0.84	—	0.58	1^b	20.9	16	18
	Mixed homogamy	0.75	0.49	0.29	1^b	9.7	15	84
Socialism	Primary phenotypic	0.76	0.71	—	1^b	15.4	16	50
	Primary + error	0.92	0.98	—	0.67	7.5	15	94
	Social homogamy	0.79	—	0.61	1^b	11.0	16	81
	Mixed homogamy	0.67	0.28	0.48	1^b	6.0	15	98
Prejudice	Primary phenotypic	0.76	0.61	—	1^b	22.9	16	12
	Primary + error	1.00	0.70	—	0.67	19.3	15	20
	Social homogamy	0.75	—	0.63	1^b	18.6	16	29
	Mixed homogamy	0.73	0.20	0.46	1^b	17.8	15	27
Permissiveness	Primary phenotypic	0.81	0.82	—	1^b	50.1	16	<0.1
	Primary + error: M	0.97	0.62	—	0.87	41.4	13	<0.1
	F	0.58	0.62	—	1.00	41.4	13	<0.1
	Social homogamy	0.85	—	0.58	1^b	67.8	16	<0.1
	Mixed homogamy	0.79	0.70	0.11	1^b	49.3	15	<0.1

[a] See text for definition of parameters.
[b] Constrained *ex hypothesi.*

not described in detail. The importance of phenotypic and social homogamy can be tested statistically by comparing the residual chi-square under the "mixed-homogamy" model with those under the corresponding reduced models (i.e. $f^* = 0$ and $s^* = 0$ respectively). A significant change in chi-square indicates that a given parameter cannot be deleted without doing violence to the data. The same procedure was used to compare the models of phenotypic assortative mating without measurement error ($r = 1$) with those in which r was allowed to take its own value.

The salient results from Feingold's analysis are summarized in Table 16.5. As we might expect from the bizarre correlations for permissiveness, none of the models can account for the pattern of spousal resemblance for that variable. For the other variables, we notice that all the models fit the data by the criterion of the goodness-of-fit test. We have shown that a sample size of 200 twin pairs and spouses is likely to be needed to distinguish models by this, less powerful, criterion (Heath and Eaves, 1985). Nevertheless, comparison of the residual chi-squares does yield some interesting (and statistically significant) contrasts. Significant improvement was achieved for the religion, authoritarianism, socialism and prejudice factors by modifications of the conventional "primary phenotypic" model of mate selection. For the first three factors, the mixed homogamy model and the addition of measurement error to the model both resulted in a significant reduction in chi-square. For the prejudice factor, the improvement from adding errors of measurement approaches significance at the 5% level. We thus conclude that either measurement error (in the broad sense of residuals uncorrelated between spouses) or social homogamy (or both) are significant components of the correlation between mates for these four attitude dimensions. If we set f^* = 0 in the mixed-homogamy model to yield the social homogamy model (see Table 16.4) then we find a significant increase in chi-square only for the authoritarianism and socialism factors. This means that there is a significant "phenotypic" component to mate selection for these variables, but for the religion and prejudice factors the social homogamy model fits as well as the mixed model.

The results, based as they are upon small samples, do not have the power to make the subtle distinctions that would lead to a more complete understanding of the mating system with respect to social attitudes. They are nevertheless significant for three reasons. First, they are important theoretically because they provide an opportunity to explore the mating system more precisely than many sociologists and geneticists have done so far. Secondly, they are important methodologically because they illustrate the basic components of a research strategy that is able to resolve effects that can only be guessed at in conventional nuclear family studies. Finally, they are important substantively because they allow us to test, and reject, for social attitudes, a model of mate selection that has dominated most research, including our own.

The analysis confirms that the process of mate selection is more complex than had been supposed previously for dimensions of similarity between spouses covering a variety of religious, moral and political values. For authoritarianism and socialism, there is fairly strong evidence that spouses select one another partly on the basis of the manifest values of the potential partner. In the case of political views this would seem to make sense; why would a conservative want to marry the radical offspring of conservative parents, or *vice versa*? In the case of authoritarianism, the statistical argument to support a model of purely phenotypic assortment seems to be strongest since the worst fit of all is given by a model that assumes that mate selection is based only on the phenotypes of the partners' parents. In no other case do we have much reason to exclude a significant component of parental influence in mate selection at this stage. For all the variables, the addition of a social-homogamy component to a "primary phenotypic" model improves the fit as much as the addition of an "error" term to the model. For the prejudice factor the addition of a social-homogamy component is marginally more successful, but the difference is not overwhelming. In the final analysis, it is tempting to believe that the phenotypes of parents contribute to the correlation between mates for religion and ethnocentrism. Unfortunately, the data do not allow us to exclude the model of "phenotypic assortment plus error", which also predicts that the phenotypic correlation between spouses is a pale reflection of a greater correlation in a latent variable. Feingold employed alpha-factor analysis to obtain estimates of the generalizability coefficients of the five factors. These coefficients were all substantially higher than the estimates of r obtained from the model-fitting analysis, suggesting that the sampling of test items alone could not account for the better fit of the "phenotypic-plus-error" model.

16.4 SUMMARY

The model of mate selection that we used in previous analyses of social attitudes assumed that spouses selected one another on the basis of the measured attitude trait. Mate selection may be more complex. Spouses may select one another for a latent variable which has only secondary effects on the measured trait. Furthermore, the latent variable may be a composite function of the spouse's own phenotypes alone ("phenotypic assortment") or reflect also the influence of their relatives' phenotypes ("social homogamy"). We show that the data gathered independently on spouses and twins are just as consistent with a mating process that matches spouses for the cultural determinants of social attitudes as they are with the model of the previous chapter in which mating is based directly on the measured phenotype for conservatism. The alternative model explains the twin and spouse data for conservatism equally well without generating a large genetic component of family resemblance. However, the data are consistent with both models for mate selection, and a study is needed that can resolve these, and several other, mechanisms of mate selection.

One design that is more informative involves the study of twins and their spouses. Preliminary data for such a study confirm that mating is not based on the measured phenotype directly, because models for assortment based on latent variables fit better. Unfortunately, sample sizes are not large enough for us to be more explicit about the relative importance of phenotypic and social homogamy.

Chapter 17

Genes, Culture and Behavior

The single most striking result to emerge from all our studies is that no one model can explain the variation and transmission of every variable we have considered. Although models involving genetic effects fit better in virtually every case than models that do not, it would be a gross oversimplification to say that the data for each variable are summarized adequately by classical heritability estimates. The usual assumptions in genetic models for behavior do not capture all the nuances of the individual differences that we have encountered.

17.1 GENETIC EFFECTS ON THE MAJOR DIMENSIONS

As the samples that we studied became larger and included more diverse kinds of relationship, we noted that the simple additive genetic model for family resemblance in personality-test scores needed to be modified in several important directions. In particular, the correlations for extraversion in large samples of twins were not consistent with the basic model, which assumed that additive genetic effects explain the similarity between twins and that random environmental effects explain all the remaining variation (see e.g. Chapter 5). We considered two possible explanations for this result, both depending on genetic effects. The first possibility is that extraversion is the result of alleles or genes displaying substantial non-additive interactions (dominance or epistasis). This result would be regarded by some population geneticists as indicating a strong relationship between the trait and fitness. A second possibility is that there are competitive social interactions between twins based on their genotypes, such that individuals who are genetically predisposed to "extraversion" create an "introverting" environment for their cotwins and *vice versa*. The precise mechanism for such interactions is still a matter for conjecture, but could involve the competition for different kinds of environmental "niche" within the family during development — "the sociable child gets the friends and leaves the books to his introverted sibling". Whatever the mechanism of such sibling interactions, they would be important theoretically for sociobiology since they indicate a kind of social interaction that is influenced by genetic effects. *If* such interactions really occur, and *if* they affect behavior that is adaptively significant, then we have a system in which the effects of "kin selection" might be expressed (see e.g. Hamilton, 1964a,b; Maynard-Smith, 1964).

It is virtually impossible to distinguish these two possibilities with the data at our disposal. The statistics on adult twins reared together could be interpreted either way. Such limited data as we have on separated twins (Chapter 6) push us slightly in favor of the postnatal competition thesis, but the evidence is flimsy. The data on juvenile extraversion (Chapter 7) seem to err on the side of the competition model. Data on "sociability" and "shyness" (Chapter 10) are also consistent with a competition/dominance model. The extraversion scales that we have employed probably give more weight to items relating to "sociability" rather than "impulsivity". There is a hint that the non-additive/competitive effects are more marked for "sociability", which might lend more support to our interpreting the low DZ correlation in social rather than purely genetic terms. Only a larger study of sociability and impulsiveness could confirm this view. In all sets of data, the amount of genetic dominance that we should have to invoke is very large indeed — much larger than has been found traditionally in experimental organisms — but Eaves (1987) has shown how duplicate gene interactions between loci can produce very low DZ correlations in randomly mating populations. The question arises: "What new data could conceivably resolve these two hypotheses?" Almost any design that we can think of is likely to require large samples. The detection of genetic non-additivity, especially the resolution of "directional" effects such as directional dominance, will require the study of inbreeding effects. Such studies are more likely to be conducted effectively in cultures where inbreeding is more common than in the West, but they are difficult to design because of the possible confounding of socio-economic, racial and cultural factors with marriage practices. The "ideal" study would control for the contaminating variables by considering close relatives discordant for whether or not they are married to a relative, and would obtain data on both offspring and their parents. The effects of sibling competition are more likely to be analyzed in an experimental study that manipulates the degree of genetic and social relatedness, the nature and availability of rewards and the opportunity for social interaction.

A second important observation is the difference in the causes of extraversion when these are compared with neuroticism. Whether we consider the individual extraversion and neuroticism items (Chapter 8) or the factor scores (Chapter 5 and 6), we find that the pattern of non-additivity/competition that we see repeatedly for extraversion is not found for neuroticism. In the case of neuroticism the bulk of the evidence from the twin studies favors an additive genetic model for family resemblance. It is commonly supposed — rightly — that factor analytic studies alone are unable to do anything but provide an arbitrary frame of reference for multiple measurements of behavior. Clearly, however, the process of refinement and experimental deduction that has been built upon the distinction between extraversion and neuroticism as indices of two distinct and independent processes underlying personality development is vindicated by the genetic analysis. The two superfactors display quite different patterns of genetic and environmental causation. We expect that rotation of the factors would obscure this fundamental difference.

The other major surprise in the genetic analysis is the strong indication of a genetic component to variation in the major dimensions of social attitudes. What kind of a model does this imply for the development of social attitudes? Certainly not a primitive model in which there are genes "coding" for particular social attitudes. The societies that our species have created, and the opportunities for expression and social interaction that they allow, are very variable and changeable. Each person is faced with a *"smorgasbord"* of attitudes and values competing for his attention, including those of his parents, his teachers, his peers and the media. Traditionally, we have viewed the development of attitudes as a matter of indoctrination, with success going to the voice that gets more of the time and offers the greatest rewards. Our analyses suggest that such a model is too simple because it ignores the unique role of the individual and his inherited preferences and sensitivities in the process of filtering, choosing and acquiring the information from the environment that helps mold his attitudes. We find that natural variation in the genotype, such as occurs within DZ twins, is correlated with the phenotypic differences in the attitudes people hold. Our model for social learning may have to accommodate the fact that each individual begins life with his/her own "agenda" expressed in differential sensitivity to particular kinds of social reward. While it would be absurd to suppose that there are genes for voting "Conservative" or "Republican", or for being "Catholic" rather than "agnostic", it is less absurd to imagine that some people are simply not "turned on" by the kinds of stimuli offered by religion or attracted by the rewards promised by a particular political party. In the context of personality, some theorists (e.g. Mischel, 1977) have suggested that the ability to "discriminate between situations" is adaptively significant. There is no reason why these individual "sensitivity parameters" should be determined socially rather than genetically. Our data suggest that, for the overall tendency to be "radical or conservative" in terms of specific issues that vary in Western cultures, is partly a function of genetic differences. The precise form and mechanism of this "function" will be a matter for the learning theorist, geneticist, physiologist and developmental psychologist to determine.

The final point to emerge from the "genetic" analysis of the principal dimensions that we have studied is that the effects of genes and environment are sometimes different between sexes. In the large samples of the Australian and Swedish studies, for example, there is compelling evidence that gene expression is dependent on sex. For example, the data on unlike-sex twin pairs suggest that there are genetic effects on neuroticism that are specific to males and females. Not all the genes affecting the personality of one sex automatically affect the other. Indeed, we find for the lie scale that the sexes not merely differ in the apparent contribution of genetic effects, but also in their sensitivity to the social environment or in the amount of variation in social pressure that is brought to bear on their behavior.

17.2 THE ENVIRONMENT

About half of the variation that we measure in personality is due to environmental effects completely specific to the individual. That is, whatever homes and teachers might do to influence behavior in a systematic way, it is clear that even twins and siblings in the same family have their own unique experiences that contribute to their personality. This is also true of the main social-attitude factors, although here the familial correlations are higher. A substantial part of this variation due to the unique environment, but by no means all, has been shown to depend on short-term environmental effects and errors of measurement. We showed that this was the case for personality in Chapters 6 and 9, and for symptoms of anxiety and depression in Chapter 11. In so far as these short-term changes are not simply errors of measurement, they may reflect the "life events" that have been studied in the context of psychiatric epidemiology but seldom, if ever, within a genetic framework.

The most striking claim to follow from our analysis of the personality variables is that there is no evidence whatever that the environment shared by family members has any impact on the development of personality differences. All four major twin studies yield estimates of the shared environmental component E_B that do not differ significantly from zero. There is nothing in the data on other relatives, adoptees or separated twins (Chapter 6) to compel us to alter this conclusion. If teachers, parents and peers influence personality development then it would seem that they do so in a manner that is highly specific to the individual, correlated with genotype and largely independent of their own personalities.

Unfortunately, we do not have any substantial body of intergenerational and adoption data for social attitudes. However, the twin data are certainly consistent with a contribution of the shared sibling environment that explains as much as 30% of the total variation in conservatism–radicalism (Chapter 14). This result is consistent over large studies in the British and Australian populations and sets social attitudes apart from personality in terms of the causes of individual differences. The main problem with this interpretation, however, when it is based purely on twins reared together, is the confounding of the environmental resemblance due to cultural inheritance with the additional genetic correlations generated by assortative mating.

17.3 ASSORTATIVE MATING

Several studies of spousal resemblance for personality show that mating is completely random for extraversion and neuroticism and only mildly assortative for psychoticism. There is a larger correlation between mates for scores on the lie scale, but the correlations for social attitudes are very high indeed (Chapters 6 and 15).

When we turn to social attitudes, the correlations between mates are often comparable to the correlations between MZ twins. Unfortunately, we still lack a large body of well-documented data on the same measures on inter-generational relationships and/or adoptees, so our interpretation of these findings has to be tentative.

At a purely empirical level we find that the primary phenotypic factors into which social attitudes are organized are also independently correlated between mates. That is, the various dimensions such as attitudes towards religion, politics and morality are kept separate in the process of mate selection or interaction between spouses. This finding for attitudes is in marked contrast to that for variables in the cognitive and educational domain, where cross-variable correlations between spouses occur, suggesting that mate selection is based on a composite phenotype rather than a collection of specific variables.

The other remarkable result is the fact that the variables that show the strongest evidence for cultural inheritance (social attitudes and the lie scale) in the twin studies are also those that show the most marked correlations between mates. This could simply be due to the fact that the apparent cultural resemblance of twins is actually no such thing but rather merely a further manifestation of the genetic consequences of assortative mating. Certainly, this hypothesis explains the data very well. A model that allows for phenotypic assortative mating, cultural inheritance and additive genetic effects yields significant estimates of the effects of genes and assortative mating, but zero estimates for the parameters of cultural inheritance. However, we considered an alternative hypothesis in which mate selection was based on the culturally transmissible determinants of social attitudes only, rather than the phenotype. Such a model is successful in ascribing all the additional resemblance of twins to non-genetic effects and leaves the genetic contribution unaffected by mate selection and resembling much more the simple "additive-genetic–random-mating" model that we found to fit many of the personality variables. This alternative model of mate selection gains some support from a small study of the spouses of twins, which suggests that models assuming phenotypic assortment alone are insufficient to predict the observed correlations for social attitudes. The model of "latent-variable assortment" predicts a very high correlation between the environments of spouses. Such extreme values are counter-intuitive, but cannot be discounted for this reason alone. Only new studies will be able to resolve these various alternatives convincingly. One thing is clear, however: researchers who hope to study social attitudes as models of cultural inheritance cannot expect to do so by concentrating their efforts only on nuclear families. We will also need data on adoptions and the relatives of twins to begin to model the process of cultural inheritance more adequately in the presence of genetic effects and assortative mating.

17.4 DEVELOPMENTAL CHANGE

We cannot begin to understand our results for adult behavior fully without consideration of the developmental processes that lead there — especially when we reflect upon the possible contribution of the family environment. Any developmental understanding of personality and attitudes has to explain why genetic differences are not "washed away" in the tide of environmental effects that accumulate during development or why the attitudes and behavior that we supposedly learn from our parents do not stay with us to create E_B in adult life.

The answers to such questions require a new kind of genotype–environmental model that reflects the way in which gene expression and social interaction change with time. Perhaps the effects of environment on personality and attitudes are simply transient. We may express shared environmental effects for as long as we live with our parents, we may display sibling interactions for a long as we live with our siblings — but when our social group changes, so do the environments that affect us. Eaves *et al.* (in preparation) show that this is apparently the case for religious affiliation. Perhaps we should not be so surprised, therefore, if there are no evident effects of the shared sibling or family environment in adults, because they are no longer engaged in the kinds of social interactions that would create them.

The data and analyses that we describe in Chapter 7 certainly indicate that developmental changes occur in whatever is measured by the EPQ in its adult and juvenile forms. The personality scores of children are generally poor predictors of the personality scores of their parents, even though for psychoticism, extraversion and neuroticism the data are fairly consistent with the existence of genetic effects in both adults and juveniles. The data for neuroticism suggest that similar genetic effects are expressed in adults and juveniles, but for extraversion and psychoticism either different genes are expressed at the two ages (or in the two generations) or there are very large non-additive components of gene action. It could be argued, for psychoticism, that the effects of the family environment are more marked in juveniles.

The juvenile data for the lie scale are especially tantalizing. In marked contrast with adults, in which there are apparently genetic effects, the twin correlations for juvenile lie scores are entirely due to the shared environment. Further examination suggests that the resemblance between juvenile twins can be explained quite well by the cultural impact of the parents' phenotypes and the reinforcing effect of the cotwin's phenotype when there is a twin. It remains to be seen whether this finding stands the test of replication, but it illustrates several important principles that have largely been ignored in genetic studies of human behavior. First, genes do not have to be expressed in childhood in order to be expressed in adult life. Genetic effects can be "switched on" when they are needed. Secondly, the effects of the shared environment created by social interaction with parents and siblings may be present in childhood, but decay once the social network changes in adulthood. Finally, the pattern of genetic and environmental effects on a trait may be substantially different at different ages, even though the behavioral content may justify giving the trait the same name throughout development.

Our analyses also extend the concept of developmental change in personality into adult life. An increase in DZ twin differences for neuroticism with age, reported in two samples, suggests that genetic effects on this trait may increase with age in adults. There are two main kinds of developmental explanation for this trend: (1) the same genetic effects are continually being reinforced by new correlated environmental effects; (2) different genes are expressed at different ages. Cross-sectional twin data cannot resolve these hypotheses. We described a developmental model, however, that allows for the decay in resemblance between relatives with increasing age that would follow if new genetic effects were continually

being expressed through adult life. There is no suggestion for either extraversion or neuroticism that such effects are important. We thus concluded
that, during adult life at least, the same genes exercise consistent and long-
term effects on personality.

17.5 MULTIPLE VARIABLES

Our analysis of the correlations between multiple variables, at the item level
and between the primary factors of personality, suggest that many of our
naive ideas about how genes and environment are expressed simply do not fit
the facts. At one stage, we had assumed that genetic effects might operate
through "generalized predisposing" variables and that the effects of the
environment would be expressed in individual idiosyncratic response
patterns because the latter may reflect the basic units of information
transmitted culturally. It turns out (see Chapters 9–11) that the genetic
contributions to specific variances in structural models for trait covariation,
to profiles of item responses within a factor, and even to individual items, are
no less specific and idiosyncratic than those of the environment. For social
attitudes, if anything, it is the cultural component that displays the most
marked generalized effects because the item loadings on the common
"conservatism" factor correlate most highly with the estimates of E_B for the
individual items.

The second main point to emerge from these studies is that we regularly
find significant differences between the covariance structure of genetic and
environmental effects. Even when we allow for trait-specific genetic and
environmental components, it is not possible, in most cases, to scale the
genetic-factor loadings as multiples of the environmental loadings. That is,
genetic and environmental effects are mediated by distinct latent variables
and cannot be viewed simply as operating through a single set of latent
"phenotypic" channels. One consequence of this result is that the factor
analysis of phenotypic correlations on random samples of unrelated
individuals will simply throw together, in a more or less arbitrary function,
processes that a more refined analysis can separate.

17.6 PERSONALITY AND PSYCHIATRIC SYMPTOMS

The analyses of Chapter 11 suggest that the same model for genetic and
environmental effects that explained items loading on the neuroticism factor
and the pattern of inheritance of the factor itself also explained the twin
resemblance for individual items related to anxiety and depression. Further-

more, the genetic correlation between the anxiety and depression items was very high indeed, and was explained almost entirely by the same genetic effects that contributed to neuroticism. There was strong evidence for a polygenic basis to liability to report symptoms characteristic of depression. Although there was some support for the large effects of a locus with a relatively rare increasing allele of large effect, the estimated contribution of such a locus to genetic variation in liability was comparatively small. Just as we found for all the personality factors, there seems to be no suggestion that variation in the family environment contributes to variation in liability to anxiety and depression. That is, family resemblance appears to be entirely genetic.

The multivariate analysis of the anxiety and depression items shows that a single common genetic factor (neuroticism) can explain much of the covariation between the items, but that there is a separate environmental factor causing correlation between the depression items. There are also separate genetic and environmental group factors and specifics. The structure of genetic and environmental covariation between the items is markedly different and points to a generalized genetic liability, with environmental effects tending to affect the development of more-specific response patterns.

This general model for the anxiety and depression symptoms is consistent with an integrated understanding of personality and specific kinds of behavioral disorder in which the personality variables in the normal population reflect the same underlying dimension of genetic liability that, in extreme cases, may give rise to psychiatric disorder.

17.7 WHERE NEXT?

17.7.1 Theory

It is clear that quantitative genetics has made significant strides over the last fifteen years in the provision of theoretical models that can capture some of the subtleties of the developmental process, cultural inheritance and mate selection that are necessary for a greater understanding of human differences. Nevertheless, there are many areas where we still do not have a clear mathematical framework to inspire data collection and analysis. This is especially true at the interface between models of social interaction and developmental change. Although we have considerable evidence that genetic and environmental parameters do not remain constant throughout life (see e.g. Chapter 7), our models for these changes are either crudely descriptive or mathematically unsophisticated. All of the conventional models for the joint effects of biological and cultural inheritance assume equilibrium conditions for the effects of assortative mating and cultural inheritance that may be sufficient empirically with the crude data at our disposal, but are difficult to justify when individual families have idiosyncratic cultural histories and patterns of social interaction. If there is genuine variability in the expression of genetic and social effects during the lifetime of the individual, for example, our ability to infer the genotype from phenotype will change during life — and therefore so will the predicted consequences of assortative mating. Similarly, the correlation between genetic and environmental effects will reflect such factors as how long children live with their parents and whether or not the cultural changes that parents engender in their offspring persist in their adult lives.

Theoretical issues of this type affect our ability to explain in a simple and elegant form the behavioral differences and changes that we see in the human species. Important though it is to understand the immediate causes and mechanisms of behavior, we should not ignore the more remote, but no less intriguing, questions that biology asks about adaptation. The danger of such sociobiological questions is that they may impart a mystical permanence to phenomena that may be revolutionized by the accidents of history. Nevertheless, we should be less than human if our curiosity did not seek for rational and simple explanations for such phenomena as the pattern of mate selection, sex differences in gene expression, the differential sensitivity of different behavioral measures to cultural inheritance and social interaction, or the changing pattern of sensitivity to the environment during development.

The problem is that our theories are constructed in ignorance of the facts. Our discussion of mate selection is only one example of this. We "know" two things about the resemblance between mates in man: (1) it is multidimensional; (2) it is positive or random but seldom negative. Whether we believe that behavioral differences are ultimately genetic or not does not detract from the simple question "Why should mate selection be like this?" For a long time it was believed that "opposites attract" on the basis of some principle of complementary needs. For individual differences in behavior, there is not the slightest evidence that this is so. We might expect that an organism that carefully optimizes all its decisions would employ a weighting function in mate selection that would tend to produce cross-variable correlations between mates and ultimately produce a common factor in the phenotypic variance. There is no hint of this for personality and attitudes. Individuals select one another on a trait-by-trait basis as if they were looking for someone just like themselves. To respond that this process merely reflects the structure of the ecological niches we find in society begs the question. We then have to ask "What explains the pattern of social organization that we see and why is it not something different?" and "Why is it that some other process of mate selection doesn't occur? Why do people select their mates from what they know? What are the risks of looking elsewhere? Why are there no *negative* correlations between mates?"

Such questions have intrinsic appeal, but we have to recognize that the number of possible explanations is large, our ability to cast them in mathematical terms is still primitive and our empirical base for deciding between the alternatives still very slender. Again, the issue of assortative mating is a case in point. We know that there are very large correlations between mates for attitudes. It is tempting to impart some sociobiological significance to this finding. However, as we examine the problem more closely, we see that there are still many basic uncertainties about the causes and consequences of mate selection. Our treatment begins by showing that two quite distinct explanations fit the data equally well. One theory is cast in purely genetic terms. The other is formulated in purely cultural terms. Some of our most recent data tend to favor some version of the second, but without more and better data on the mechanism of mate selection "all is but a woven web of guesses" (Xenophanes).

17.7.2 Data

Our book suggests that we need to enter a new generation of data collection. The days of cross-sectional studies of a large collection of variables are numbered. It seems that models have a hard time explaining all the patterns we see in terms of one or two global parameters. Genetic and environmental parameters vary significantly with sex, age, and variable. Studies will need to focus on a more selected set of problems.

(a) Longitudinal studies

Some of the most compelling questions in behavior genetics now relate to the processes of change. The ordinary longitudinal study of parents and offspring can only describe change — it cannot begin to explain it because social interactions and biological changes are confounded. Such developmental genetic studies as have been conducted have shown that the area is ripe with possibilities, but they have tended to be too broad in their scope. All the considerations of power and sample size apply as much to genetic studies of developmental change as they do to cross-sectional studies. It is unlikely that these will be met without carefully targeted studies of development in particular age ranges. The constraints of sample size may necessitate cooperation between centers and compromises about the "intensity" of measurement to which individuals can be subjected.

(b) Assortative mating and cultural inheritance

Our studies suggest that the processes of mate selection and cultural inheritance are closely intertwined and are not captured with conventional studies of nuclear families or twins. What we say about mate selection is dependent on our assumptions about cultural inheritance and *vice versa*. The data that could resolve these issues are collectable, and the domain of social attitudes comprises an eminently tractable system from the standpoint of data collection that does not have all the technical and cost ramifications of cognitive testing. In Chapter 16 we described one strategy for the resolution of different mechanisms of mate selection. The power of such a study would be enhanced enormously by the collection of data on adult adoptees.

(c) Environmental measurements

The weakest area in behavioral genetics today is the treatment of genotype–environmental interaction. It will stay this way as long as genetic studies fail to measure the environmental factors that may contribute to $G \times E$. There are problems in measuring the environment, as we have seen, because many so-called "environmental indices" behave just like any other inherited personality variable in twin studies. More careful study of the "environment" is needed to assess those variables that might be genuine antecedents of behavioral change from those that are the consequence of genetic differences. The resolution of such different types of environmental variable is not just of theoretical interest, though the "genetic environment" is theoretically important, but has implications for intervention. It is tempting to ascribe the onset of certain kinds of psychiatric disorder to adverse "life events", but such ascriptions would be far more compelling if the life events had no apparent genetic component. A recent paper (Heath *et al.*, 1986) has shown how social support, indexed by marital status, moderates the expression of genetic effects on symptoms of anxiety and depression.

If such environmental factors can be identified then it becomes possible to study their interaction with genotype. Experience has shown that the analysis of $G \times E$ is very weak when the environment cannot be measured, but simulation studies (see e.g. Eaves, 1984) suggest that, once the environment can be measured, it becomes feasible to detect the genetic control of sensitivity to particular environmental changes.

17.8 CONCLUSION

There are consistent patterns in the causes of variation in personality and attitudes. The consistencies occur in the different effects of genes and environment on different attitude and personality measures, and in the similarities that we find between studies of different populations. The causes of variation in personality and attitudes are more complex than we had thought fifteen years ago. The main reasons for our changing view have been the collection of more extensive data, better theoretical models, and more detailed analysis that, we hope, will lay a strong foundation for new investigations to resolve the uncertainties that our own work has created.

Appendices

APPENDIX A

EXAMPLE OF AN SAS PROGRAM FOR ITERATIVE WEIGHTED LEAST-SQUARES ANALYSIS OF TWIN MEAN SQUARES

```
1       PROC MATRIX PRINT;
2       *   READ OBSERVED MEAN SQUARES, D.F. AND MODEL (X, D AND A);
3       *   EXAMPLE DATA FROM NMSQT NEUROTICISM SCORES;
4       X = 41.9/14.5/34.2/21.7;
5       D = 266/267/175/176;
6       * MODEL GIVES COEFFICIENTS FOR EW AND VA;
7       A = 1 2/1 0/1 1.5/1 0.5;
8       * SET INITIAL VALUES FOR ACCURACY AND CHISQUARE;
9       CHIOLD = - 100; ACC = 100;
10      * PUT EXPECTED = OBSERVED FOR FIRST ITERATION;
11      E = X;
12      * THEN ITERATE AS LONG AS NEW CHISQUARE EXCEEDS OLD BY >0.01;
13      DO WHILE (ACC > 0.01);
14      * OBTAIN WEIGHTS;
15              T = 2#E##2;
16              W = DIAG(D#/T);
17      * GET RIGHT HAND SIDE OF NORMAL EQUATIONS;
18              R = A'*W*X;
19      * GET LEFT HAND SIDE ("INFORMATION MATRIX");
20              S = A'*W*A;
21      * INVERT INFORMATION MATRIX AND SOLVE FOR ESTIMATES;
22              C = INV(S);
23              B = C*R;
24      * OBTAIN NEW EXPECTED VALUES;
25              E = A*B;
26      * GET DEVIATIONS AND COMPUTE CHISQUARE;
27              F = X - E;
28              CHISQ = F'*W*F;
29              ACC = ABS(CHISQ-CHIOLD);
30              CHIOLD = CHISQ;
31      END;
32      STOP;
```

APPENDIX B
ADULT EYSENCK PERSONALITY QUESTIONNAIRE

Occupation ....

Age .... Sex ....

Instructions   Please answer each question by putting a circle around the "YES" or the "NO" following the question. There are no right or wrong answers, and no trick questions. Work quickly and do not think too long about the exact meaning of the questions.

PLEASE REMEMBER TO ANSWER EACH QUESTION

1	Do you have many different hobbies?	YES	NO
2	Do you stop to think things over before doing anything?	YES	NO
3	Does your mood often go up and down?	YES	NO
4	Have you ever taken the praise for something you knew someone else had really done?	YES	NO
5	Are you a talkative person?	YES	NO
6	Would being in debt worry you?	YES	NO
7	Do you ever feel "just miserable" for no reason?	YES	NO
8	Were you ever greedy by helping yourself to more than your share of anything?	YES	NO
9	Do you lock up your house carefully at night?	YES	NO
10	Are you rather lively?	YES	NO
11	Would it upset you a lot to see a child or an animal suffer?	YES	NO
12	Do you often worry about things you should not have done or said?	YES	NO
13	If you say you will do something, do you always keep your promise no matter how inconvenient it might be?	YES	NO
14	Can you usually let yourself go and enjoy yourself at a lively party?	YES	NO
15	Are you an irritable person?	YES	NO
16	Have you ever blamed someone for doing something you knew was really your fault?	YES	NO
17	Do you enjoy meeting new people?	YES	NO
18	Do you believe insurance schemes are a good idea?	YES	. NO
19	Are your feelings easily hurt?	YES	NO
20	Are *all* your habits good and desirable ones?	YES	NO
21	Do you tend to keep in the background on social occasions?	YES	NO
22	Would you take drugs which may have strange or dangerous effects?	YES	NO
23	Do you often feel "fed-up"?	YES	NO

24	Have you ever taken anything (even a pin or button) that belonged to someone else?	YES	NO
25	Do you like going out a lot?	YES	NO
26	Do you enjoy hurting people you love?	YES	NO
27	Are you often troubled about feelings of guilt?	YES	NO
28	Do you sometimes talk about things you know nothing about?	YES	NO
29	Do you prefer reading to meeting people?	YES	NO
30	Do you have enemies who want to harm you?	YES	NO
31	Would you call yourself a nervous person?	YES	NO
32	Do you have many friends?	YES	NO
33	Do you enjoy practical jokes that can sometimes really hurt people?	YES	NO
34	Are you a worrier?	YES	NO
35	As a child did you do as you were told immediately and without grumbling?	YES	NO
36	Would you call yourself happy-go-lucky?	YES	NO
37	Do good manners and cleanliness matter much to you?	YES	NO
38	Do you worry about awful things that might happen?	YES	NO
39	Have you ever broken or lost something belonging to someone else?	Yes	NO
40	Do you usually take the initiative in making new friends?	YES	NO
41	Would you call yourself tense or "highly strung"?	YES	NO
42	Are you mostly quiet when you are with other people?	YES	NO
43	Do you think marriage is old-fashioned and should be done away with?	YES	NO
44	Do you sometimes boast a little?	YES	NO
45	Can you easily get some life into a rather dull party?	YES	NO
46	Do people who drive carefully annoy you?	YES	NO
47	Do you worry about your health?	YES	NO
48	Have you ever said anything bad or nasty about anyone?	YES	NO
49	Do you like telling jokes and funny stories to your friends?	YES	NO
50	Do most things taste the same to you?	YES	NO
51	As a child were you ever cheeky to your parents?	YES	NO
52	Do you like mixing with people?	YES	NO
53	Does it worry you if you know there are mistakes in your work?	YES	NO
54	Do you suffer from sleeplessness?	YES	NO
55	Do you always wash before a meal?	YES	NO
56	Do you nearly always have a "ready answer" when people talk to you?	YES	NO
57	Do you like to arrive at appointments in plenty of time?	YES	NO
58	Have you often felt listless and tired for no reason?	YES	NO
59	Have you ever cheated at a game?	YES	NO
60	Do you like doing things in which you have to act quickly?	YES	NO
61	Is (or was) your mother a good woman?	YES	NO
62	Do you often feel life is very dull?	YES	NO

63   Have you ever taken advantage of someone? ....YES   NO

64   Do you often take on more activities than you have time for? ....YES   NO

65   Are there several people who keep trying to avoid you? ....YES   NO

66   Do you worry a lot about your looks? ....YES   NO

67   Do you think people spend too much time safeguarding their future
     with savings and insurances? ....YES   NO

68   Have you ever wished that you were dead? ....YES   NO

69   Would you dodge paying taxes if you were sure you could never
     be found out? ....YES   NO

70   Can you get a party going? ....YES   NO

71   Do you try not to be rude to people? ....YES   NO

72   Do you worry too long after an embarrassing experience? ....YES   NO

73   Have you ever insisted on having your own way? ....YES   NO

74   When you catch a train do you often arrive at the last minute? ....YES   NO

75   Do you suffer from "nerves"? ....YES   NO

76   Do your friendships break up easily without it being your fault? ....YES   NO

77   Do you often feel lonely? ....YES   NO

78   Do you always practice what you preach? ....YES   NO

79   Do you sometimes like teasing animals? ....YES   NO

80   Are you easily hurt when people find fault with you or the work
     you do? ....YES   NO

81   Have you ever been late for an appointment or work? ....YES   NO

82   Do you like plenty of bustle and excitement around you? ....YES   NO

83   Would you like other people to be afraid of you? ....YES   NO

84   Are you sometimes bubbling over with energy and sometimes very
     sluggish? ....YES   NO

85   Do you sometimes put off until tomorrow what you ought to do
     today? ....YES   NO

86   Do other people think of you as being very lively? ....YES   NO

87   Do people tell you a lot of lies? ....YES   NO

88   Are you touchy about some things? ....YES   NO

89   Are you always willing to admit it when you have made a mistake? ...YES   NO

90   Would you feel very sorry for an animal caught in a trap? ....YES   NO

PLEASE CHECK TO SEE THAT YOU HAVE ANSWERED ALL THE QUESTIONS

Weight matrices for *published* EPQ (90 items)

Index Number	Factor	EPQ Item Number							
		P		E		N		L	
		Yes	No	Yes	No	Yes	No	Yes	No
1			2	1		3			4
2			6	5		7			8
3			9	10		12		13	
4			11	14		15			16
5			18	17		19		20	
6		22			21	23			24
7		26		25		27			28
8		30			29	31		35	
9		33		32		34			39
10			37	36		38			44
11		43		40		41			48
12		46			42	47			51
13		50		45		54		55	
14			53	49		58			59
15			57	52		62			63
16			61	56		66			69
17		65		60		68			73
18		67		64		72		78	
19			71	70		75			81
20		74		82			77		85
21		76		86			80	89	
22		79					84		
23		83					88		
24		87							
25			90						

Note: The numbers in the body of the table refer to the item numbers on the EPQ. The "index numbers" refer to the order with which each item is identified within a scale. For example, item L21 is item 89 on the published EPQ and a "Yes" answer adds 1 to the raw lie score.

JUNIOR EYSENCK PERSONALITY QUESTIONNAIRE

Age .... Sex ....

Instructions Please answer each question by putting a circle around the "YES" or the "NO" following the question. There are no right or wrong answers, and no trick questions. Work quickly and do not think too long about the exact meaning of the questions.

REMEMBER TO ANSWER EACH QUESTION

---

1 Do you like plenty of excitement going on around you? ....YES NO

2 Are you moody? ....YES NO

3 Do you enjoy hurting people you like? ....YES NO

4 Were you ever greedy by helping yourself to more than your share of anything? ....YES NO

5 Do you nearly always have a quick answer when people talk to you? ....YES NO

6 Do you very easily feel bored? ....YES NO

7 Would you enjoy practical jokes that could sometimes really hurt people? ....YES NO

8 Do you always do as you are told at once? ....YES NO

9 Would you rather be alone instead of meeting other children? ....YES NO

10 Do ideas run through your head so that you cannot sleep? ....YES NO

11 Have you ever broken any rules at school? ....YES NO

12 Would you like other children to be afraid of you? ....YES NO

13 Are you rather lively? ....YES NO

14 Do lots of things annoy you? ....YES NO

15 Would you enjoy cutting up animals in Science class? ....YES NO

16 Did you ever take anything (even a pin or button) that belonged to someone else? ....YES NO

17 Have you got lots of friends? ....YES NO

18 Do you ever feel "just miserable" for no good reason? ....YES NO

19 Do you sometimes like teasing animals? ....YES NO

20 Did you ever pretend you did not hear when someone was calling you? ....YES NO

21 Would you like to explore an old haunted castle? ....YES NO

22 Do you often feel life is very dull? ....YES NO

23 Do you seem to get into more quarrels and scraps than most children? YES NO

24 Do you always finish your homework before you play? ....YES NO

25 Do you like doing things where you have to act quickly? ....YES NO

26 Do you worry about awful things that might happen? ....YES NO

27 When you hear children using bad language do you try to stop them? .YES NO

28 Can you get a party going? ....YES NO

29 Are you easily hurt when people find things wrong with you or the
work you do? ....YES NO

30 Would it upset you a lot to see a dog that has just been run over? ....YES NO

31 Do you always say you are sorry when you have been rude? ....YES NO

32 Is there someone who is trying to get their own back for what they
think you did to them? ....YES NO

33 Do you think water ski-ing would be fun? ....YES NO

34 Do you often feel tired for no reason? ....YES NO

35 Do you rather enjoy teasing other children? ....YES NO

36 Are you always quiet when older people are talking? ....YES NO

37 When you make new friends do you usually make the first move? .....YES NO

38 Are you touchy about some things? ....YES NO

39 Do you seem to get into a lot of fights? ....YES NO

40 Have you ever said anything bad or nasty about anyone? ....YES NO

41 Do you like telling jokes or funny stories to your friends? ....YES NO

42 Are you in more trouble at school than most children? ....YES NO

43 Do you generally pick up papers and rubbish others throw on the
classroom floor? ....YES NO

44 Have you many different hobbies and interests? ....YES NO

45 Are your feelings rather easily hurt? ....YES NO

46 Do you like playing pranks on others? ....YES NO

47 Do you always wash before a meal? ....YES NO

48 Would you rather sit and watch than play at parties? ....YES NO

49 Do you often feel "fed-up"? ....YES NO

50 Is it sometimes rather fun to watch a gang tease or bully a small child? YES NO

51 Are you always quiet in class, even when the teacher is out of the
room? ....YES NO

52 Do you like doing things that are a bit frightening? ....YES NO

53 Do you sometimes get so restless that you cannot sit still in a chair
for long? ....YES NO

54 Would you like to go to the Moon on your own? ....YES NO

55 At prayers or assembly, do you always sing when the others are
singing? ....YES NO

56 Do you like mixing with other children? ....YES NO

57 Are your parents far too strict with you? ....YES NO

58 Would you like parachute jumping? ....YES NO

59 Do you worry for a long while if you feel you have made a fool of
yourself? ....YES NO

60 Do you always eat everything you are given at meals? ....YES NO

61 Can you let yourself go and enjoy yourself a lot at a lively party? ......YES NO

62 Do you sometimes feel life is just not worth living? ....YES NO

63 Would you feel very sorry for an animal caught in a trap? ....YES NO

64 Have you ever been cheeky to your parents? ....YES NO

65 Do you often make up your mind to do things suddenly? ....YES NO

66 Does your mind often wander off when you are doing some work? ....YES NO

67 Do you enjoy diving or jumping into the sea or a pool? ....YES NO

68 Do you find it hard to get to sleep at night because you are worrying
about things? ....YES NO

69 Did you ever write or scribble in a school or library book? ....YES NO

70 Do other people think of you as being very lively? ....YES NO

71 Do you often feel lonely? ....YES NO

72 Are you always specially careful with other people's things? ....YES NO

73 Do you always share all the sweets you have? ....YES NO

74 Do you like going out a lot? ....YES NO

75 Have you ever cheated at a game? ....YES NO

76 Do you find it hard to really enjoy yourself at a lively party? ....YES NO

77 Do you sometimes feel specially cheerful and at other times sad
without any good reason? ....YES NO

78 Do you throw waste paper on the floor when there is no waste
paper basket handy? ....YES NO

79 Would you call yourself happy-go-lucky? ....YES NO

80 Do you often need kind friends to cheer you up? ....YES NO

81 Would you like to drive or ride on a fast motor bike? ....YES NO

PLEASE MAKE SURE YOU HAVE ANSWERED ALL THE QUESTIONS

Weight matrices for *published* JEPQ (81 items)

P	YES:	3, 7, 12, 15, 19, 23, 32, 35, 39, 42, 46, 50, 54, 57
	NO:	30, 63, 72
E	YES:	1, 5, 13, 17, 21, 25, 28, 33, 37, 41, 44, 52, 56, 58, 61, 65, 67, 70, 74, 79, 81
	NO:	9, 48, 76
N	YES:	2, 6, 10, 14, 18, 22, 26, 29, 34, 38, 45, 49, 53, 59, 62, 66, 68, 71, 77, 80
L	YES:	8, 24, 27, 31, 36, 43, 47, 51, 55, 60, 73
	NO:	4, 11, 16, 20, 40, 64, 69, 75, 78

APPENDIX D

PUBLIC OPINION INVENTORY

It is hoped you will be interested in this survey of public opinion. Below are given 60 statements which represent widely held opinions on various social questions, selected from speeches, books, newspapers and other sources. They were chosen in such a way that most people are likely to agree with some, and to disagree with others.

After each statement, you are requested to record your personal opinion regarding it. You should use the following system of marking:

+ +   if you strongly agree with the statement
+   if you agree on the whole
O   if you can't decide for or against, or if you think the question is worded in such a way that you can't give an answer
—   if you disagree on the whole
— — if you strongly disagree

Please answer frankly. Remember this is not a test; there are no "right" or "wrong" answers. The answer required is your own personal opinion. Be sure not to omit any questions. The questionnaire is anonymous, *so please do not sign your name.*

Do not consult any other person while you are giving your answers.

	Opinion statements	Your opinion	
1	The nation exists for the benefit of the individuals composing it, not the individuals for the benefit of the nation.		
2	Coloured people are innately inferior to white people.		
3	War is inherent in human nature.		
4	Ultimately, private property should be abolished and complete socialism introduced.	R +	
5	Persons with serious hereditary defects and diseases should be compulsorily sterilized.		
6	In the interests of peace, we must give up part of our national sovereignty.		
7	Production and trade should be free from government interference.	R −	
8	Divorce laws should be altered to make divorce easier.		T −
9	The so-called underdog deserves little sympathy or help from successful people.		T −
10	Crimes of violence should be punished by flogging.	R −	T −
11	The nationalization of the great industries is likely to lead to inefficiency, bureaucracy and stagnation.	R −	
12	Men and women have the right to find out whether they are sexually suited before marriage (e.g. by trial marriage.)	R +	T −

13 "My country right or wrong" is a saying which expresses a
     fundamentally desirable attitude.   R −

14 The average man can live a good enough life without
     religion.   T −

15 It would be a mistake to have coloured people as foremen
     over whites.

16 People should realize that their greatest obligation is to
     their family.

17 There is no survival of any kind after death.   T −

18 The death penalty is barbaric, and should be abolished.   R +   T +

19 There may be a few exceptions, but in general, Jews are
     pretty much alike.   T −

20 The dropping of the first atom bomb on a Japanese city,
     killing thousands of innocent women and children, was
     morally wrong and incompatible with our kind of
     civilization.   T +

21 Birth control, except when recommended by a doctor,
     should be made illegal.   T +

22 People suffering from incurable diseases should have the
     choice of being put painlessly to death.   T −

23 Sunday observance is old-fashioned, and should cease to
     govern our behaviour.

24 Capitalism is immoral because it exploits the worker by
     failing to give him full value for his productive labour.   R +

25 We should believe without question all that we are taught
     by the Church.   R −

26 A person should be free to take his own life, if he wishes to
     do so, without any interference from society.   T −

27 Free love between men and women should be encouraged
     as a means towards mental and physical health.   R +   T −

28 Compulsory military training in peacetime is essential for
     the survival of this country.   T −

29 Sex crimes such as rape and attacks on children, deserve
     more than mere imprisonment; such criminals ought to be
     flogged or worse.   R −

30 A white lie is often a good thing.   T −

31 The idea of God is an invention of the human mind.   T −

32 It is wrong that men should be permitted greater sexual
     freedom than women by society.

33 The Church should attempt to increase its influence on the
     life of the nation.   T +

34 Conscientious objectors are traitors to their country, and
     should be treated accordingly.

35 The laws against abortion should be abolished.

36 Most religious people are hypocrites.   T −

37 Sex relations except in marriage are always wrong.   R −   T +

38	European refugees should be left to fend for themselves.		T −
39	Only by going back to religion can civilization hope to survive.		
40	It is wrong to punish a man if he helps another country because he prefers it to his own.	R +	
41	It is just as well that the struggle of life tends to weed out those who cannot stand the pace.		T −
42	In taking part in any form of world organization, this country should make certain that none of its independence and power is lost.	R −	
43	Nowadays, more and more people are prying into matters which do not concern them.		T −
44	All forms of discrimination against the coloured races, the Jews, etc., should be made illegal, and subject to heavy penalties.		
45	It is right and proper that religious education in schools should be compulsory.		
46	Jews are as valuable citizens as any other group.		T +
47	Our treatment of criminals is too harsh; we should try to cure them, not punish them.	R +	T +
48	The Church is the main bulwark opposing the evil trends in modern society.		T +
49	There is no harm in travelling occasionally without a ticket, if you can get away with it.		T −
50	The Japanese are by nature a cruel people.		
51	Life is so short that a man is justified in enjoying himself as much as he can.		T −
52.	An occupation by a foreign power is better than war.	R +	T +
53.	Christ was divine, wholly or partly in a sense different from other men.		T +
54	It would be best to keep coloured people in their own districts and schools, in order to prevent too much contact with whites.		
55	Homosexuals are hardly better than criminals, and ought to be severely punished.		
56	The universe was created by God.		T +
57	Blood sports, like fox hunting for instance, are vicious and cruel, and should be forbidden.		T +
58	The maintenance of internal order within the nation is more important than ensuring that there is complete freedom for all.		T −
59	Every person should have complete faith in some supernatural power whose decisions he obeys without question.		
60	The practical man is of more use to society than the thinker.		

APPENDIX E

ITEM MEANS BY TWIN GROUP FOR THE POI ITEMS

Item	MZ_f $n = 650$	MZ_m $n = 240$	DZ_f $n = 388$	DZ_m $n = 118$	DZ_{fm} $n = 122$	DZ_{mf} $n = 132$
1	2.45	2.30	2.49	2.24	2.24	2.22
2	4.14	4.29	4.26	4.15	4.37	4.10
3	2.61	2.47	2.66	2.29	2.73	2.68
4	4.32	4.16	4.23	4.10	4.09	3.94
5	2.92	3.16	2.91	3.12	3.15	3.34
6	3.38	3.18	3.25	2.95	3.24	3.22
7	2.85	2.97	2.93	3.40	2.96	3.33
8	3.17	2.80	3.20	3.00	3.18	2.94
9	3.92	4.13	3.94	4.01	4.05	3.98
10	2.96	3.07	3.30	3.48	3.39	3.44
11	2.74	2.70	2.74	2.83	2.79	2.87
12	2.96	2.31	2.89	2.27	2.60	2.36
13	3.14	3.35	3.27	3.41	3.46	3.50
14	2.67	2.32	2.63	2.26	2.47	2.18
15	3.54	3.66	3.61	3.88	3.60	3.38
16	2.35	2.23	2.37	2.39	2.59	2.43
17	3.47	2.93	3.56	3.02	3.45	3.09
18	3.16	3.12	3.02	2.91	2.67	2.98
19	3.15	3.25	3.32	3.34	3.47	3.40
20	2.04	2.72	2.04	2.53	1.94	2.57
21	4.36	4.50	4.55	4.54	4.44	4.46
22	2.38	2.21	2.34	2.14	2.38	2.31
23	2.95	2.39	2.94	2.49	2.91	2.34
24	3.17	3.26	3.26	3.33	3.12	3.19
25	4.12	4.39	4.20	4.45	4.43	4.47
26	3.19	3.48	3.20	2.81	3.17	3.21
27	3.94	3.41	3.99	3.35	3.89	3.43
28	3.03	3.50	3.18	3.86	3.43	3.70
29	2.52	2.81	2.79	3.30	2.85	3.03
30	2.09	2.19	2.06	2.25	2.07	1.99
31	3.61	3.04	3.65	3.00	3.39	2.95
32	2.25	2.56	2.32	2.34	2.17	2.54
33	3.04	3.39	2.88	3.33	3.16	3.29
34	3.86	4.18	4.01	4.22	4.17	4.12
35	3.30	3.11	3.34	2.86	3.39	3.07
36	3.61	3.25	3.64	3.32	3.73	3.52
37	3.52	4.20	3.68	4.13	3.79	4.09
38	4.03	3.97	4.13	4.02	4.21	3.91
39	3.31	3.62	3.30	3.81	3.40	3.61

Item	MZ$_f$ $n = 650$	MZ$_m$ $n = 240$	DZ$_f$ $n = 388$	DZ$_m$ $n = 118$	DZ$_{fm}$ $n = 122$	DZ$_{mf}$ $n = 132$
40	2.91	2.85	2.98	3.05	2.98	2.88
41	3.29	3.38	3.24	3.25	3.50	3.32
42	2.35	2.64	2.49	2.88	2.57	2.65
43	2.49	2.36	2.46	2.57	2.64	2.39
44	2.45	2.32	2.37	2.53	2.28	2.59
45	2.89	3.31	2.88	3.34	3.17	3.37
46	1.77	1.65	1.68	1.70	1.57	1.68
47	3.18	3.13	3.17	2.88	2.94	2.91
48	3.07	3.37	3.11	3.42	3.19	3.45
49	3.57	3.25	3.52	3.39	3.35	3.45
50	3.37	3.46	3.35	3.73	3.53	3.57
51	2.29	2.04	2.31	2.31	2.39	2.28
52	3.53	3.98	3.51	3.65	3.43	3.56
53	2.33	2.75	2.31	2.77	2.47	2.57
54	3.95	4.20	4.19	4.32	4.21	4.18
55	4.17	4.19	4.39	4.14	4.48	4.13
56	2.45	2.98	2.46	3.19	2.73	3.04
57	2.18	2.44	2.34	2.50	2.16	2.47
58	2.73	2.77	2.84	2.77	2.80	2.70
59	3.79	4.06	3.79	4.18	3.91	4.06
60	3.35	3.48	3.54	3.57	3.50	3.49

APPENDIX F

ITEM VARIANCES BY TWIN GROUP FOR THE POI ITEMS

Item	MZ_f $n = 650$	MZ_m $n = 240$	DZ_f $n = 388$	DZ_m $n = 118$	DZ_{fm} $n = 122$	DZ_{mf} $n = 132$
1	1.14	1.18	1.14	1.26	0.87	1.46
2	1.16	0.93	1.05	1.34	1.03	1.39
3	1.25	1.39	1.38	1.27	1.40	1.59
4	0.99	1.44	1.06	1.25	1.41	1.72
5	1.87	1.77	1.85	1.93	1.90	1.69
6	1.48	1.48	1.45	1.67	1.53	1.60
7	1.43	1.59	1.51	1.46	1.42	1.75
8	1.81	1.70	1.82	1.63	2.02	1.71
9	1.09	0.89	1.16	1.00	1.24	1.40
10	2.23	2.43	2.22	2.25	2.02	2.14
11	1.42	1.68	1.51	1.54	1.36	1.89
12	1.97	1.44	1.89	1.48	1.75	1.57
13	1.14	1.21	1.30	1.49	1.15	1.38
14	1.65	1.39	1.60	1.55	1.46	1.17
15	1.53	1.35	1.53	1.17	1.51	1.78
16	1.32	1.12	1.33	1.42	1.20	1.29
17	1.73	1.65	1.69	1.95	1.80	1.83
18	2.13	2.39	2.27	2.75	2.32	2.52
19	1.38	1.51	1.45	1.47	1.55	1.78
20	1.22	1.79	1.32	1.48	1.17	1.85
21	1.17	0.94	0.76	0.80	1.23	1.03
22	1.64	1.38	1.59	1.49	1.57	1.52
23	1.64	1.73	1.73	1.77	1.78	1.54
24	1.10	1.67	1.22	1.78	1.36	1.88
25	1.04	0.77	0.89	0.76	0.60	0.78
26	1.75	1.52	1.89	1.71	1.69	1.68
27	1.24	1.63	1.11	1.51	1.11	1.53
28	1.57	1.75	1.75	1.49	1.63	1.48
29	2.17	2.34	2.53	2.23	2.42	2.35
30	0.72	0.72	0.94	0.74	1.08	0.55
31	1.59	1.71	1.47	1.95	1.51	1.93
32	1.43	1.39	1.56	1.03	1.20	1.30
33	1.59	1.64	1.56	1.85	1.58	1.65
34	1.09	0.81	1.09	1.22	0.95	1.12
35	1.73	1.67	1.74	1.81	1.72	1.72
36	1.35	1.45	1.29	1.40	1.23	1.47
37	1.66	1.03	1.56	1.05	1.66	1.32
38	0.83	0.89	0.81	1.08	0.80	1.01
39	1.53	1.46	1.49	1.47	1.48	1.53

Item	MZ_f $n = 650$	MZ_m $n = 240$	DZ_f $n = 388$	DZ_m $n = 118$	DZ_{fm} $n = 122$	DZ_{mf} $n = 132$
40	1.17	1.43	1.23	1.36	1.31	1.52
41	1.15	1.25	1.09	1.34	1.17	1.32
42	1.27	1.66	1.33	1.61	1.31	1.55
43	1.26	1.23	1.27	1.51	1.36	1.21
44	1.38	1.44	1.31	1.69	1.46	1.76
45	1.57	1.67	1.57	1.87	1.72	1.57
46	0.59	0.43	0.45	0.38	0.36	0.49
47	1.74	1.99	1.80	1.74	1.81	2.02
48	1.27	1.41	1.27	1.54	1.13	1.47
49	1.61	1.55	1.57	1.66	1.48	1.82
50	1.20	1.00	1.22	1.22	1.20	1.36
51	1.26	1.13	1.24	1.36	1.72	1.43
52	1.20	1.00	1.22	1.22	1.20	1.36
53	1.26	1.32	1.24	1.32	1.24	1.67
54	1.31	1.10	1.04	0.82	0.95	1.05
55	0.96	0.86	0.63	1.44	0.47	1.18
56	1.55	1.58	1.61	1.69	1.64	2.08
57	1.55	1.80	1.81	2.06	1.77	2.11
58	1.18	1.40	1.22	1.27	1.18	1.39
59	1.15	1.04	0.123	0.93	0.94	0.87
60	1.37	1.33	1.26	1.26	1.30	1.24

References

Adorno, T.W., Frenkel-Brunswick, E., Levinson, P.J. and Sanford, R.N. (1950). *The Authoritarian Personality*. New York: Harper & Row.

Allport, G.W. (1937). *Personality: A Psychological Interpretation*. New York: Holt.

Altemeyer, B. (1981). *Right-wing Authoritarianism*. Manitoba: University of Manitoba Press.

Argyle, M. and Cook, M. (1976). *Gaze and Mutual Gaze*. Cambridge University Press.

Axelrod, R. and Hamilton, W.D. (1981). The evolution of cooperation. *Science* 211, 1390–1396.

Ayala, F. (1982). *Evolutionary and Population Genetics: A Primer*. Reading, Mass: Addison-Wesley.

Baker, R.J. and Nelder, J.A. (1978). *The GLIM System, Release 3: Generalised Linear Interactive Modelling*. Oxford: Numerical Algorithms Group.

Baltes, P.B. and Nesselroade, J.R. (1973). The developmental analysis of individual differences on multiple measures. *Life-Span Developmental Psychology: Methodological Issues* (ed. J.R. Nesselroade and H.W. Reese). New York: Academic Press.

Barker, E.N. (1958) Authoritarianism of the political right, centre and left. Doctoral dissertation. Columbia University, New York.

Barrett, P. and Eysenck, S.B.G. (1984). The assessment of personality factors across 25 countries. Personality and individual differences. To be published.

Bartlett, M.S. (1947). The use of transformations. *Biometrics* 3, 39–52.

Bateson, W. and Punnett, R.C. (1908). *Experimental Studies in the Physiology of Heredity. Reports to the Evolution Committee of the Royal Society*, II, III and IV. London: Harrison & Sons.

Bedford, A., Foulds, G.A. and Sheffield, B.F. (1976). A new personal disturbance scale (DSSI/sAD). *British Journal of Social and Clinical Psychology* 15, 387–394.

Behrman, J.R., Hrubec, A., Taubman, P. and Wales, T.J. (1980). *Socioeconomic Success: A Study of the Effects of Genetic Endowments, Family Environment, and Schooling*. Amsterdam: Elsevier.

Berki, R.N. (1975). *Socialism*. London: Dent.

Bieber, F.R., Nance, W.E., Morton, C.C., Brown, J.A., Redwine, F.O., Jordan, R.L. and Mohanakumar, T. (1981). Genetic studies of an acardiac monster: Evidence of polar body twinning in man. *Science* 213, 775–777.

Birnbaum, A. (1968). Some latent trait models and their use in inferring an examinee's ability. *Statistical Theories of Mental Test Scores* (ed. F.M. Lord and M.R. Novick), pp. 397–479. Reading, Mass.: Addison-Wesley.

Bock, R. D. and Aitkin, M. (1981). Marginal maximum-likelihood estimation of item parameters: Application of an EM algorithm. *Psychometrika* 46, 443–459.

Bock, R. D. and Lieberman, M. (1970). Fitting a response model for *n* dichotomously scored items. *Psychometrika* 35, 179–197.

Boller, P. F. (1981) *American Thought in Transition. The Impact of Evolutionary Naturalism, 1865–1900.* Lanham, Maryland: University Press of America.

Boltomore, T. (1984). *Sociology and Socialism.* Brighton: Wheatsheaf.

Boorman, S. A. and Levitt, P. R. (1980). *The Genetics of Altruism.* New York: Academic Press.

Borecki, I. B. and Ashton, G. C. (1984). Evidence for a major gene influencing performance on a vocabulary test. *Behavior Genetics* 14, 63–79.

Bouchard, T., Heston, L. Eckert, E., Keger, M. and Resnick, S. (1981). The Minnesota study of twins reared apart: Project description and sample results in the developmental domain. *Twin Research 3: Part B: Intelligence, Personality and Development* (ed. L. Gedda, P. Parisi and W. E. Nance), pp. 227–284. New York: Alan Liss.

Braithwaite, V. (1982). The structure of social values: Validation of Rokeach's two-value model. *British Journal of Social Psychology* 21, 203–211.

Breese, E. L. and Mather, K. (1957). The organisation of polygenic activity with a chromosone in Drosophila, I. Hair characteristics. *Heredity* 11, 373–395.

Breese, E. L. and Mather, K. (1960). The organisation of polygenic activity within a chromosome in Drosophila. II. Viability. *Heredity* 14, 375–399.

Brengelmann, J. C. and Brengelmann, L. (1960a). Deutsche validierung von frage-bogen der extraversion, neurotischen, tendenz und rigiditat. *Zeitschrift für experimentelle und angewandte Psychologie* 7, 291–331.

Brengelmann, J. C. and Brengelmann, L. (1960b) Deutsche validierung von frage-bogen dogmatischer und intoloranter haltungen. *Zeitschrift für experimentelle und angewandte Psychologie* 7, 451–471.

Brittan, S. (1968). *Left or Right: The Bogus Dilemma.* London: Secker & Warburg.

Broadhurst, P. L. (1975). The Maudsley reactive and non-reactive strains of rats: A survey. *Behavior Genetics* 5, 299–319.

Broadhurst, P. L. and Jinks, J. L. (1966). Stability and change in the inheritance of behavior in rats: A further analysis of statistics from a diallel cross. *Proceedings of the Royal Society of London* B165, 450–472.

Bucio-Alanis, L. (1966). Environmental and genotype–environmental components of variability I. Inbred lines. *Heredity* 21, 387–397.

Bulmer, M. G. (1970). *The Biology of Twinning in Man.* Oxford University Press.

Bulmer, M. G. (1980). *The Mathematical Theory of Quantitative Genetics.* Oxford: Clarendon Press.

Burt, C. (1962). Francis Galton and his contribution to psychology. *British Journal of Statistical Psychology* 15, 1–49.

Caligari, P. D. S. and Mather, K. (1975). Genotype–environment interactions. *Proceedings of the Royal Society of London* B, 191, 387–411.

Cantor, R. M., Nance, W. E., Eaves, L. J., Winter, P. M. and Blanchard, M. M. (1983). Analysis of the covariance structure of digital ridge counts in the offspring of monozygotic twins. *Genetics* 103, 495–512.

Carey, G. (1986). Sibling imitation and contrast effects. *Behavior Genetics* 16, 319–342.

Carey, G. (1986). A general multivariate approach to linear modeling in human genetics. *American Journal of Human Genetics* 39, 775–786.

Carrigan, P.M. (1960). Extraversion–introversion as a dimension of personality: A reappraisal. *Psychological Bulletin* 57, 329–360.

Carter, H.D. (1925). Twin similarities in emotional traits. *Character and Personality* 4, 61–78.

Caten, C.E. (1979). Quantitative genetic variation in fungi. In *Quantitative Genetic Variation* (ed. J.N. Thompson and J.M. Thoday), pp. 35–60. New York: Academic Press.

Cattell, R.B., Blewett, D.B. and Beloff, J.R. (1955). The inheritance of personality: A multiple variance analysis determination of approximate nature–nurture ratios for primary personality factors in Q-data. *American Journal of Human Genetics* 7, 122–146.

Cattell, R.B. (1960). The multiple abstract variance analysis equations and solutions: For nature–nurture research on continuous variables. *Psychological Review* 67, 353–372.

Cattell, R.B. (1963a). The interaction of hereditary and environmental influences. *British Journal of Statistical Psychology* 16, 191–210.

Cattell, R.B. (1963b). The theory of fluid and crystalised intelligence: A critical experiment. *Journal of Educational Psychology* 54, 1–22.

Cattell, R.B. (1965). Methodological and conceptual advances in evaluating hereditary and environmental influences and their interaction. *Methods and Goals in Human Behavior Genetics* (ed. S.G. Vandenberg). New York: Academic Press.

Cattell, R.B. (1982). *The Inheritance of Personality and Ability: Research methods and findings*. New York: Academic Press.

Cattell, R.B., Eber, H.W. and Tatsuoka, M. (1970). *Handbook for the Sixteen Personality Factor Questionnaire*. Champaign, Ill:

Cavalli-Sforza, L.L. and Feldman, M. (1973a). Models for cultural inheritance 1: Group mean and within group variation. *Theoretical Population Biology* 4, 42–55.

Cavalli-Sforza, L.L. and Feldman, M.W. (1973b). Cultural versus biological inheritance: Phenotypic transmission from parents to children (a theory of direct effects of parental phenotypes on children's phenotypes). *American Journal of Human Genetics* 25, 618–637.

Cavalli-Sforza, L.L. and Feldman, M.W. (1981). *Cultural Transmission and Evolution: A Quantitative Approach*. Princeton University Press.

Cavalli-Sforza, L.L., Feldman, M.W., Chen, K.H. and Dornbusch, S.M. (1982). Theory and observation in cultural transmission. *Science* 218, 19–27.

Cederlof, R., Friberg, L., Jonsson, E. and Kaij, L. (1961). Studies on similarity diagnosis in twins with the aid of mailed questionnaires. *Acta Genetica et Statistica Medica* 11, 338–362.

Chamove, A.S., Eysenck, H.J. and Harlow, H.F. (1972). Personality in monkeys: Factor analysis of rhesus social behavior. *Quarterly Journal of Experimental Psychology* 24, 496–504.

Christie, R. (1956). Eysenck's treatment of the personality of communists. *Psychological Bulletin* 53, 411–430.

Claridge, G., Canter, S. and Hume, W.I. (1973). *Personality and Biological Variations: A Study of Twins*. New York: Pergamon Press.

Cloetta, B. (1983). Der Fragebogen zur Erfassung von Machiavellismus und Konservatismus MK. *Schweizer Aeitschrift fur Psychologische Anwendung* 42, 127–159.

Cloninger, C. R. (1980). Interpretation of intrinsic and extrinsic structural relations by path analysis: Theory and applications to assortative mating. *Genetical Research* 36, 135–145.

Cloninger, C. R., Rice, T. and Reich, T. (1979) Multifactorial inheritance with cultural transmission and assortative mating. *American Journal of Human Genetics* 31, 178–188.

Cochrane, R., Billig, M. and Hogg, M. (1979). Politics and values in Britain: A test of Rokeach's two-value model. *British Journal of Social and Clinical Psychology* 18, 159–169.

Conley, J. J. (1984). The hierarchy of consistency: A review and model of longitudinal findings on adult individual differences in intelligence, personality, and self-opinion. *Personality and Individual Differences* 5, 11–26.

Conte, S. D. and deBoor, C. (1980). *Elementary Numerical Analysis: An Algorithmic Approach*, (3rd edn). Tokyo: McGraw-Hill Kogakusha.

Corey, L. A., Kang, K. W., Christian, J. C., Norton, J. A., Harris, R. E. and Nance, W. E. (1976). Effects of chorion type on variation in cord blood cholesterol of monozygotic twins. *American Journal of Human Genetics* 28, 433–441.

Corey, L. A., Eaves, L. J., Mellen, B. G. and Nance, W. E. (1986). Testing for developmental changes in gene expression on resemblance for quantitative traits in kinships of monozygotic twins. *Genetic Epidemiology* 3, 73–83.

Costa, P. T., McCrae, R. R. and Norris, A. H. (1983). Recent longitudinal research on personality and aging. *Longitudinal Studies in Aging* (ed. K. W. Shaie). New York: Guilford Press.

Craik, I. (1984). *Modern Social Theory: From Parsons to Habermas*. Brighton: Wheatsheaf.

Cronbach, L. J., Gleser, G. C. and Rajaratnain, N. (1963). Theory of generalizability: A liberalization of reliability theory. *British Journal of Statistical Psychology* 16, 137–163.

Crow, J. F. and Kimura, M. (1970). *An Introduction to Population Genetics Theory*. New York: Harper & Row.

Darwin, C. (1859). *On the Origin of Species*. London: Murray.

Davis, P. F. and Polonsky, I. (1965). Numerical interpolation, differentiation, and integration. *Handbook of Mathematical Functions* (ed. M. Abramowitz and I. A. Stegun). New York: Dover.

Duncan, O. D. (1966). Path analysis: Sociological examples. *American Journal of Sociology* 72, 1–16.

East, E. M. (1915). Studies on size inheritance in *Nicotiana*. *Genetics* 1, 164–176.

Eaves, L. J. (1969). The genetic analysis of continuous variation: A comparison of experimental designs applicable to human data. *British Journal of Mathematical and Statistical Psychology* 22, 131–147.

Eaves, L. J. (1970). Aspects of Human Psychogenetics. PhD thesis, University of Birmingham.

Eaves, L. J. (1972). Computer simulation of sample size and experimental design in human psychogenetics. *Psychological Bulletin*, 77, 144–152.

Eaves, L. J. (1973a). The structure of genotypic and environmental covariation of personality measurements: An analysis of the PEN. *British Journal of Social and Clinical Psychology* 12, 275–282.

Eaves, L. J. (1973b). Assortative mating and intelligence: An analysis of pedigree data. *Heredity* 30, 199–210.

Eaves, L. J. (1976a). A model for sibling effects in man. *Heredity* 36, 205–214.

Eaves, L.J. (1976b). The effect of cultural transmission on continuous variation. *Heredity* 37, 41–57.

Eaves, L.J. (1977). Inferring the causes of human variation. *Journal of the Royal Statistical Society* A, 140, 324–355.

Eaves, L.J. (1980a). Twins as a basis for the causal analysis of personality. *Twin Research: Psychology and Methodology* (ed. W.E. Nance), pp. 151–174. New York: Alan R. Liss.

Eaves, L.J. (1980b). The use of twins in the analysis of assortative mating. *Heredity* 43, 399–409.

Eaves, L.J. (1982). The utility of twins. *Genetic Basis of the Epilepsies* (ed. V.E. Anderson), pp. 249–276. New York: Raven.

Eaves, L.J. (1983). Errors of inference in the detection of major gene effects on psychological test scores. *American Journal of Human Genetics* 35, 1179–1189.

Eaves, L.J. (1984). The resolution of genotype × environment interaction in segregation analysis of nuclear families. *Genetic Epidemiology* 1, 215–228.

Eaves, L.J. (1987). Dominance alone is not enough. *Behavior Genetics* 8, 27–33.

Eaves, L.J. and Brumpton, R.J. (1972). Factors of covariation in *Nicotiana Rustica*. *Heredity* 29, 151–175.

Eaves, L.J. and Eysenck, H.J. (1974). Genetics and the development of social attitudes. *Nature* 249, 288–289.

Eaves, L.J. and Eysenck, H.J. (1975). The nature of extraversion: A genetical analysis. *Journal of Personality and Social Psychology* 32, 102–112.

Eaves, L.J. and Eysenck, H.J. (1976a). Genotype × age interaction for neuroticism. *Behavior Genetics* 6, 359–362.

Eaves, L.J. and Eysenck, H.J. (1976b). Genetic and environmental components of inconsistency and unrepeatability in twins' responses to a neuroticism questionnaire. *Behavior Genetics* 6, 145–160.

Eaves, L.J. and Eysenck, H.J. (1977). A genotype–environmental model for psychoticism. *Advances in Behavior Research and Therapy* 1, 5–26.

Eaves, L.J. and Eysenck, H.J. (1980). The genetics of smoking. *The Causes and Effects of Smoking* (ed. H.J. Eysenck), pp. 158–314. London: Temple-Smith.

Eaves, L.J. and Gale, J.S. (1974). A method for analysing the genetic basis of covariation. *Behavior Genetics*, 4, 253–267.

Eaves, L.J. and Heath, A.C. (1981a). Detection of the effects of asymmetric assortative mating. *Nature* 289, 205–206.

Eaves, L.J. and Heath, A.C. (1981b) Sex limitation and "asymmetric" assortative mating. *Twin Research 3: Intelligence, Personality, and Development* (ed. L.L. Gedda, P. Parisi and W.E. Nance), pp. 73–86. New York: Alan Liss.

Eaves, L.J. and Jinks, J.L. (1972). Insignificance of evidence for differences in heritability of IQ between races and social classes. *Nature* 240, 84–88.

Eaves, L.J. and Young, P.A. (1981). Genetical theory and personality differences. *Dimensions of Personality* (ed. R. Lynn), pp.129–180. Oxford: Pergamon Press.

Eaves, L.J., Last, K.A., Martin, N.G. and Jinks, J.L. (1977). A progressive approach to non-additivity and genotype-environmental covariance in the analysis of human differences. *British Journal of Mathematical and Statistical Psychology* 30, 1–42.

Eaves, L.J., Last, K.A, Young, P.A. and Martin, N.G. (1978). Model-fitting approaches to the analysis of human behavior. *Heredity* 41, 249–320.

Eaves, L.J., Heath, A.C. and Martin, N.G. (1984). A note on the generalized effects of assortative mating. *Behavior Genetics* 14, 371–376.

Eaves, L.J., Kendler, K.S. and Schulz, S.C. (1986a). The familial vs. sporadic classification: Its power for the resolution of genetic and environmental etiologic factors. *Journal of Psychiatric Research* **20**, 115–130.

Eaves, L.J., Long, J. and Heath, A.C. (1986b). A theory of developmental change in quantitative phenotypes applied to cognitive development. *Behavior Genetics* **16**, 143–162.

Eaves, L.J., Martin, N.G., Heath, A.C. and Kendler, K.S. (1987). Testing genetic models for multiple systems: An application to the genetic analysis of liability to depression. *Behavior Genetics* **17**, 331–342.

Eaves, L.J., Martin, N.G. and Heath, A.C. (in preparation). Religious affiliation in twins and their parents: Testing a model of cultural inheritance. For *Behavior Genetics*.

Edwards, J.H. (1969). Familial predisposition in man. *British Medical Bulletin* **25**, 58–63.

Effron, B. (1982). *The Jackknife, the Bootstrap and other Resampling Plans*. Philadelphia: SIAM.

Eichorn, D.H., Clausen, J.A., Haan, N., Honzik, M.P. and Mussen, P. (1981). *Present and Past in Middle Life*. New York: Academic Press.

Elston, R.C. and Bocklage, C.E. (1978). An examination of fundamental assumptions of the twin method. *Twin Research: Psychology and Methodology* (ed. W.E. Nance), pp. 189–200. New York: Alan Liss.

Elston, R.C. and Stewart, J. (1971). A general model for the genetic analysis of pedigree data. *Human Heredity* **21**, 523–542.

Emmerick, W. (1964). Continuity and stability in early social development. *Child Development* **35**, 311–332.

Emmerick, W. (1967). Stability and change in early personality development. *The Young Child: Review of Research* (ed. W.W. Hartrup and N.L. Smothergill). Washington: National Association for Education.

Emmerick, W. (1968). Personality development and concepts of structure. *Child Development* **39**, 671–690.

Endler, N.S. and Hunt, J.M. (1976). S–R inventories of hostility and comparisons of the proportions of variance from persons, responses, and situations for hostility and anxiousness. *Interactional Psychology and Personality* (ed. N.S. Endler and D. Magnusson). New York: Wiley.

Eysenck, H.J. (1951). Primary social attitudes as related to social class and political party. *The British Journal of Sociology* **2**, 198–209.

Eysenck, H.J. (1952a). *The Scientific Study of Personality*. London: Routledge & Kegan Paul.

Eysenck, H.J. (1952b). Uses and abuses of factor analysis. *Applied Statistics* **1**, 45–47.

Eysenck, H.J. (1953). The logical basis of factor analysis. *American Psychologist* **8**, 105–119.

Eysenck, H.J. (1954). *The Psychology of Politics*. London: Routledge & Kegan Paul.

Eysenck, H.J. (1956a). The inheritance of extraversion–introversion. *Acta Psychologica* **12**, 95–110.

Eysenck, H.J. (1956b). The psychology of politics: A reply. *Psychological Bulletin* **53**, 177–182.

Eysenck, H.J. (1956c). The psychology of politics and the personality similarities between fascists and communists. *Psychological Bulletin*, **53**, 431–438.

Eysenck, H.J. (1960). *Experiments in Personality*. London: Routledge & Kegan Paul.

Eysenck, H. J. (1967a). *Dimensions of Personality*. London: Routledge & Kegan Paul.

Eysenck, H. J. (1967b). *The Biological Basis of Personality*. Springfield: C. C. Thomas.

Eysenck, H. J. (1970). A dimensional system of psychodiagnosis. *New Approaches in Personality Classification* (ed. S. H. Mahrer). New York: Columbia University Press.

Eysenck, H. J. (1971a). *Readings in Extraversion–Introversion*, Vols. 1–3. London: Staples Press.

Eysenck, H. J. (1971b). Social attitudes and social class. *British Journal of Social and Clinical Psychology* 10, 201–212.

Eysenck, H. J. (1972). The paradox of socialism: Social attitude and social class. *Psychology is About People* (ed. H. J. Eysenck), pp. 200–235. London: Allen Lane.

Eysenck, H. J. (1973). *Handbook of Abnormal Psychology*. London: Pitman.

Eysenck, H. J. (1975a). *Manual of the Eysenck Personality Questionnaire*. San Diego: Digits.

Eysenck, H. J. (1975b). The structure of social attitudes. *British Journal of Social and Clinical Psychology* 14, 323–331.

Eysenck, H. J. (1976a). *The Measurement of Personality*. Lancaster: Medical & Technical Publishing.

Eysenck, H. J. (ed.) (1976b). *Sex and Personality*. London: Open Books.

Eysenck, H. J. (1976c). Structure of social attitudes. *Psychological Reports*, **39**, 463–466.

Eysenck, H. J. (1977). *Crime and Personality*, 3rd edn. London: Routledge & Kegan Paul.

Eysenck, H. J. (1980). Man as a biological animal: Comments in the sociobiology debate. *International Sociobiology* 2, 43–51.

Eysenck, H. J. (1980a). The biological nature of man. *Journal of Social and Biological Structures* 3, 125–134.

Eysenck, H. J. (ed.) (1981). *A model for Personality*. New York: Springer-Verlag.

Eysenck, H. J. (1983a). Is there a paradigm in personality research? *Journal of Personality in Research* 17, 369–397.

Eysenck, H. J. (1983b). Psychopharmacology and personality. In *Response Variability to Psychotropic Drugs* (ed. W. J. Janke). New York: Pergamon Press.

Eysenck, H. J. (1984). The theory of intelligence and the psychophysiology of cognition. *Advances in the Psychology of Human Intelligence* (ed. R. J. Sternberg). Hillsdale: Lawrence Erlbaum.

Eysenck, H. J. (1985). Personality theory and the problem of criminality. *Theory and Practice: Applying Psychology to Imprisonment* (ed. B. McGurk, D. Thorton and M. Williams). London.

Eysenck, H. J. and Broadhurst, P. L. (1965). Experiments with animals. *Experiments in Motivation* (ed. H. J. Eysenck). Oxford: Pergamon Press.

Eysenck, H. J. and Coulter, T. T. (1972). The personality and attitudes of working class British communists and fascists. *The Journal of Social Psychology* 87, 59–73.

Eysenck, H. J. and Eysenck, M. W. (1985). *Personality and Individual Differences*. New York: Plenum Press.

Eysenck, H. J. and Eysenck, S. B. G. (1967). On the unitary nature of extraversion. *Acta Psychologica* **26**, 383–390.

Eysenck, H. J. and Eysenck, S. B. G. (1969). *Personality Structure and Measurement*. London: Routledge & Kegan Paul.

Eysenck, H. J. and Eysenck, S. B. G. (1975). *Manual of the Eysenck Personality Questionnaire*. San Diego: Digits.

Eysenck, H. J. and Eysenck, S. B. G. (1976). *Psychoticism as a Dimension of Personality*. London: Hodder & Stoughton.

Eysenck, H. J. and Gudjonsson, G. H. (1988). *The Causes and Cures of Criminality*. New York: Plenum Press.

Eysenck, H. J. and Prell, D. B. (1951). The inheritance of neuroticism: An experimental study. *Journal of Mental Science* 97, 441–465.

Eysenck, H. J. and Wakefield, J. A. (1981). Psychological factors as predictors of marital satisfaction. *Advances in Behaviour Research and Therapy*, 3, 151–192.

Eysenck, H. J. and Wilson, G. (1978). *The Psychological Basis of Ideology*. Lancaster: Medical & Technical Publishing.

Eysenck, M. W. and Eysenck, H. J. (1980). Mischel and the concept of personality. *British Journal of Psychology* 71, 191–204.

Eysenck, S. B. G. and Eysenck, H. J. (1963). On the dual nature of extraversion. *British Journal of Social and Clinical Psychology* 2, 46–55.

Eysenck, S. B. G. and Eysenck, H. J. (1969). Scores on three personality variables as a function of age, sex, and social class. *British Journal of Social and Clinical Psychology*, 8, 69–76.

Eysenck, S. B. G. and Eysenck, H. J. (1977). The place of impulsiveness in a dimensional scheme of personality. *British Journal of Social and Clinical Psychology*, 16, 57–68.

Eysenck, S. B. G., Nias, D. K. B. and Eysenck, H. J. (1971). The interpretation of children's Lie Scale scores. *British Journal of Educational Psychology* 41, 23–31.

Eysenck, S. G. and Eysenck, H. J. (1976). Personality and mental illness. *Psychological Reports* 39, 1011–1022.

Falconer, D. S. (1960). *Quantitative Genetics*. Edinburgh: Oliver & Boyd.

Falconer, D. S. (1963). Quantitative inheritance. *Methodology in Mammalian Genetics* (ed. W. J. Burdette). San Francisco: Holden-Day.

Falconer, D. S. (1965). The inheritance of liability to certain diseases estimated from the incidence among relatives. *Annals of Human Genetics* 29, 51–76.

Falconer, D. S. (1981). *Introduction to Quantitative Genetics*, 2nd edn. London: Longman.

Feather, N. T. (1975). Factor structure of the conservatism scale. *Australian Psychologist* 10, 179–184.

Feather, N. T. (1978). Family resemblance in conservatism: Are daughters more similar to parents than their sons are? *Journal of Personality* 46, 260–278.

Feingold, L. (1984). Genetic and environmental determinants of social attitudes. Doctoral dissertation, University of Oxford.

Feldman, M., Franklin, I. and Thompson, G. (1974). Selection in complex genetic systems I: The symmetric equilibria of the three locus symmetric viability model. *Genetics* 76, 135–162.

Feldman, M. W. and Lewontin, R. C. (1975). The heritability hang-up. *Science* 190, 1163–1168.

Ferguson, L. W. (1939). Primary social attitudes. *Journal of Psychology* 8, 217–223.

Fischbein, S. (1981). Heredity–environment influences on growth and environment during adolescence. *Twin research 3: Intelligence, personality, and development* (ed. L. L. Gedda, P. Parisi and W. E. Nance), pp. 211–226. New York: Alan R. Liss.

Fisher, R.A. (1918). The correlation between relatives on the supposition of Mendelian inheritance. *Transactions of the Royal Society of Edinburgh* **52**, 399–433.

Fisher, R.A. (1930). *The Genetical Theory of Natural Selection.* London: Oxford University Press.

Fisher, R.A. (1935). The detection of linkage with 'Dominant' abnormalities. *Annals of Eugenics* **6**, 187–201.

Fisher, R.A., Immer, F.R. and Tedin, O. (1932). The genetical interpretation of statistics of the third degree in the study of quantitative inheritance. *Genetics* **17**, 107–124.

Forrest, D.W. (1974). *Francis Galton: The life and work of a Victorian Genius.* London: Paul Elek.

Freeman, D. (1983). *Margaret Mead and Samoa: The Making and Unmaking of an Anthropological Myth.* Cambridge, Mass.: Harvard University Press.

Fulker, D.W. (1975). Review of "The science and politics of IQ." *American Journal of Psychology* **88**, 505–519.

Fuller, J.L. and Thompson, W.R. (1978). *Foundations of Behavior Genetics.* Saint Louis: C.V. Mosby.

Galton, F. (1869). *Hereditary Genius: An Inquiry into its Laws and Consequences.* London: Macmillan.

Galton, F. (1883). *Inquiries into Human Faculty.* London: Macmillan.

Garcia, I. and Sevilla, L. (1984). Extraversion and neuroticism in rats. *Personality and Individual Differences* **5**, 511–532.

Gottesman, I.I. (1963). Heritability of personality: A demonstration. *Psychological Monographs* **77**, No. 572.

Gottesman, I.I. (1965). Personality and natural selection. *Methods and Goals in Human Behavior Genetics* (ed. S.G. Vandenberg). New York: Academic Press.

Gottesman, I.I. (1966). Genetic variance in adaptive personality traits. *Journal of Child Psychology and Child Psychiatry* **7**, 199–208.

Gottesman, I.I. and Shields, J. (1967). A polygenic theory of schizophrenia. *Proceedings of the National Academy of Sciences of the USA* **58**, 199–205.

Gottesman, I.I. and Shields, J. (1968). In pursuit of the schizophrenic genotype. *Progress in Human Behavior Genetics* (ed. S.G. Vandenberg), pp. 67–104. Baltimore: John Hopkins.

Gottesman, I.I. and Shields J. (1982). *Schizophrenia: The Epigenetic Perspective.* New York: Cambridge University Press.

Gotz, K.O. and Gotz, K. (1979a). Personality characteristics of professional artists. *Perceptual and Motor Skills,* **49**, 327–334.

Gotz, K.O. and Gotz, K. (1979b). Personality characteristics of successful artists. *Perceptual and Motor Skills* **49**, 919–924.

Gray, J.A. (1970). The psychophysiological basis of introversion–extraversion. *Behaviour Research and Therapy* **8**, 249–266.

Gray, J.A. (1973). Causal theories of personality and how to test them. *Multivariate Analysis and Psychological Theory* (ed. J.R. Royce). New York: Academic Press.

Gray, J.A. (1981). A critique of Eysenck's theory of personality. *A Model for Personality* (ed. H.J. Eysenck). New York: Springer-Verlag.

Greig, D.M. (1980). *Optimisation.* London: Longman.

Grossarth-Maticek, R. (1975). *Revolution der Gestorten?* Heidelberg: Quelle und Meyer.

Guiganino, B.M. and Hindley, C.B. (1982). Stability of individual differences in personality structure from three to five years. *Personality and Individual Differences* **3**, 287–301.

Haley, C. S., Jinks, J. L. and Last, K. A. (1981). The monozygotic twin half-sib method for analysing maternal effects and sex linkage in humans. *Heredity* **46**, 227–238.

Hall, C. S. and Lindzey, G. (1970). *Theories of Personality*, 2nd edn. New York: Wiley.

Hamilton, W. D. (1964a). The genetical evolution of social behavior I. *Journal of Theoretical Biology*, **7**, 1–16.

Hamilton, W. D. (1964b). The genetical evolution of social behavior II. *Journal of Theoretical Biology* **7**, 17–52.

Hay, D. A. and O'Brien P. (1981). The interaction of family attitudes and cognitive abilities in the La Trobe twin study of behavioral and biological development. *Twin research 3: Intelligence, Personality, and Development* (ed. L. L. Gedda, P. Parisi, and W. E. Nance), pp. 235–250. New York: Alan R. Liss.

Hayek, F. A. (1960). *The Constitution of Liberty*. London: Routledge & Kegan Paul.

Heath, A. C. (1983). Human quantitative genetics: Some issues and applications. Doctoral dissertation, University of Oxford.

Heath, A. C. and Eaves, L. J. (1985). Resolving the effects of phenotype and social background on mate selection. *Behavior Genetics* **15**, 15–30.

Heath, A. C., Martin, N. G., Eaves, L. J. and Loesch, D. (1984). Evidence for polygenic epistatic interactions in man? *Genetics* **106**, 719–727.

Heath, A. C., Berg, K., Eaves, L. J., Solaas, M. H., Corey, L. A., Sundet, H. M., Magnus, P. and Nance, W. E. (1985a). Educational policy and the heritability of educational attainment. *Nature* **314**, 734–736.

Heath, A. C., Berg, K., Eaves, L. J., Solaas, M. H., Sundet, J., Nance, W. E., Corey, L. A. and Magnus, P. (1985b). No decline in assortative mating for educational level. *Behavior Genetics* **15**, 349–369.

Heath, A. C., Kendler, K. S., Eaves, L. J. and Markell, D., (1986). The resolution of cultural and biological inheritance: Informativeness of different relationships. *Behavior Genetics* **15**, 439–465.

Heavens, P. C. C. (1983). A factorial study of a new measure of authoritarianism. *Personality and Individual Differences* **4**, 693–694.

Hewitt, J. K. (1974). An analysis of data from a twin study of social attitudes. Masters thesis, University of Birmingham.

Hewitt, J. K. (1984) Normal components of personality variation. *Journal of Personality and Social Psychology* **47**, 671–675.

Hewitt, J. K., Eysenck, H. J. and Eaves, L. J. (1977). Structure of social attitudes after twenty five years: A replication. *Psychological Reports* **40**, 183–188.

Hindley, C. B. and Guiganino, B. M. (1982). Stability of individual differences in personality structures from 3 to 5 years. *Personality and Individual Differences* **3**, 287–301.

Holzinger, K. J. (1929). The relative effect of nature and nurture influences on twin differences. *Journal of Educational Psychology* **20**, 241–248.

Imel, P. (1974). Maternal effects on personality. *Behavior Genetics* **4**, 133–144.

IMSL (1979) *International Mathematical and Statistical Library*, Edition 7. Houston, Texas: IMSL.

Jablon, S., Neel, J. V., Gershowitz, H. and Atkinson, G. F. (1967). The NAS–NRC Twin Panel: Methods of construction of the panel, zygosity diagnosis, and proposed use. *American Journal of Human Genetics* **19**, 133–161.

Jacquard, A. (1984). *In Praise of Difference: Genetics and Human Affairs*. New York: Columbia University Press.

Jaensch, E. R. (1938). *Der Gegentypus*. Leipzig: J. A. Barth.

Jardine, R. (1985). A twin study of personality, social attitudes, and drinking behavior. PhD thesis, Australian National University.

Jardine, R. and Martin, N. G. (1984). No evidence for sex-linked or sex-limited gene expression influencing spatial orientation. *Behavior Genetics* 14, 345–354.

Jardine, R., Martin, N. G. and Henderson, A. S. (1984). Genetic covariation between neuroticism and the symptoms of anxiety and depression. *Genetic Epidemiology* 1, 89–107.

Jaspars, J. and Furnham, A. (1983). The evidences for interactionism in psychology. A critical analysis of the situation–response inventories. *Personality and Individual Differences* 4, 627–644.

Jencks, C., Smith, M., Acland, H., Bane, M. J., Cohen, D., Gintis, H. and Heyns, B. (1973). *Inequality: A Reassessment of Effect of Family and Schooling in America*. New York: Basic Books.

Jensen, A. R. (1978). Genetic and behavioral effects of non-random mating. *Human Variation: The Biopsychology of Age, Race, and Sex*. (ed. R. T. Osborne, C. E. Noble, and N. Wegl). New York: Academic Press.

Jinks, J. A. and Towey, P. (1976). Estimating the number of genes in a polygenic system by genotype assay. *Heredity* 37, 69–81.

Jinks, J. L. (1977) Comments on "Inferring the causes of human variation". *Journal of the Royal Statistical Society* 140, 352–353.

Jinks, J. L. and Fulker, D. W. (1970). Comparison of the biometrical, genetical, MAVA, and classical approaches to the analysis of human behavior. *Psychological Bulletin* 73, 311–349.

Jinks, J. L. and Mather, K. (1955). Stability in development of heterozygotes and homozygotes. *Proceedings of the Royal Society of London* B, 143, 561–578.

Johannsen, W. (1909). *Elemente der exakten Erblichkditslehre*. Jena: Fisher.

Jöreskog, K. (1973). Analysis of covariance structures. *Multivariate analysis III. Proceedings of the Third International Symposium on Multivariate Analysis* (ed. P. R. Krishnarah). New York: Academic Press.

Jöreskog, K. (1978). The structural analysis of covariance matrices. *Psychometrika* 43, 443–477.

Jöreskog, K. and Sörbom, D. (1981). *LISREL V Users Guide*. University of Uppsala.

Jupp, J. (1968). *Political Parties*. London: Routledge & Kegan Paul.

Kamin, L. (1972). *The Science and Politics of IQ*. Potomac: Erlbaum.

Karlin, S. and Feldman, M. (1970). Linkage and selection: Two locus symmetric viability models. *Theoretical Population Biology* 3, 210–238.

Kasriel, J. and Eaves, L. J. (1976). The zygosity of twins: Further evidence on the agreement between diagnosis by blood groups and written questionnaires. *Journal of Biosocial Science* 8, 263–266.

Kawasaki, T., Delea, C. S., Bartter, F. C. and Smith, H. (1978). The effect of high-sodium and low-sodium intakes on blood pressure and other related variables in human subjects with idiopathic hypertension. *American Journal of Medicine* 64, 193–198.

Kearsey, M. J. and Barnes, B. W. (1970). Variation for metrical characters in *Drosophila* populations II. Natural selection. *Heredity* 25, 11–21.

Kearsey, M. J. and Kojima, K. (1967). The genetic architecture of body weight and hatchability in *Drosophila melanogaster*. *Genetics* 56, 23–37.

Kendall, M.G. and Stuart, A. (1977). *The Advanced Theory of Statistics*, 4th edn. New York: Macmillan.

Kendler, K.S. and Eaves, L.J. (1986). Models for the joint effects of genotype and environment on liability to psychiatric illness. *American Journal of Psychiatry* **143**, 279–289.

Kendler, K.S., Heath, A.C., Martin, N.G. and Eaves, L.J. (1986). Symptoms of anxiety and depression in a volunteer twin population: The etiologic role of genetic and environmental factors. *Archives of General Psychiatry* **43**, 213–221.

Kendler, K.S., Heath, A.C., Martin, N.G. and Eaves, N.G. (1987). Symptoms of anxiety and symptoms of depression: Same genes, different environments? *Archives of General Psychiatry* **44**, 451–457.

Kerlinger, F.N. (1970). A social attitude scale: Evidence on reliability and validity. *Psychological Reports* **26**, 379–383.

Kerlinger, F.N. (1972). The structure and content of social attitudes referents: A preliminary study. *Educational and Psychological Measurement* **32**, 613–630.

Klein, T.W., DeFries, J.C. and Finkbeiner, C.T. (1973). Heritability and genetic correlation: Standard errors of estimates and sample size. *Behavior Genetics* **5**, 355–364.

Kool, V.K. and Ray, J.Z. (1983). *Authoritarianism across Cultures*. Bombay: Himalaya Publishing House.

Kreml, W.P. (1977). *The Anti-Authoritarian Personality*. Oxford: Pergamon Press.

Kuhn, T.S. (1962). *The Structure of Scientific Revolutions*. University of Chicago Press.

Kuhn, T.S. (1970). Logic of discovery or psychology of research. In *Criticism and the Growth of Knowledge* (ed. I. Lakatos and A. Musgrave), pp. 1–24. Cambridge University Press.

Kuhn, T.S. (1974). Second thoughts on paradigms. *The Structure of Scientific Theories* (ed.), pp. 459–982. London: University of Illinois Press.

Lalouel, J.M., Rao, D.C., Morton, N.E. and Elston, R.C. (1983). A unified model for complex segregation analysis. *American Journal of Human Genetics* **35**, 816–826.

Lange, K.L., Westlake, J. and Spence, M.A. (1976). Extensions to pedigree analysis III. Variance components by the scoring method. *Annals of Human Genetics* **39**, 485–491.

Last, K.A. (1978). Genetical aspects of human behavior. PhD thesis, University of Birmingham, UK.

Lathrop, G.M., Lalouel, J.M. and Jacquard, A. (1984). Path analysis of family resemblance and gene–environment interaction. *Biometrics* **40**, 611–625.

Lathrop, G.M., Lalouel, J.M., Julier, C. and Ott, J. (1985). Multilocus linkage analysis in humans: Detection of linkage and estimation of recombination. *American Journal of Human Genetics* **37**, 482–498.

Lee, S.Y. and Jennrich, R.I. (1979). A study of algorithms for covariance structure analysis with special comparisons using factor analysis. *Psychometrika* **44**, 99–113.

Lewontin, R.C. (1974). *The Genetic Basis of Evolutionary Change*. New York: Columbia University Press.

Lewontin, R.C., Rose, S. and Kamin, L.J. (1984). *Not in our Genes: Biology, Ideology, and Human Nature*. New York: Pantheon Books.

Li, C.C. (1976). *First Course in Population Genetics*. Pacific Grove, California: The Boxwood Press.

Loehlin, J. C. (1979). Combining data from different groups in human behavior genetics. *Theoretical Advances in Behavior Genetics* (ed. J. R. Royce and L. P. Mos.), pp. 303–336 Rockville: Sijthoff & Noordhoff.

Loehlin, J. C. (1981). Personality resemblance in adoptive families. *Behavior Genetics* 11, 309–330.

Loehlin, J. C. (1982). Personality resemblances between unwed mothers and their adopted away offspring. *Journal of Personality and Social Psychology* 42, 1089–1099.

Loehlin, J. C. and Nichols, R. C. (1976). *Heredity, Environment, and Personality: A Study of 850 Sets of Twins.* Austin: University of Texas Press.

Loehlin, J. C. and Vandenberg, S. C. (1968). Genetic and environmental components in the covariation of cognitive abilities: An additive model. In *Progress in Human Behavior Genetics* (ed. S. G. Vandenberg), pp. 261–286. Baltimore: Johns Hopkins.

Lord, F. M. and Novick, M. R. (1968). *Statistical Theories of Mental Test Scores.* Reading, Massachusetts: Addison-Wesley.

Lumsden, C. J. and Wilson, E. O. (1981). *Genes, Mind, and Culture: The Coevolutionary Process.* Cambridge Massachusetts: Harvard University Press.

Lykken, D. T. (1978). Volunteer bias in twin research: The rule of two thirds. *Social Biology* 25, 1–9.

Lykken, D. T. (1982). Research with twins: The concept of emergenesis. *The Society for Psychophysiological Research* 19, 361–373.

Lytton, H. (1977). Do parents create or respond to differences in twins? *Developmental Psychology* 13, 456–459.

McCrae, R. R. and Costa, P. T. (1984). *Emerging Lives, Enduring Dispositions.* Boston: Little, Brown.

McGue, M., Wette, R. and Rao, D. C. (1984). Evaluation of path analysis through computer simulation: Effect of incorrectly assuming independent distributions of familial correlations. *Genetic Epidemiology* 1, 255–270.

McClearn, G. E. and Johnson, R. C. (1982). Galton's data a century later: the laboratory, the measures and the subjects. *Behavior Genetics* 12, 591.

Magnus, P., Berg, K. and Nance, W. E. (1983). Predicting zygosity in Norwegian twin pairs born 1915–1960. *Clinical Genetics* 24, 103–112.

Magnusson, D. and Endler, N. S. (1977). *Personality at the Crossroads: Current Issues in Interactional Psychology.* Hillsdale: Lawrence Erlbaum.

Mannheim, K. (1936). *Ideology and Utopia.* London: Routledge & Kegan Paul.

Manning, D., O'Sullivan, N. and Raden, P. (1976a). *Conservatism.* London: Dent.

Manning, D., O'Sullivan, N. and Raden, P. (1976b) *Liberalism.* London: Dent.

Manning, D., O'Sullivan, N. and Raden, P. (1980). Authoritarianism and overt aggression. *Psychological Reports* 47, 452–454.

Martin, N. G. and Eaves, L. J. (1977). The genetical analysis of covariance structure. *Heredity* 38, 79–95.

Martin, N. G. and Eysenck, H. J. (1976). Genetic factors in sexual behavior. *Sex and Personality* (ed. H. J. Eysenck), pp. 192–219. London: Open Books.

Martin, N. G. and Jardine, R. (1986). Eysenck's contribution to behavior genetics. *Hans Eysenck: Consensus and Controversy* (ed. S. Modgil and C. Modgil), pp. 13–62. Lewes, Sussex: Falmer Press.

Martin, N. G. and Martin, P. G. (1975). The inheritance of scholastic ability in a sample of twins. I. Ascertainment of the sample and diagnosis of zygosity. *Annals of Human Genetics* 39, 213–218.

Martin, N. G., Eaves, L. J. and Eysenck, H. J. (1977). Genetical, environmental, and personality factors influencing the age of first sexual intercourse in twins. *Journal of Biosocial Science* **9**, 91–97.

Martin, N.G., Eaves, L.J., Kearsey, M.J. and Davies, P. (1978). The power of the classical twin study. *Heredity* **40**, 97–116.

Martin, N.G., Eaves, L.J. and Loesch, D.Z. (1982). A genetical analysis of covariation between finger ridge counts. *Annals of Human Biology* **9**, 539–552.

Martin, N.G., Jardine, R. and Eaves, L.J. (1984). Is there only one set of genes for different abilities? A reanalysis of the National Merit Scholarship Qualifying Test (NMSQT) data. *Behavior Genetics* **14**, 355–370.

Martin, N.G., Eaves, L.J., Jardine, R., Heath, A.C., Feingold, L.F. and Eysenck, H.J. (1986). Transmission of social attitudes. *Proceedings of the National Academy of Sciences of the USA* **83**, 4364–4368.

Martin, N.G., Eaves, L.J. and Heath, A.C. (1987). Prospects for detecting genotype and environment interactions in twins with breast cancer. *Acta Geneticae Medicae et Gemellologiae* **36**, 5–20.

Maruyama, T. and Yasuda, N. (1970). Use of graph theory in computation of inbreeding and kinship correlations. *Biometrics* **26**, 209–220.

Mather, K. (1949). *Biometrical Genetics*, 1st edn. London: Methuen.

Mather, K. (1953). Genetical control of stability in development. *Heredity* **7**, 297–336.

Mather, K. (1966). Variability and selection. *Proceedings of the Royal Society of London* **B164**, 328–340.

Mather, K. (1967). Complementary and duplicate gene interaction in biometrical genetics. *Heredity* **22**, 97–103.

Mather, K. (1974). Non-allelic interactions in continuous variation of randomly breeding populations. *Heredity* **32**, 414–419.

Mather, K. (1975). Genotype × environment interactions II. Some genetical considerations. *Heredity* **35**, 31–53.

Mather, K. and Caligari, P.D.S. (1976). Genotype × environment interaction IV. The effect of the background genotype. *Heredity* **36**, 41–48.

Mather, K. and Jinks, J.L. (1982). *Biometrical Genetics*, 3rd edn. London: Chapman and Hall.

Maynard-Smith, J. (1964). Group selection and kin selection *Nature* **201**, 1145–1147.

Meszaros, I. (1985). *Ideology and Social Science*. Brighton: Wheatsheaf.

Mischel, W. (1968). *Personality and Assessment*. London: Wiley.

Mischel, W. (1971). *Introduction to Personality*. New York: Holt, Rinehart & Wilson.

Mischel, W. (1977). The interaction of person and situation. *Personality at the Crossroads: Current Issues in Interaction Psychology* (ed. D. Magnusson and N.S. Endler). Hillsdale, New Jersey: Lawrence Erlbaum.

Mischel, W. and Peake, P.K. (1982). Beyond *déjà vu* in search for cross-situational consistency. *Psychological Review* **89**, 730–755.

Mislevy, R.J. (1984). Estimating latent distributions. *Psychometrika* **49**, 359–381.

Moran, P.A.P. (1973). A note on heritability and the correlation between relatives. *Annals of Human Genetics* **37**, 217.

Morton, N.E. (1974). Analysis of family resemblance. I. Introduction. *American Journal of Human Genetics* **26**, 318–330.

Morton, N.E. and MacLean, C.J. (1974). Analysis of family resemblance. III.

Complex segregation of quantitative traits. *American Journal of Human Genetics* **26**, 489–503.

Morton, N. E. and Rao, D. C. (1979). Causal analysis of family resemblance. *Genetic analysis of Common Diseases* (ed. C. F. Sing and M. Skolnick), pp. 431–452. New York: Alan R. Liss.

Murphy, E. A. and Williams, R. A. (1984). The dynamics of quantifiable homeostasis. IV: Zero-order homeostasis. *American Journal of Medical Genetics* **18**, 99–113.

NAG (1982). *Fortran Library Manual*, Mark 9. Oxford: Numerical Algorithms Group.

Nance, W. E. and Corey, L. A. (1976). Genetic models for the analysis of data from the families of identical twins. *Genetics* **83**, 811–825.

Nance, W. E., Kramer, A. A., Corey, L. A., Winter, P. M. and Eaves, L. J. (1983). A causal analysis of birth weight in the offspring of monozygotic twins. *American Journal of Human Genetics* **35**, 1211–1223.

Nelder, J. A. (1975). *General Linear Interactive Modelling. GLIM Manual Release* 2. Oxford: Numerical Algorithms Group.

Nelder, J. A. and Wedderburn, R. W. M. (1972). Generalized linear models. *Journal of the Royal Statistical Society* A. **135**, 370–384.

Newman, H. H., Freeman, F. N. and Holzinger, K. J. (1937). *Twins: A Study of Heredity and Environment*. University of Chicago Press.

Nichols, R. C. (1969). The resemblance of twins in personality and interests. *Behavioral Genetics: Method and Research* (ed. M. Manosevitz, G. Lindzey and D. D. Theissen). New York: Appleton-Century-Crofts.

Nichols, R. C. and Bilbro, W. C. (1966). The diagnosis of twin zygosity. *Acta Genetica et Statistica Medica* **16**, 265–275.

Nilsson-Ehle, H. (1909). Kreuzung untersuchungen an Hafer und Weizen. Lund.

Numerical Algorithms Group (NAG) (1985). *The General Linear Interactive Modeling System*, Release 3.77. Oxford: NAG.

Numerical Algorithms Group (NAG) (1987). *NAG Fortran Library Manual*. Mark 12. Oxford: NAG.

O'Donald, P. (1980). *Genetic Models of Sexual Selection*. Cambridge University Press.

O'Sullivan, N. (1984). *Revolutionary Theory and Political Reality*. Brighton: Wheatsheaf.

Olsson, U. (1979). Maximum likelihood estimation of the polychoric correlation coefficient. *Psychometrika* **44**, 443–460.

Owen, D. R. and Sines, J. O. (1970). Heritability of personality in childre. *Behavior Genetics* **1**, 235–247.

Partanen, J., Braun, K., and Markkanen, T. (1966). Inheritance of drinking behavior: A study on intelligence, personality and use of alcohol of adult twins. Helsinki: Finnish Foundation for Alcohol Study.

Paxman, G. J. (1956). Differentiation and stability in the development of *Nicotiana rustica*. *Annals of Botany* **20**, 331–347.

Pearson, E. S. and Hartley, E. O. (1972). *Biometrika Tables for Statisticians*. Cambridge University Press.

Pearson, K. (1900). On the correlations of characters not quantitatively measurable. *Philosophical Transactions of the Royal Society of London* **A195**, 1–47.

Pearson, K. (1903). On a generalized theory of alternative inheritance with special reference to Mendel's laws. *Philosophical Transactions of the Royal Society of London* **A203**, 53–87.

Pearson, K. (1904). On the laws of inheritance in man II. *Biometrika* **3**, 131–190.

Pearson, K. (1973). *The Life, Letters, and Labours of Francis Galton*. Cambridge University Press.

Pearson, K. and Lee, A. (1903). On the laws of inheritance in man. I. Inheritance of physical characters. *Biometrika* **2**, 357–462.

Penrose, L. S. (1951). Measurement of pleiotropic effects in phenylketonuria. *Annals of Eugenics* **16**, 134–141.

Phelan, M. C., Pellock, J. M. and Nance, W. E. (1982). Discordant expression of fetal hydantoin syndrome in heteropaternal dizygotic twins. *New England Journal of Medicine*, **307**, 99–101.

Plomin, R. and DeFries, J. C. (1985). *Origins of Individual Differences in Infancy: The Colorado Adoption Project*. Orlando, Florida: Academic Press.

Popper, K. R. (1960). *The Poverty of Historicism*, 2nd edn. London: Routledge & Kegan Paul.

Price, G. R. and Maynard-Smith, J. (1973). The logic of animal conflict. *Nature* **15**, 246.

Price, R. A., Vandenberg, S. G., Iyer, H. and Williams, J. S., (1982). Components of variation in normal personality. *Journal of Personality and Social Psychology* **43**, 328–340.

Rao, D. C. and Morton, N. E. (1978). IQ as a paradigm in genetic epidemiology. *Genetic Epidemiology* (ed. N. E. Morton and C. E. Chung), p. 145. New York: Academic Press.

Rao, D. C., Morton, N. E. and Yee, S. (1974). Analysis of family resemblance. II. A linear model for family correlation. *American Journal of Human Genetics* **26**, 331–359.

Rao, D. C., Morton, N. E. and Yee, S. (1976). Resolution of cultural and biological inheritance by path analysis. *American Journal of Human Genetics* **26**, 331–359.

Rao, D. C., Morton, N. E., Elston, R. C. and Yee, S. (1977). Causal analysis of academic performance. *Behavior Genetics* **7**, 147–159.

Rao, D. C., Morton, N. E., Gottesman, I. I. and Lew, R. (1981). Path analysis of qualitative data on pairs of relatives: Application to schizophrenia. *Human Heredity* **31**, 325–333.

Ray, A. A. and Sall, J. P. (1982). *SAS User's Guide: Statistics*. Cary, North Caroline: SAS Institute.

Ray, J. J. and Bozek, R. S. (1981). Authoritarianism and Eysenck's P scale. *The Journal of Social Psychology* **113**, 231–234.

Record, R. G., McKeown, T. and Edwards, J. H. (1970). An investigation of the difference in measured intelligence between twin and single births. *Annals of Human Genetics* **34**, 11–20.

Reich, T., Rice, J., Cloninger, C. R., Wette, R. and James, J. (1978). The use of multiple thresholds and segregation analysis in analyzing the phenotypic heterogeneity of multifactorial traits. *Annals of Human Genetics* **42**, 371–390.

Reznikoff, M. and Honeyman, M. S. (1967). MMPI profiles of monozygotic and dizygotic twin pairs. *Journal of Consulting Psychology* **31**, 100.

Rice, J., Cloninger, R. C. and Reich, T. (1978). Multifactorial inheritance with cultural transmission and assortative mating I: Description and basic properties of unitary models. *American Journal of Human Genetics* **30**, 618–643.

Rice, J., Cloninger, C. R. and Reich, T. (1980). Analysis of behavioral traits in the presence of cultural transmission and assortative mating: Applications to IQ and SES. *Behavior Genetics* 10, 73–92.

Rice, J., Nichols, P. L. and Gottesman, I. I. (1981). Assessment of sex differences for multifactorial traits using path and analysis: Application to learning difficulties. *Psychiatry Research* 4, 301–312.

Rokeach, M. (1960). *The Open and Closed Mind.* New York: Basic Books.

Rokeach, M. (1973). *The Nature of Human Values.* New York: Free Press.

Rokeach, M. (1979). The two-value model of political ideology and British politics. *British Journal of Social and Clinical Psychology* 18, 169–172.

Rokeach, M. and Hanley, C. (1956). Eysenck's tender-mindedness dimension: A critique. *Psychological Bulletin* 53, 169–176.

Rose, R. J. and Ditto, W. B. (1983). A developmental–genetic analysis of common fears from early childhood to adolescence. *Child Development* 54, 361–368.

Roth, E. (1972). *Der Werteinstellungs-Test.* Berlin: Huber.

Royce, J. R. and Powell, S. (1983). *Theory of Personality and Individual Differences: Factors, Systems, and Processes.* Englewood Cliffs, New Jersey: Prentice-Hall.

Ruch, W. and Hehl, F. (1985). Conservatism as a predictor of responses to humour I. A comparison of four scales. *Personality and Individual Differences* 7, 1–14.

Runyan, W. M. (1982). *Life Histories and Psychobiography.* Oxford University Press.

Russell, B. (1961). *History of Western Philosophy,* 2nd edn. London: George Allen & Unwin.

SAS (1982). SAS User's Guide: Statistics, 1982 edition, SAS Institute Inc, Cary, NC.

SAS (1985a). SAS/IML User's Guide, Version 5 edition. SAS Institute Inc, Cary, NC.

SAS (1985b). SAS User's Guide, Statistics, Version 5 edition. SAS Institute Inc, Cary, NC.

Scarr, S. (1969a). Environmental bias in twin studies. In *Behavioral Genetics: Method and Research* (ed. M. Manosevitz, G. Lindzey and D. Thiessen). New York: Appleton Century Crofts.

Scarr, S. (1969b). Social introversion–extraversion as a heritable response. *Child Development* 40, 823–832.

Scarr, S., Webber, P. L., Weinberg, R. A. and Wittig, M. A. (1981). Personality resemblance among adolescents and their parents in biologically related and adoptive families. *Twin research 3: Intelligence, Personality, and Development* (ed. L. L. Gedda, P. Parisi and W. E. Nance), pp. 99–120. New York: Alan Liss.

Schaie, K. and Strother, C. R. (1968). A cross-sequential study of age changes in cognitive behavior. *Psychological Bulletin* 70, 671–680.

Scheinfeld, A. (1967). *Twins and Supertwins.* Philadephia: J. B. Lippincott.

Schneewind, K. A., Schroder, G. and Cattell, R. B. (1983). *Der 16-personlichkeits-faktoren-test.* Bern: Huber.

Schneider, J. and Minkumar, H. (1972). Deutsche neukonstruktion einer konser-vativismusskala. *Diagnostica* 18, 37–48.

Schuerger, J. M., Taid, E. and Taveruelli, M. (1982). Temporal stability of personality by questionnaire. *Social Psychology* 43, 176–182.

Shields, J. (1962). *Monozygotic Twins: Brought Up Apart and Brought Up Together.* Oxford University Press.

Simonov, P. V. (1981). Role of limbic structures in individual characteristics of behaviour. *Acta Neurobiologica* 41, 573–582.

Simpson, I. N. and Caten, C. E. (1979). Recurrent mutation and selection for increased penicillin titre in *Aspergillus nidulans. Journal of General Microbiology* 113, 209–217.

Smith, C. (1971). Discriminating between different modes of inheritance in genetic disease. *Clinical Genetics* 2, 303–313.

Smith, G., Falconer, D.S. and Duncan, L.P.J. (1972). A statistical and genetical study of diabetes II. Heritability and liability. *Annals of Human Genetics* 35, 281–299.

Smithers, A.G. and Lobley, D.M. (1978). Dogmatism, social attitudes, and personality. *British Journal of Social and Clinical Psychology* 17, 135–142.

Snedecor, G.W. and Cochran, W.G. (1980) *Statistical Methods* 7th edn. Ames, Iowa: Iowa State University Press.

Software Arts (1982). *TK! Solver* (microcomputer program). Software Arts.

Sparrow, N.H. and Ross, J. (1964). The dual nature of extraversion: A replication. *Australian Journal of Psychology* 16, 216–218.

Stellar, M. and Hunze, D. (1984). Zur Selbstbeschreibung von Delinquenten im Freiburger Personlichkeitsinventar FPE—eine Sekundaranalyse empirisher untersuchungen. *Zeitsschrift für Differentialle und Diagnostische Psychologie* 5, 87–109.

Stelmack, R.M. (1981). The psychophysiology of extraversion and neuroticism. *A model for Personality* (ed. H.J. Eysenck) New York: Springer-Verlag.

Stone, W.F. (1974). *The Psychology of Politics.* London: Free House.

Stone, W.F. and Russ, R.C. (1976). Machiavellianism as tough-mindedness. *Journal of Social Psychology* 98, 213–220.

Sutton, W.S. (1903). The chromosomes in heredity. *Biological Bulletin, Marine Biology Laboratory, Woods Hole* 4, 231–248.

Tellegen, A., Lykken, D.T., Bouchard, T.J.J.R. and Rich, S. (1984). Heritability of personality: An analysis using twins reared apart and together. *Proceedings of Symposium at the 14th Annual Meeting of the Behavior Genetics Association. Bloomington,* Indiana. Unpublished.

Thomas, W. (1979). *The Philosophic Radicals.* Oxford: Clarendon Press.

Thouless, R.N. (1935). The tendency to certainty in religious beliefs. *British Journal of Psychology* 26, 16–31.

Torgersen, S. (1979). The nature and origin of common phobic fears. *British Journal of Psychiatry* 134, 343–351.

Torgersen, W.S. (1958). *Theory and Methods of Scaling.* New York: Wiley.

Trojak, J.E. and Murphy, E.A. (1981). Recurrence risks for autosomal, epistatic two locus sytems: The effects of linkage disequilibrium. *American Journal of Medical Genetics* 9, 219–229.

Urbach, P. (1974). Progress and degeneration in the 'IQ debate', I and II. *British Journal of Philosophical Science* 25, 99–135.

Vandenberg, S.G. (1972). Assortative mating, or who marries who? *Behavior Genetics* 2, 127–157.

Vandenberg, S.G. (1962). The hereditary abilities study: Hereditary components in a psychological test battery. *American Journal of Human Genetics* 14, 220–237.

Vandenberg, S.G. (1966). Contribution of twin research to psychology. *Psychological Bulletin* 66, 327–352.

Walford, G. (1979). *Ideologies and their Functions.* London: Villien Publications.

Watson, J.D. and Crick, F.C. (1953). Molecular structure of nucleic acids. A structure of deoxyribose nucleic acids. *Nature* 171, 737–738.

Wilde, G.J.S. (1964). Inheritance of personality traits. *Acta Psychologica* 22, 37–51.

Wilde, O. (1899). *The Importance of Being Earnest*.

Williams, J. S. and Iyer, H. (1981). A statistical model and analyses for genetic and environmental effects in responses from twin-family studies. *Acta Geneticae Medicae et Gemellologiae: Twin Research* **30**, 9–38.

Wilson, G. D. (1981). Personality and social behaviour. *A Model for Personality* (ed. H. J. Eysenck). New York: Springer.

Wilson, G. D. (ed.) (1973) *The Psychology of Conservatism*. London: Academic Press.

Wilson, G. D. and Patterson, J. R. (1970). *Manual for the Conservatism Scale*. London:

Wilson, R. S. (1972). Twins: Early mental development. *Science* **175**, 914–917.

Woody, E. and Claridge, G. (1977). Psychoticism and thinking. *British Journal of Social and Clinical Psychology* **16**, 241–248.

Wright, S. (1921). Correlation and causation. *Journal of Agriculture Research* **20**, 551.

Wright, S. (1934). The method of path coefficients. *Annals of Mathematical Statistics* **5**, 161–215.

Wright, S. (1968). *Evolution and the Genetics of Populations*. University of Chicago Press.

Young, J. P. R., Fenton, G. W. and Lader, M. H. (1971). The inheritance of neurotic traits: A twin study of the Middlesex Hospital Questionnaire. *British Journal of Psychiatry* **119**, 393–398.

Young, P. A., Eaves, L. J. and Eysenck, H. J. (1980). Intergenerational stability and change in the causes of variation in personality. *Personality and Individual Differences* **1**, 35–55.

Zajonc, R. B. and Markus, G. B. (1975). Birth order and intellectual development. *Psychological Review* **82**, 74–88.

Zuckerman, M. (1979). *Sensation seeking: Beyond the Optimal Level of Arousal*. Hillsdale, New Jersey: Lawrence Erlbaum.

Zuckerman, M., Ballinger, J. C. and Post, R. M. (1984). The neurobiology of some dimensions of personality. *International Review of Neurobiology* **25**, 391–436.

Index

455